I0010789

# Game Development with Unity for .NET Developers

The ultimate guide to creating games with
Unity and Microsoft Game Stack

**Jiadong Chen**

BIRMINGHAM—MUMBAI

# Game Development with Unity for .NET Developers

**Group Product Manager**: Pavan Ramchandani
**Publishing Product Manager**: Aaron Tanna
**Senior Editor**: Aamir Ahmed
**Content Development Editor**: Feza Shaikh
**Technical Editor**: Joseph Aloocaran
**Copy Editor**: Safis Editing
**Project Coordinator**: Manthan Patel
**Proofreader**: Safis Editing
**Indexer**: Tejal Daruwale Soni
**Production Designer**: Aparna Bhagat
**Marketing Coordinator**: Anamika Singh

First published: May 2022

Production reference: 2130522

Published by Packt Publishing Ltd.
Livery Place
35 Livery Street
Birmingham
B3 2PB, UK.

ISBN 978-1-80107-807-8

www.packt.com

*This book is dedicated to my wife, Yi Liang, for her encouragement and support, especially during the COVID-19 pandemic; her positive attitude toward life has inspired me and helped me through difficult times.*

*– Jiadong Chen*

# Foreword

*"A delayed game is **eventually good**, but a rushed game is **forever bad**."*

*– Industry catchphrase*

You now hold in your hands the tools to make great games. If this was a Mario game, then you just punched a question mark block, and then a glowing, flashing, smoking mushroom just rose out of it. If this was a Zelda game, then you just navigated a boss dungeon, found a treasure chest, pulled this book out of it, and raised the book high above your head (with some suitable music playing). In other words, you just acquired a key powerup for your quest.

You can feel that dedication to excellence (which the industry quote refers to) when you play Nintendo games, Halo, Minecraft, and Sonic the Hedgehog (and pretty much whatever your favorite game is). Now you have the same opportunity: you can use your knowledge and your new Unity skills (granted, you probably haven't learned them yet) to make a powerful statement through your work.

*"Nintendo's philosophy is to **never go the easy path**; it's always to challenge ourselves and **try to do something new**."*

*– Shigeru Miyamoto (2005)*

So that's why Nintendo makes some great games (and some really weird ones)! Regardless, here's your chance to make something that's unique and that really shows who you are. With these tools, your passion opens limitless opportunities.

And that brings us to a Microsoft quote (you knew it was coming).

*"Learning to fly is **not pretty**, but **flying is**."*

*– Satya Nadella, Hit Refresh*

You learn tools to create masterpieces (or to at least get a job, or a better one). It's not easy to learn these new tools, but after you gain this knowledge, you can make great games.

*Game Development with Unity for .NET Developers* makes that process a whole lot easier! Unity is the world's most widely used, real-time 3D development platform for a reason! If you approach Unity with the .NET Framework, then you can leverage the power of C#, Microsoft Game Dev tools, Microsoft Azure Cloud services, and Azure PlayFab. You'll also see how these resources work seamlessly with Visual Studio and GitHub.

Our author, Jiadong Chen, used to work at Unity Technologies as a field engineer. He's been working in this .NET and Unity gaming stack for over 9 years. He's a member of the .NET Foundation, and has been a Microsoft MVP (Most Valuable Professional) for 6 years. (That means he's been awarded the title every year since 2015, based on his impact on the Microsoft developer community.) As you can imagine, his MVP award category is Developer Technologies. In other words, Jiadong is the perfect person to teach you how to use the .NET Framework and Microsoft developer stack to learn how to develop with the Unity game engine and how to take your games to the next level!

Make no mistake: as the title implies, this is a book for .NET developers to learn Unity. First, Jiadong starts with the basics of the Unity game engine. Then, you'll dig into scripting, using Unity to build your game UI, animating your game graphics, building physics, and adding audio and video (the basic components of building your game). Next, Jiadong gets into game math, while also using a rendering pipeline, data-oriented tech, a serialization system, and assets management. Finally, he shows you how to leverage the Microsoft Game Dev suite of technologies, Azure Cloud, and Azure PlayFab as you develop with the Unity engine.

In *Game Development with Unity for .NET Developers*, Jiadong Chen takes you on a tour, explaining the concepts with images and examples, so that you can fully understand each topic. He then takes you through the process, using real code, so that you can implement your own solutions. The process of learning how to develop games isn't easy, but this book will make it a lot easier.

After you're done reading the book, if life was a Mario game, you'll have grabbed the flag at the top of the flag pole and entered a sewer pipe! If this was a Zelda game, you'll have collected the Triforce. And if it's a Sonic game, then you just beat Dr. Robotnik in a giant robot mech that looks like him. As Satya Nadella implied, learning to fly won't be easy, but once you're done, you're going to be able to really take off and do some great things with Unity!

*Ed Price*

*Senior Program Manager of Architectural Publishing*
*Microsoft, Azure Architecture Center (*`http://aka.ms/Architecture`*) Co-author of seven books, including Meg the Mechanical Engineer, The Azure Cloud Native Architecture Mapbook (Packt), and ASP.NET Core 5 for Beginners (Packt)*

# Contributors

## About the author

**Jiadong Chen** is one of 3,000 international Microsoft® **Most Valuable Professional** (**MVP**) award winners, recognized by Microsoft as one of the technology industry's best and brightest six years in a row, and is currently working as a senior software developer at Company-X, based in Hamilton, New Zealand.

He specializes in the Microsoft Azure cloud, Unity and XR development, and .NET/C#. He is a Microsoft Certified Azure Solutions Architect Expert, a Microsoft Certified Azure Developer, a Microsoft Certified Azure AI Engineer, and a Microsoft Certified Trainer. He is also a member of the .NET Foundation.

Before joining Company-X, Jiadong worked for Unity, the creator of the world's most widely used real-time 3D development platform, as a field engineer.

*After this long adventure, I would first like to thank my wife for her support. Writing a book is not an easy task, especially during the COVID-19 pandemic, and my wife's support has given me the psychological comfort to overcome the difficulties and finish the book.*

*A special thanks to the Packt team (especially Aaron, Manthan, and Feza) for being with me on this journey.*

# About the reviewer

**Simon Jackson** is a long-time software engineer and architect with many years of Unity game development experience, as well as an author of several Unity game development titles. He loves to both create Unity projects and lend a hand to help educate others, whether it's via a blog, vlog, user group, or major speaking event.

He currently works at a mixed-reality research lab called xRealityLabs, building the future of digital solutions for the construction and medical industries.

His primary focus at the moment is on the Reality Toolkit project, which is aimed at building a cross-platform mixed-reality framework to enable both VR and AR developers to build efficient solutions in Unity.

# Table of Contents

# Part 2: Using C# Scripts to Work with Unity's Built-In Modules

## 3
## Developing UI with the Unity UI System

## 4
## Creating Animations with the Unity Animation System

# Part 3: Advanced Scripting in Unity

# 9

# The Data-Oriented Technology Stack in Unity

# 10

# Serialization System and Assets Management in Unity and Azure

# 11

## Working with Microsoft Game Dev, Azure Cloud, PlayFab, and Unity

## Index

## Other Books You May Enjoy

# Preface

As one of the most widely used game engines in the world, Unity provides easy-to-use and powerful game development tools, which undoubtedly attracts many developers to choose it to develop their own games. However, the tools needed in modern game development are not limited to game engines; other tools and services such as the cloud are increasingly used in game development. In this book, we will explore how to use the Unity game engine and the Microsoft Game Dev, including the Microsoft Azure Cloud and Microsoft Azure PlayFab services, to create games.

Starting by understanding the fundamentals of the Unity game engine, you will gradually become familiar with the Unity Editor and the key concepts of writing Unity scripts in C#, which will get you ready to make your own game.

Then, you'll learn how to work with Unity's built-in modules, such as the UI system, animation system, physics system, and how to integrate video and audio in your game to make your game more interesting.

As you progress through the chapters, I'll take you through advanced topics, such as the math involved in computer graphics, how to create post-processing effects in Unity with the new Scriptable Render Pipeline, how to use Unity's C# Job System to implement multithreading, and how to use Unity's **Entity Component System** (**ECS**) to write game logic code in a data-oriented way to improve game performance.

Along the way, you'll also learn about the Microsoft Game Dev, the Azure cloud services, Azure PlayFab, and using the Unity3D PlayFab SDK to access the PlayFab API to save and load data from the cloud.

By the end of this book, you'll be familiar with the Unity game engine, have a high-level understanding of the Azure cloud, and be ready to develop your own games.

## Who this book is for

The book is for developers with intermediate .NET and C# programming experience who are interested in learning game development with Unity. Basic experience in C# programming is assumed.

# What this book covers

*Chapter 1, Hello Unity*, explains the fundamentals of the Unity game engine. Beginning with the Unity installation process and then exploring the editor, you will also learn about .NET profiles and the scripting backend offered by Unity, and finally, you will have a broad understanding of Unity.

*Chapter 2, Scripting Concepts in Unity*, continues from the previous chapter and introduces scripting in Unity in detail. It begins by introducing the most commonly used classes in Unity scripting and then explains the life cycle of scripts. It also covers how to create a new script in Unity and attach a script as a component to a GameObject, and demonstrates how to add or remove packages through the Unity Package Manager.

*Chapter 3, Developing UI with the Unity UI System*, covers the different types of UI elements commonly used in Unity. Additionally, this chapter also discusses how to develop UI in Unity by using a **Model View ViewModel** (**MVVM**) architectural pattern. It ends by exploring some optimization tips for Unity UI.

*Chapter 4, Creating Animations with the Unity Animation System*, covers the most important concepts of the Unity animation system, such as animation clips, Animator Controller, Avatar, and the Animator component. Here, you will implement 3D and 2D animations using the animation system. It ends by exploring some optimization tips for the animation system in Unity.

*Chapter 5, Working with the Unity Physics System*, presents an overview of the physics solutions provided by Unity, including two built-in physics solutions, the NVIDIA PhysX engine and the Box2D engine. It also covers key concepts in the Unity physics system, such as Collider and Rigidbody. Here, you will implement a physics-based ping-pong game. It ends by exploring some optimization tips for the physics system in Unity.

*Chapter 6, Integrating Audio and Video in a Unity Project*, covers key concepts in the Unity audio system and video system, such as audio clip assets, Audio Source components, Audio Listener components, and Video Player components. It ends by exploring some optimization tips for the audio system in Unity.

*Chapter 7, Understanding the Mathematics of Computer Graphics in Unity*, covers the mathematics related to computer graphics, such as coordinate systems, vectors, matrices, and quaternions.

*Chapter 8, The Scriptable Render Pipeline in Unity*, presents an overview of three ready-made render pipelines to choose from in Unity, namely the legacy built-in render pipeline and two pre-made render pipelines based on the Scriptable Render Pipeline, the Universal Render Pipeline and the High Definition Render Pipeline. It also covers how to use the Universal Render Pipeline Asset to configure the render pipeline and how to use the Volume framework to apply post-processing effects to a game. It ends by exploring some optimization tips for the Universal Render Pipeline.

*Chapter 9, Using Data-Oriented Technology Stack in Unity*, covers what data-oriented design is and the difference between data-oriented design and traditional object-oriented design. It also explores the **Data-Oriented Technology Stack** (**DOTS**) in Unity and the three technology modules that make it up – namely, the C# Job System, ECS, and the Burst compiler.

*Chapter 10, Serialization System and Assets Management in Unity and Azure*, discusses binary serialization, YAML serialization, and JSON serialization in Unity. It also covers the assets workflow in Unity and ends by exploring how to create an Azure Blob storage service in the Azure cloud and load addressable content from Azure into a Unity project.

*Chapter 11, Working with Microsoft Game Dev, Azure Cloud, PlayFab, and Unity*, discusses what Microsoft Game Dev, Microsoft Azure Cloud, and Azure PlayFab are and why you should consider using them in game development. Here, you will implement the registration, login, and leaderboard functions in a Unity project through the API of Azure PlayFab.

# To get the most out of this book

This book assumes that you have some familiarity with .NET and C#. This book covers basic concepts, advanced topics of the Unity game engine, and also other technologies such as the Microsoft Azure cloud and Azure PlayFab.

You'll also need a **Long-Term Support** (**LTS**) version of Unity installed on your computer – 2020 or later is recommended. You can find out how to install Unity on your computer in *Chapter 1, Hello Unity*. All code examples have been tested with Unity 2020.3.24 on a Windows OS. However, they should work with future version releases too.

You'll also need a Microsoft Azure cloud subscription, and you can apply for a free Azure account at the following link: `https://azure.microsoft.com/en-in/free/`.

| Software/hardware covered in the book | Operating system requirements |
| --- | --- |
| Unity 2020.3+ | Windows, macOS, or Linux |
| The Microsoft Azure portal | Windows, macOS, or Linux |
| Visual Studio 2019 Community | Windows or macOS |

If you wish to download sample projects from our GitHub repository, you will need a Git client; we recommend GitHub Desktop as it is the easiest to use. You can download it at the following link: `https://desktop.github.com`.

If you are using the Windows OS, you can also consider using Git for Windows. It can be downloaded at the following link: `https://git-scm.com/download/win`.

**If you are using the digital version of this book, we advise you to type the code yourself or access the code from the book's GitHub repository (a link is available in the next section). Doing so will help you avoid any potential errors related to the copying and pasting of code.**

## Download the example code files

The code bundle for the book is also hosted on GitHub at `https://github.com/PacktPublishing/Game-Development-with-Unity-for-.NET-Developers`. If there's an update to the code, it will be updated on the existing GitHub repository.

We also have other code bundles from our rich catalog of books and videos available at `https://github.com/PacktPublishing/`. Check them out!

## Download the color images

We also provide a PDF file that has color images of the screenshots and diagrams used in this book. You can download it here: `https://static.packt-cdn.com/downloads/9781801078078_ColorImages.pdf`.

## Conventions used

There are a number of text conventions used throughout this book.

`Code in text`: Indicates code words in text, database table names, folder names, filenames, file extensions, pathnames, dummy URLs, user input, and Twitter handles. Here is an example: "If some content is generated at the beginning of the object collision via `OnCollisionEnter`, and you want to destroy them when the object collision ends, then you should consider using `OnCollisionExit`."

A block of code is set as follows:

```
using UnityEngine;

public class TriggerTest : MonoBehaviour
{
    private void OnTriggerStay(Collider other)
    {
        Debug.Log($"{this} stays {other}");
    }
}
```

When we wish to draw your attention to a particular part of a code block, the relevant lines or items are set in bold:

```
using UnityEngine;

public class PingPongBall : MonoBehaviour
{
    [SerializeField] private Rigidbody _rigidbody;
    [SerializeField] private Vector3 _initialImpulse;

    private void Start()
    {
        _rigidbody.AddForce(_initialImpulse,
            ForceMode.Impulse);
    }
}
```

**Bold**: Indicates a new term, an important word, or words that you see onscreen. For instance, words in menus or dialog boxes appear in **bold**. Here is an example: "Select **3D Object | Plane** to create a new **Plane** object in the editor."

---

**Tips or Important Notes**
Appear like this.

---

# Get in touch

Feedback from our readers is always welcome.

**General feedback**: If you have questions about any aspect of this book, email us at customercare@packtpub.com and mention the book title in the subject of your message.

**Errata**: Although we have taken every care to ensure the accuracy of our content, mistakes do happen. If you have found a mistake in this book, we would be grateful if you would report this to us. Please visit www.packtpub.com/support/errata and fill in the form.

**Piracy**: If you come across any illegal copies of our works in any form on the internet, we would be grateful if you would provide us with the location address or website name. Please contact us at copyright@packt.com with a link to the material.

**If you are interested in becoming an author**: If there is a topic that you have expertise in and you are interested in either writing or contributing to a book, please visit authors.packtpub.com.

# Share Your Thoughts

Once you've read, we'd love to hear your thoughts! Scan the QR code below to go straight to the Amazon review page for this book and share your feedback.

https://packt.link/r/1801078076

Your review is important to us and the tech community and will help us make sure we're delivering excellent quality content.

# Part 1: Basic Unity Concepts

In this section of the book, we'll explore the fundamentals of the Unity game engine and introduce some of the key concepts of scripting in Unity to get you ready to make your own games.

This section includes the following chapters:

- *Chapter 1, Hello Unity*
- *Chapter 2, Scripting Concepts in Unity*

# 1
# Hello Unity

Before we get started with using Unity to develop games, I think it's good to first understand Unity itself. Many people, especially those who are interested in games and game development, know that Unity is a widely used game engine, and you may have played many games developed with Unity. But you may not be familiar with how to use Unity to develop games. For example, there are many different Unity versions available, so how do you choose the version that suits you? Unity provides different subscription plans, but which subscription plan is right for your situation?

If you have never used Unity before, it is necessary for you to learn how to use the Unity Editor first. In addition to the Unity Editor, what features does the Unity engine provide to help game developers develop games? It is also important to know the features in Unity. If you are a .NET developer, then it's likely that you are familiar with Visual Studio. You need to know how to use Visual Studio to develop a Unity game. But developing a Unity game is different from developing a .NET application.

Am I asking too many questions? Don't worry – this chapter will help you answer them.

In this chapter, we will introduce how to choose the right release of Unity and provide an overview of how to download and install Unity via the Unity Hub or the Unity installer. Then, we will choose the right subscription plan for your situation. At this point, you should have installed Unity and opened the Unity Editor.

If you have only just started using the Unity Editor, you may not know how to use it. We will first explore the Unity Editor and then discuss the different features provided by Unity. We will then introduce the .NET profiles in Unity and the scripting backend offered by Unity. Finally, we will present how to use Visual Studio to develop Unity games.

We will cover the following key topics in this chapter:

- Getting started with the Unity Editor
- Working with different features in Unity
- .NET/C# and scripting in Unity
- Building Unity games with Visual Studio

# Technical requirements

Before starting, I highly recommend you first check whether your system can run the Unity Editor. The following table gives the minimum requirements to run the Unity Editor:

| Operating system | Operating system version | CPU | Graphics API |
| --- | --- | --- | --- |
| Windows | Windows 7 (SP1+) and Windows 10 – 64-bit versions only | x64 architecture with SSE2 instruction set support | DX10-, DX11-, and DX12-capable GPUs |
| macOS | High Sierra 10.13+ | x64 architecture with SSE2 instruction set support | Metal-capable Intel and AMD GPUs |
| Linux | Ubuntu 20.04, Ubuntu 18.04, and CentOS 7 | x64 architecture with SSE2 instruction set support | OpenGL 3.2+ or Vulkan-capable, NVIDIA, and AMD GPUs |

# Getting started with the Unity Editor

Whether you are an independent game developer or work in a team for a company, you need to do two things before installing or even downloading Unity:

- Choose the right Unity release for you.
- Choose the right subscription plan for you.

Therefore, before introducing how to install Unity and exploring the Unity Editor, let's first introduce the Unity release and subscription plans. We hope that by reading these contents, you can find the right release for you and choose a suitable subscription plan.

# Choosing the right Unity release for you

Nowadays, Unity offers two different release versions each year. They are as follows:

- Tech Stream releases
- **Long-Term Support** (**LTS**) releases:

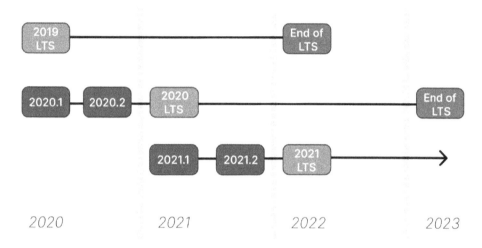

Figure 1.1 – Unity releases

You may not be quite sure which version of Unity is best to use in your project, so I will explain these two different releases so that you can get an idea of how to choose the right release for you.

LTS releases provide developers with maximum stability and full support for their projects, and they are the last Tech Stream releases of each year. With LTS releases, there are no new features or API changes. The updates of LTS releases address crashes, and fix bugs and any minor issues. As I mentioned at the beginning of this section, each year, Unity releases new versions of the LTS release, and each one is supported for 2 full years from the date of the announcement.

Therefore, if you are looking for performance and stability, or your project is already in production or in the middle of development, it is a good idea to use the latest LTS release version to ensure best performance and stability.

> **Note**
> At the time of writing (April 2022), there are two LTS releases, namely Unity 2020 LTS and Unity 2019.4. Unity 2020 LTS is the latest LTS release and has the same feature set as the Unity 2020.2 Tech Stream release. Alternatively, Unity 2019.4 is the legacy LTS release now.

The Tech Stream releases give developers who want to explore the latest in-progress features an option to use them to prepare for future projects. Unlike the LTS releases, a Tech Stream release will be released twice a year (typically published in the first and last quarters) and will only be supported until the next Tech Stream release is officially published.

Therefore, if you are preparing for your next project or working on prototyping and experimentation, you should try the Tech Stream releases.

> **Note**
> At the time of writing (April 2022), the latest Tech Stream release is Unity 2021.2.

By reading this section, I hope you have gained an understanding of the Unity releases, and you should be able to choose the right Unity release according to your situation.

When writing this book, I chose the latest LTS version, Unity 2020.3.

# Choosing the right subscription plan for you

Unity is a widely used game engine, and many independent game developers use Unity to develop their games. But technically speaking, Unity is not a free game engine. In this section, I will introduce several different subscription plans offered by Unity. I hope that after reading this section, you can choose a subscription plan that suits your situation.

Unity offers a range of plans, from the free Personal plan for individual learners to Enterprise plans used by large organizations:

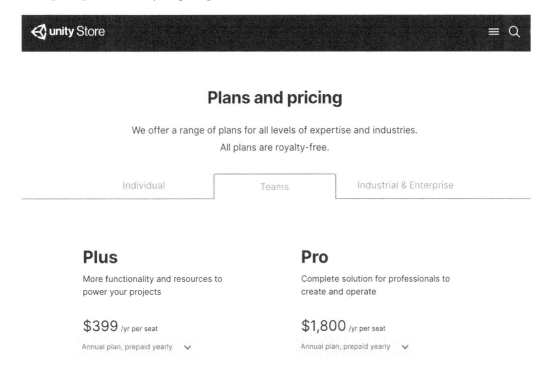

Figure 1.2 – The Plans and pricing page

Because each Unity plan has different eligibility requirements, you should choose the right plan for your project. Next, I will introduce the subscription plans:

- The **Personal plan** is free and includes all the basic functionality of Unity. You can choose this plan if you work as an individual and have earned less than $100,000 in revenue or funding for your Unity project in the past 12 months. In addition, if you are a student or educator, you can get additional benefits, but before that, you need to join the **GitHub Student Developer Pack** to be verified.

- The **Plus plan** is a paid plan and offers more functionality and training resources, such as advanced cloud diagnostics and splash screen customization. If you have earned more than $100,000 but less than $200,000 in revenue from using Unity in the past 12 months, you should choose this plan.

- The **Pro plan** is also a paid plan. Compared with the Plus plan, you can get more technical support from Unity by using the Pro plan. If your organization has earned more than $200,000 in the last 12 months from any source, you must use the Pro plan or the **Enterprise plan**.

- The **Enterprise plan** is specifically for teams with at least 20 members and provides more support than the Pro plan. For example, a customer success manager from Unity will be assigned to your organization to provide guidance, orchestrate resources, and serve as an internal advocate.

I hope this section was helpful for you in choosing the right Unity plan for your situation. Next, let's download and install the Unity Editor!

## Downloading and installing the Unity Editor

There are two different ways to download and install the Unity Editor. The first and recommended way to download and install Unity is to use the **Unity Hub**.

The Unity Hub is a management tool that can be used to manage all your Unity projects and Unity installations. We can take the following steps to install the Unity Hub and the Unity Editor:

1.  To install the Unity Hub, visit the **Download Unity** page at `https://unity3d.com/get-unity/download`:

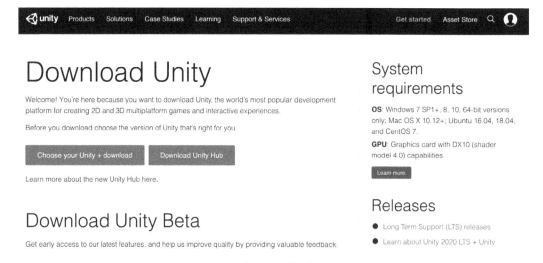

Figure 1.3 – The Download Unity page

As you can see in the **System requirements** section in the preceding screenshot, the Unity Hub supports **Windows**, **Mac OS X**, **Ubuntu**, and **CentOS**.

2.  Installing the Unity Hub is very easy; you just need to choose the folder where the Unity Hub is installed. Then, click on the **Install** button:

Figure 1.4 – Unity Hub Setup

3.  After installing the Unity Hub, select the **Run Unity Hub** option and click on the **Finish** button to launch the Unity Hub:

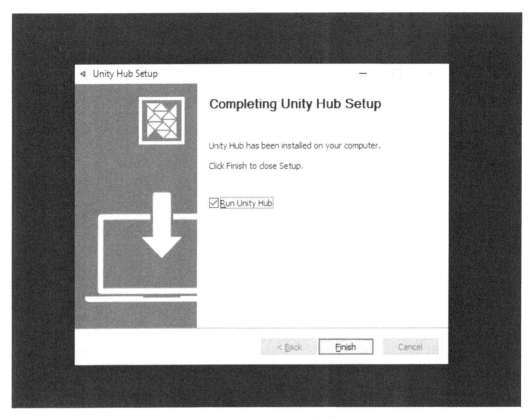

Figure 1.5 – Completing Unity Hub Setup

I am using the latest version of Unity Hub (*version 3.0.0*) at the time of writing. If you have used previous versions of Unity Hub, you will find that the launch page of the new version of Unity Hub is completely different.

4.  You need a Unity account to access the Unity Editor and the Unity Hub. If you don't have a Unity account yet, then you need to create a new one:

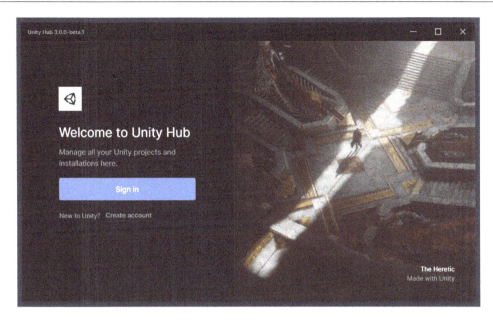

Figure 1.6 – Unity Hub

5.  When you sign into the Unity Hub for the first time, you will be asked to add an active license, as you can see at the top of the following screenshot. Click on the **Manage licenses** button to open the **Licenses** setting panel:

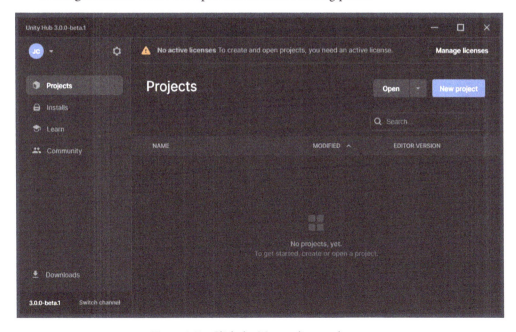

Figure 1.7 – Click the Manage licenses button

6.   There are two buttons available for you to add a new license. You can click either the **Add** button at the top-right corner or the **Add license** button:

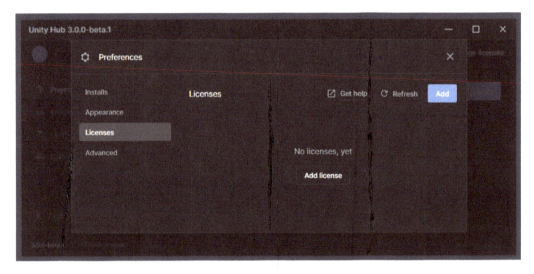

Figure 1.8 – The Licenses setting panel

Then, you have different options to activate the license. We have discussed the different Unity subscription plans in the previous section:

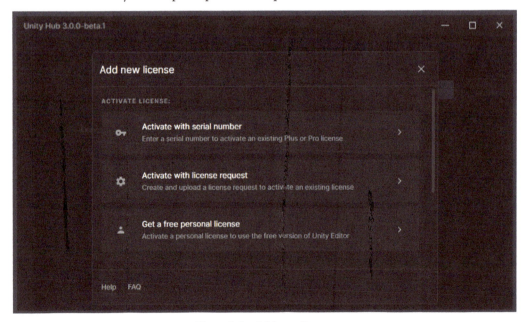

Figure 1.9 – Add new license

7.  After adding the new license, we can start exploring the Unity Hub. From the **Projects** view, you can find a list of Unity projects that are tracked by the Unity Hub. You can also create a brand-new project by clicking the **New project** button at the upper-right corner of the **Projects** view, or you can import an existing project by clicking the **Open** button:

Figure 1.10 – The Projects view

8.  To install the Unity Editor, open the **Installs** view, where you can manage the installation of multiple versions of the Unity Editor:

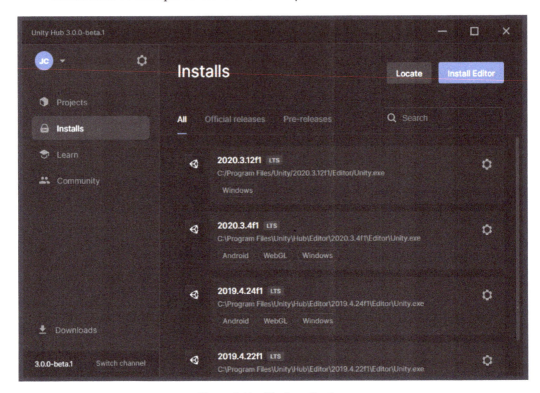

Figure 1.11 – The Installs view

There is a list of the Unity Editors that are installed and managed by the Unity Hub. Similar to the **Projects** view, you can download and install a new Unity Editor, or you can import an existing Unity Editor that is not managed by the Unity Hub, such as the Unity Editor that we installed using the Unity installer.

9.  Open the **Install Unity Editor** panel by clicking the **Install Editor** button on the **Installs** view. Then, you will see the latest version of each release:

Figure 1.12 – Install Unity Editor

> **Note**
> Unity 2018 LTS has reached the end of its support cycle, so you should not install it.

We will use the latest version of the Unity 2020 LTS release, so we need to install Unity **2020.3.13f1** here:

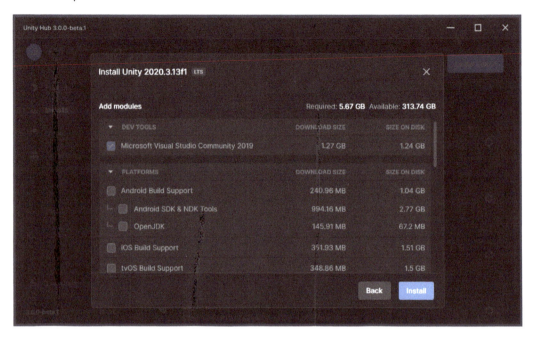

Figure 1.13 – Install Unity 2020.3.13f1

Then, we need to select the modules that need to be installed. As you can see in the preceding screenshot, **Microsoft Visual Studio Community 2019** will be installed by default, which will be our **Integrated Development Environment** (**IDE**) to develop games in Unity.

> **Note**
>
> If you want to change the installation location, you can change it in the **Installs** settings of the **Preferences** panel.

Once it has been downloaded and installed, we are ready to start exploring the Unity Editor!

Sometimes, you may need a specific version that is not available through the Unity Hub, such as some older Unity versions. At this point, you can also install the Unity Editor a second way, which is through the **Unity installer**. You can use the Unity installer to download the previous versions of Unity:

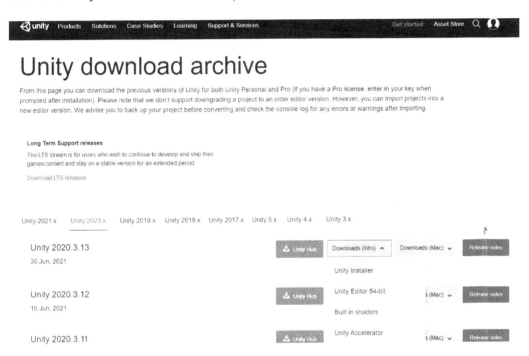

Figure 1.14 – The Unity download archive page

Now, follow these steps to install the Unity Editor through the Unity installer:

1. To download a previous version of Unity, you should access the Unity download archive page at `https://unity3d.com/get-unity/download/archive`.

2. Click on the **Next** button and choose the components of Unity that you want to download and install. The Unity installer should resemble the following screenshot:

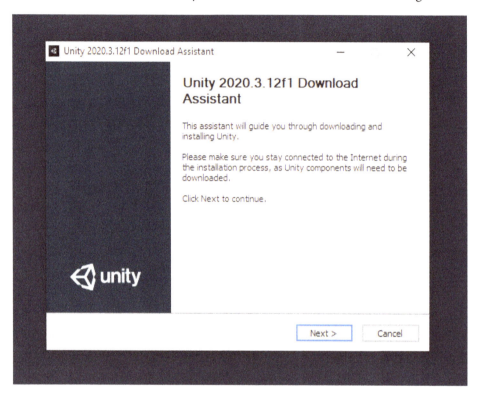

Figure 1.15 – The Unity installer

3. The Unity Editor is selected by default; in order to build games for different platforms, you also need to select the corresponding **build support** components. For example, if you want to build an Android game running on Android devices, you need to download and install the **Android Build Support** component:

Figure 1.16 – Choosing components

4. Click the **Next** button, and then you need to choose **download and install locations**:

Figure 1.17 – Choose Download and Install locations

5.  After specifying where to download and install these files, click the **Next** button to download Unity:

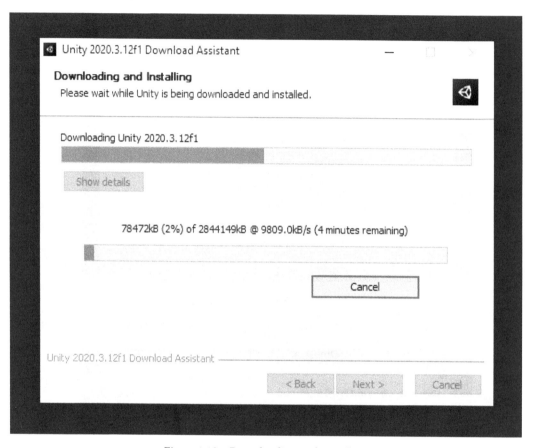

Figure 1.18 – Downloading and Installing

After the download and installation are complete, the Unity Editor icon will appear on your desktop.

# Exploring the Unity Editor

The first thing we need to do is use the Unity Hub to create a new Unity project. As I mentioned in the previous section, we will create a brand-new project by clicking the **New Project** button at the upper-right corner of the **Projects** view.

Figure 1.19 – Creating a new project

As shown in the preceding screenshot, we can choose different Unity Editor versions for this new project, and Unity provides us with some built-in project templates, such as the **2D**, **3D**, **HDRP**, and **URP** templates. You can also download and install more templates from Unity, such as the **VR** template and the **AR** template. In the **PROJECT SETTINGS** section, you can set the name of the project and the location of the project.

Here, we will choose the default **3D** project template and name our project `UnityBook`. Then, click on the **Create project** button. After that, the Unity Editor you previously selected will launch and open a new project for you:

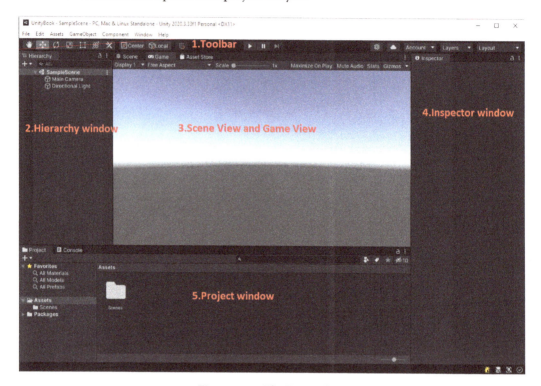

Figure 1.20 – The Unity Editor

As you can see in the preceding screenshot, the Unity Editor layout organizes the most important windows for you. Specifically, the default layout divides the editor interface into five key areas. From top to bottom, they are as follows:

1. **Toolbar**
2. **Hierarchy window**
3. **Scene view and Game view**
4. **Inspector window**
5. **Project window**

Next, I will introduce these UI areas in order.

## The Toolbar

The Toolbar is always at the top of the Unity Editor interface, and it consists of several groups of controls:

Figure 1.21 – The Toolbar

From left to right, the first tool in the Toolbar is the **transform tools set**. The transform tools are used in the **Scene** view and allow you to pan around the Scene and move, rotate, and scale individual GameObjects in the Scene:

Figure 1.22  – The transform tools set

The next tool is the **Gizmo handle position toggles set,** which is used to define the position of any transform tool Gizmo in the **Scene** view:

Figure 1.23 – The Gizmo handle position toggles set

Then, you can find the **Play, Pause, and Step buttons** in the center. You can use these buttons in the **Game** view:

Figure 1.24 – The Play, Pause, and Step buttons

On the right side, let's take a look at the **Unity Plastic SCM** button first, which allows you to access the Plastic SCM version control and source code management tool in the Unity Editor directly. You can click the **cloud** button to open the **Unity Services** window, where you are able to access a lot of cloud services provided by Unity, such as the **Cloud Build** service, the **Analytics** service, and the **Ads** service.

You can also access your Unity account from the **Account** drop-down menu. There are two other drop-down menus on the right, namely **Layers** and **Layout**; you can control which objects in the **Scene** view appear using the **Layers** drop-down menu and change or create a new layout of your Unity Editor using the **Layout** drop-down menu:

Figure 1.25 – The Unity Collaborate and Unity Services buttons and
the Unity Account, Layers, and Layout dropdowns

## The Hierarchy window

The second area is the **Hierarchy** window. As you can see in the following screenshot, the **Hierarchy** window in the Unity Editor displays everything in a **Scene**; the things in the Scene, such as **Main Camera**, **Directional Light**, and the 3D cube, are called **GameObjects**.

We can also organize all the objects in the game world in the **Hierarchy** window:

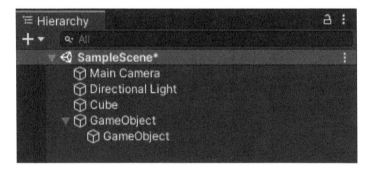

Figure 1.26 – The Hierarchy window

It is very easy to create a new GameObject in a Scene. You only need to right-click on the **Hierarchy** window, and a menu will pop up where you can select the object you want to create:

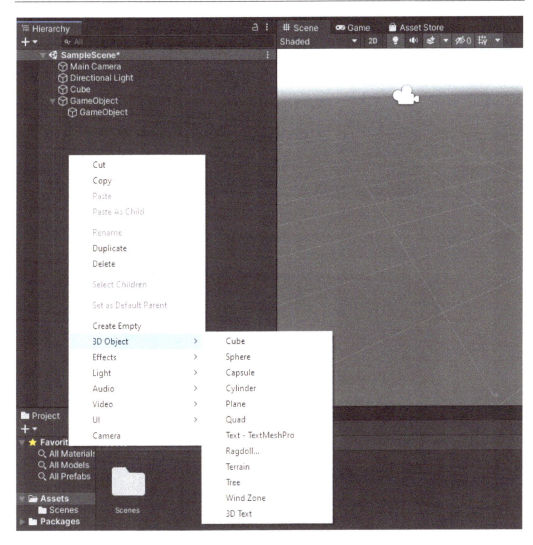

Figure 1.27 – Creating a new GameObject

It is worth noting that Unity uses parent-child hierarchies to organize GameObjects, so you can create one object as a child of another. If you want to create a new GameObject as a child of another GameObject, then you only need to select the parent GameObject first and then right-click to create the child GameObject:

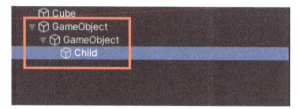

Figure 1.28 – The parent-child hierarchy

Another way to create a parent-child hierarchy is to directly drag an existing GameObject onto the parent GameObject in the **Hierarchy** window:

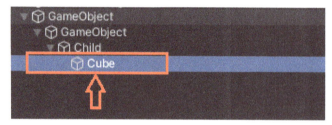

Figure 1.29 – The parent-child hierarchy

As you can see in the preceding screenshot, we dragged the GameObject named **Cube** onto the GameObject named **Child** to create a parent-child hierarchy.

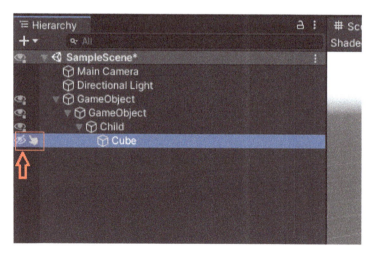

Figure 1.30 – Hiding and showing GameObjects

Another feature of the **Hierarchy** window is that it allows you to hide and show GameObjects in the **Scene** view without changing their visibility in the **Game** view or the final application.

## The Scene view and the Game view

The center of the default Unity Editor layout is the **Scene** view and the **Game** view, which is the most important window in the Unity Editor. The **Scene** view is an interactive view of the game world you are creating:

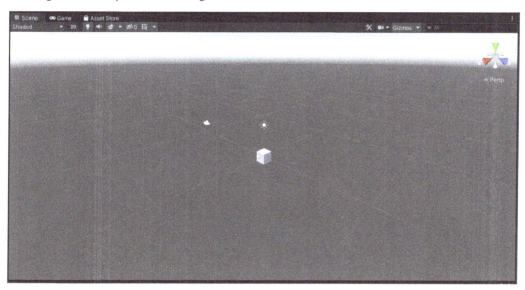

Figure 1.31 – The Scene view

You can use the **Scene** view to manipulate GameObjects and view them from various angles. Also, there are some useful tools available in the **Scene** view, such as the **Scene** Gizmo tool at the upper-right corner of the **Scene** view:

Figure 1.32 – The Scene Gizmo tool

It shows the current orientation of the **Scene** view camera and allows you to quickly modify the angle of the view and projection mode.

If you want to modify the settings of the **Scene** view camera, you can click the **Camera** button next to the **Gizmos** button to open the **Scene Camera** settings window:

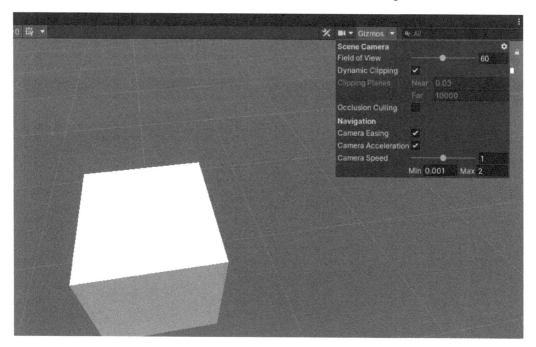

Figure 1.33 – The Scene Camera settings

Here, you can adjust some settings of the **Scene** view camera, such as **Field of View** and **Camera Speed**.

A visual grid is another useful tool that you can use in the **Scene** view to help you align GameObjects by moving them to the nearest grid location:

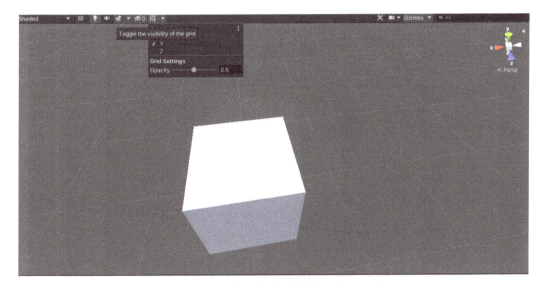

Figure 1.34 – Toggle the visibility of the grid

As you can see in the preceding screenshot, you can also move a GameObject to a grid projected along the **X**, **Y**, or **Z** axes.

The last useful tool in the **Scene** view that I want to introduce is **draw mode** used in the Scene:

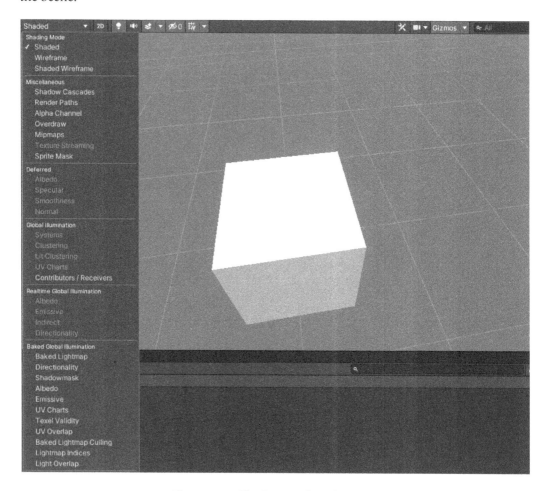

Figure 1.35 – The draw mode in the Scene

This is useful if your project uses the built-in render pipeline of Unity because a different draw mode in the Scene can help you understand and debug the lighting in it.

In the default layout, the **Game** view also appears in the same area as the **Scene** view. You can click the **Game** button to switch to the **Game** view from the **Scene** view:

Figure 1.36 – Click the Game button to switch to the Game view

The **Game** view represents your final published game. The content of the **Game** view is rendered from the camera(s) in your game. In the **Game** view, you cannot modify the viewing angle and projection mode at will, as with the **Scene** view. You need to modify the settings of the camera object to achieve this function:

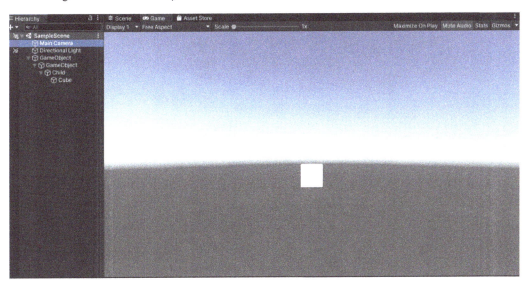

Figure 1.37 – The Game view

You can run your game directly inside the **Game** view by clicking the **Play** button on the Toolbar. It's important to note that in **Play** mode, any changes you make are temporary and will be reset when you exit it; therefore, it is not a good idea to make lots of changes in play mode.

I want to introduce three tools in the **Game** view, namely **Aspect**, **Maximize On Play**, and **Stats**.

The **Aspect** drop-down menu is very useful when you develop games for different screens with different aspect ratios. You can select different values to test how your game looks on these screens, and you even can add custom values by clicking the **plus** button at the bottom of the menu:

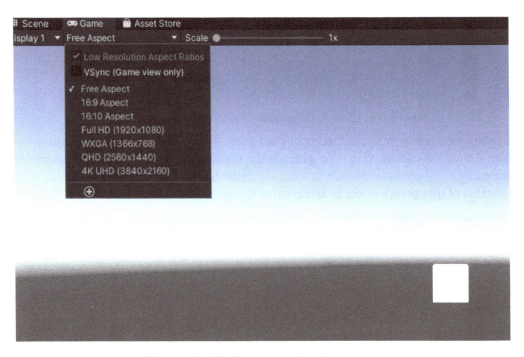

Figure 1.38 – The Free Aspect drop-down menu

The second feature is called **Maximize On Play**, which can maximize the **Game** view for a full-screen preview when you enter play mode:

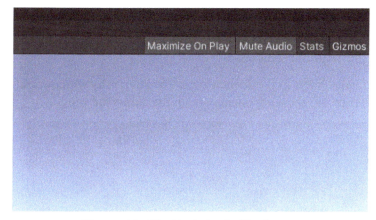

Figure 1.39 – The Maximize On Play button

The third feature is called **Stats**. This feature is useful because it can display the **rendering statistics** about your game's audio and graphics. Therefore, you can use it for monitoring the performance of your game while in play mode:

Figure 1.40 – The Stats window

In the **Scene** view, you can view and adjust the game world you are creating. In the **Game** view, you can see your final game. So, this area is very important in the Editor. Next, let's take a look at the UI area related to a specific GameObject in the Scene.

## The Inspector window

If you want to modify the properties of a GameObject or a component on a GameObject, you need to use the **Inspector** window.

You can select a GameObject in the **Scene** view or the **Hierarchy** window, and then you will see the properties and the components of it in the **Inspector** window:

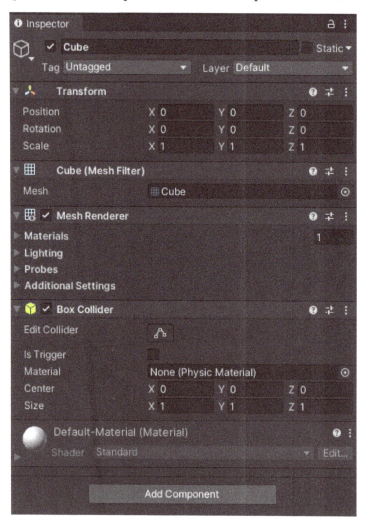

Figure 1.41 – The Inspector window of a GameObject

You can modify these properties directly in the **Inspector** window, which also provides some useful tools that can help you modify your GameObjects.

For example, if you want to copy the values of a component on a GameObject, you can right-click on the component, and then a menu will pop up; from there, you can select the **Copy Component** command:

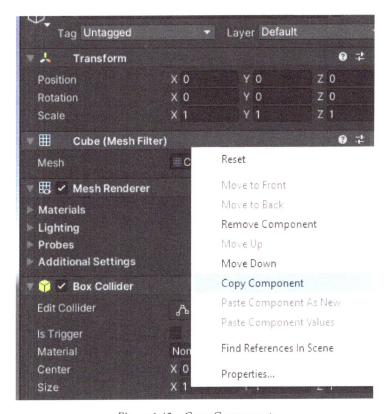

Figure 1.42 – Copy Component

Not only can the GameObjects in the **Scene** view be inspected but also the digital assets in the **Project** window. You can select a digital asset in the **Project** window, and the **Inspector** window will display the settings that control how Unity imports and uses the asset at runtime:

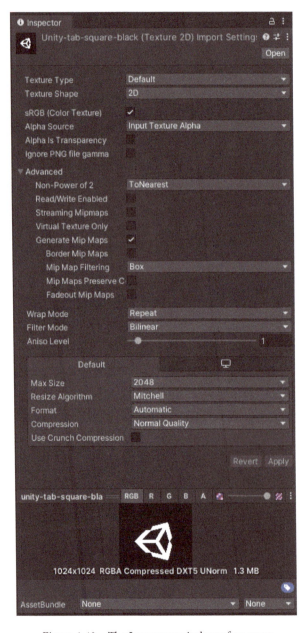

Figure 1.43 – The Inspector window of an asset

In this section, we learned how to view and modify the properties of a GameObject and asset through the **Inspector** window.

## The Project window

The final window I will introduce is the **Project** window. In the **Project** window, you can find all the digital assets of your project. The **Project** window works like a file browser, organizing assets files in folders:

Figure 1.44 – The Project window

The **Project** window is the main way to navigate and find assets in your game. It provides two ways to search assets, by type or label:

Figure 1.45 – Searching assets by type

It is very easy to import external digital assets or create an asset inside the Unity Editor directly. You just need to right-click on the Project window and a menu will pop up where you can create a new asset or import an existing asset:

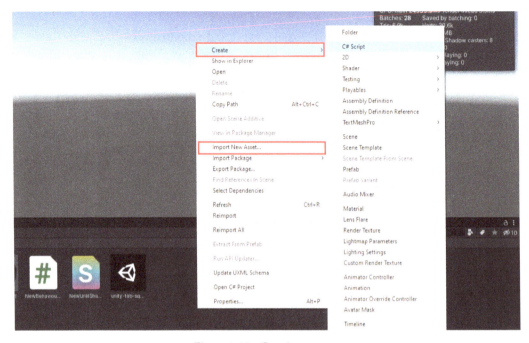

Figure 1.46 – Creating an asset

I hope you now have a good understanding of the Unity Editor by reading this section. Next, I will introduce what a game engine is and what important features Unity provides as a game engine.

# Working with different features in Unity

Nowadays, Unity is no longer just a game engine but also a creative tool widely used in various industries. However, Unity still retains its game engine roots, and it remains one of the most popular game engines. To learn how to use Unity to develop games, you must first understand what features Unity provides for game developers as a game engine.

In fact, almost all game engines provide similar functional modules to Unity to game developers. So, the first question is, what exactly is a game engine?

# What is a game engine?

The term **game engine** is widely used in the game industry, but not everyone knows what this term means, especially new game developers. So, I will explain what a game engine is and, at the same time, introduce the corresponding functions in Unity.

A game engine is not just a computer graphics renderer. Of course, rendering is an important function of a game engine, but the process of creating a game is much more complicated than just rendering.

As a game developer, you need to import different types of **digital assets**, such as 3D models, 2D textures, and audio, and most of these digital assets are not created inside a game engine. Therefore, a game engine should provide the function of managing digital assets. In addition to digital assets, you also need to use **scripts** to add game logic to guide these assets to perform correct behaviors, such as character interactions.

**UI** is another integral part of a game, and even some gameplay is based on UI. Therefore, a good game engine should provide an easy-to-use and powerful UI toolkit to develop user interfaces for games.

You can use other software to develop animation files and import them into a game engine, but in order for animation files to be played and controlled correctly in the game, the game engine needs to provide an **animation** feature.

At the same time, a **physical** effect is a common function in modern games, so a powerful game engine should provide a physical function so that game developers do not need to implement a physical effect from scratch.

There is no doubt that adding **video and audio** to your game can make your game livelier and more interesting. With audio especially, suitable background music and some appropriate sound effects can make your game feel completely different. Even if it is just a prototype, background music and sound effects can make the game more complete and more professional. Therefore, although many people often ignore the functions of video and sound when talking about game engines, I don't think that a game engine without video and audio functions is a good one.

As you can see, there are many features in a game engine for game developers to develop their games. A game engine integrates all aspects of creating a game to create a complete game user experience. So, in game development, you will deal with different functions. For example, you may need to properly manage digital assets and create appropriate digital assets for your game engine to optimize performance at runtime, or you may need to know how to use the scripting function provided by the game engine you are using to develop logic for your game.

As one of the most popular game engines, Unity also provides the aforementioned functions. In the following subsection, I will introduce these functions in Unity.

# Features in Unity

Like other excellent game engines, Unity also provides many functions for game developers. You will be introduced to these functions in the following sections.

## Graphics

The first feature I want to introduce is **graphics** in Unity. You can use Unity's graphics features to create beautiful, optimized graphics on various platforms:

Figure 1.47 – A Unity HDRP template Scene

A render pipeline performs a series of operations that render the contents of a Scene on a screen. There are three render pipelines available in Unity:

- The **Built-in Render Pipeline**, which is the default render pipeline in Unity. You cannot modify this render pipeline.

- The **Universal Render Pipeline** (**URP**), which allows developers to customize and create optimized graphics for different platforms.

- The **High Definition Render Pipeline** (**HDRP**), which focuses on cutting-edge, high-fidelity graphics on high-end platforms.

In addition, you can also create your own render pipeline by using the **Scriptable Render Pipeline API** in Unity. We will introduce it in detail in *Chapter 8*, *The Scriptable Render Pipeline in Unity*.

## Scripting

**Scripting** is another essential feature of Unity. You need scripts to implement the game logic in your games.

The Unity engine is built with native C/C++ internally, but it offers scripting APIs in C#, so you do not have to learn C/C++ to create a game. In the following sections and chapters, you will learn more about the concepts of scripting.

## UI

UI is very important for a game, and Unity offers three different UI solutions for game developers:

- The **Immediate Mode Graphical User Interface (IMGUI)**
- The **Unity UI (uGUI)** package
- The **UI Toolkit**

The IMGUI is a relatively old UI solution in Unity, and it is not recommended for building a runtime UI. The UI Toolkit is the latest UI solution; however, it is still missing some features that you can find in the uGUI package and the IMGUI. The uGUI package is a mature UI solution in Unity, which is widely used in the game industry. We will introduce the uGUI package in detail in *Chapter 3, Developing UI with the Unity UI System*.

## Animation

Animation can make your game more vivid. Unity provides a powerful animation feature called **Mecanim** that allows you to retarget an animation, control the weight of it at runtime, and call events from the animation playback.

We will introduce Unity's animation system in *Chapter 4, Creating Animations with the Unity Animation System*.

## Physics

**Physical simulation** is an indispensable feature in certain types of games, and some gameplays are even based entirely on physical simulation. There are different physics engine implementations in Unity, and you can select one according to your game needs.

We will introduce Unity's physics engine implementations in *Chapter 5, Working with the Unity Physics System*.

## Video and audio

Good background music, sound effects, and video can make your game stand out. This is a feature that cannot be ignored. Unity provides **video and audio** features, allowing your game to play videos on different platforms, and supports real-time mixing and full 3D spatial sound effects.

We will discuss video and audio more in *Chapter 6, Integrating Audio and Video in a Unity Project*.

## Assets

You can import your digital asset files into the Unity Editor, such as 3D models and 2D textures. Unity offers an Asset Import Pipeline to process these imported assets. You can also customize the import settings to control how Unity imports and uses the assets at runtime.

We will introduce **assets management** and **serialization** in *Chapter 10, Serialization System and Assets Management In Unity and Azure*.

We've briefly introduced the functions that a game engine needs to provide and the functions provided by Unity. Next, let's introduce .NET/C# and scripting in Unity.

# .NET/C# and scripting in Unity

Unity is a game engine written in C/C++, but in order to make it easier for game developers to develop games, Unity provides C# (*pronounced C-sharp*) as a scripting programming language to write game logic in Unity. This is because compared with C/C++, C# is easier to learn. In addition, it is a "managed language," which means that it will automatically manage memory for you – allocate release memory, cover memory leaks, and so on.

In this section, we will introduce .NET/C# and scripting in Unity.

## .NET profiles in Unity

The Unity game engine uses **Mono**, an open source ECMA CLI, C#, and .NET implementation, for scripting. You can follow the development of Unity's fork of Mono on GitHub: `https://github.com/Unity-Technologies/mono/tree/unity-master-new-unitychanges`.

Unity provides different .NET profiles. If you are using the legacy version of Unity, which is before Unity 2018, you may find that it provides two API compatibility levels in the **Player** settings panel (**Edit | Project Settings | Player | Other Settings**), which are **.NET 2.0 Subset** and **.NET 2.0**. First of all, if you are using a legacy version of Unity, then I strongly recommend that you update your Unity version. Secondly, both the **.NET 2.0 Subset** and **.NET 2.0** profiles in Unity are closely aligned with the .NET 2.0 profile from Microsoft.

If you are using a modern version of Unity, which is Unity 2019 or later, you will find another two .NET profiles supported by Unity, which are **.NET Standard 2.0** and **.NET 4.x**:

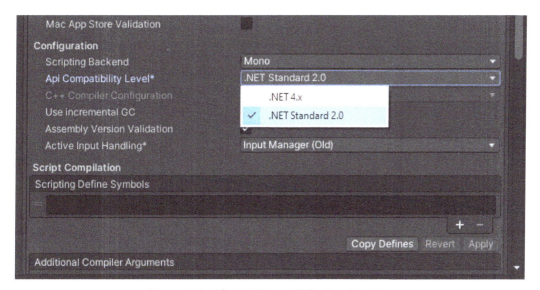

Figure 1.48 – The Api Compatibility Level settings

> **Note**
> The name of the .NET Standard 2.0 profile can be a bit misleading because it is not related to the **.NET 2.0** and **.NET 2.0 Subset** profiles from the legacy versions of Unity.

.NET Standard is a formal specification of the .NET APIs that all .NET platforms have to implement. These .NET platforms include .NET Framework, .NET Core, Xamarin, and Mono. You can find the .NET Standard repository on GitHub: `https://github.com/dotnet/standard`.

On the other hand, the .NET 4.x profile in Unity matches the .NET 4 series (.NET 4.5, .NET 4.6, .NET 4.7, and so on) of profiles from the .NET Framework.

Therefore, it is a good idea to use the .NET Standard 2.0 profile in Unity, and you should choose the .NET 4.x profile only for compatibility reasons.

## Scripting backends in Unity

In addition to the .NET profiles, Unity also provides two different **scripting backends**, which are **Mono** and **IL2CPP** (which stands for **Intermediate Language To C++**):

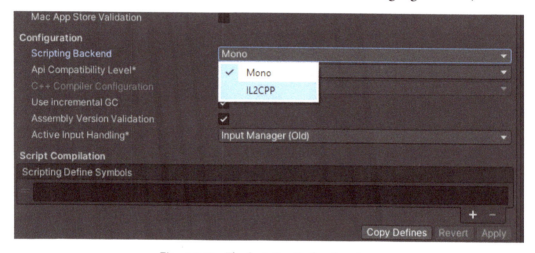

Figure 1.49 – The Scripting Backend settings

You can change the scripting backend of your project in the same settings panel, which can be found by going to **Edit | Project Settings | Player | Other Settings**.

The key difference between the two scripting backends is how they compile your Unity scripting API code (C# code):

- The Mono scripting backend uses **Just-in-Time** (**JIT**) compilation and compiles code on demand at runtime. It will compile your Unity scripting API code to regular .NET DLLs. And, as I mentioned in the previous sections, Unity uses an implementation of the standard Mono runtime for scripting that natively supports C#.

- Alternatively, the IL2CPP scripting backend uses **Ahead-of-Time** (**AOT**) compilation and compiles your entire application before it is run. And it not only compiles your Unity scripting API code into .NET DLL but also converts all managed assemblies into standard C++ code. In addition, the runtime of IL2CPP is developed by Unity, which is an alternative to the Mono runtime:

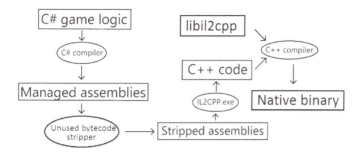

Figure 1.50 – The IL2CPP scripting backend

As shown in *Figure 1.50*, IL2CPP not only compiles C# code into managed assemblies but also further converts assemblies into C++ code, and then compiles the C++ code into the native binary format.

Clearly, IL2CPP takes more time to compile code compared to Mono, so why do we still need IL2CPP?

Well, first, IL2CPP uses AOT compilation, which takes longer to compile, but when you ship the game for a specific platform, the binary files are fully specified, which means that compared to Mono, code generation is greatly improved.

Second, it is worth noting that IL2CPP is the only scripting backend available when building for **iOS** and **WebGL**. In addition to iOS and WebGL, Unity has added support for **Android 64-bit** in Unity 2018.2 to comply with the **Google Play Store** policy, which requires that starting from August 1, 2019, your apps published on Google Play need to support 64-bit architectures:

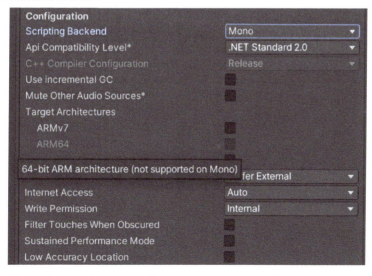

Figure 1.51 – The Android 64-bit ARM architecture is not supported on the Mono scripting backend

And as you can see in the preceding screenshot, the Android 64-bit ARM architecture is not supported on the Mono scripting backend. In this situation, you have to choose the IL2CPP scripting backend.

So, whether we use IL2CPP for better code generation or some specific platforms or architectures, spending more compilation time is still a disadvantage of IL2CPP. So, how should we optimize the compilation time of IL2CPP? I think the following tips will help:

- Don't delete the previous `build` folder, and build your project with the IL2CPP scripting backend at the same location as the folder. This is because we can use incremental building, which means the C++ compiler only recompiles files that have changed since the last build.

- Store your project and target build folder on a **Solid-State Drive** (**SSD**). This is because when IL2CPP is selected, the compilation process will convert the IL code into C++ and compile it, which involves a lot of read/write operations. A faster storage device will speed up this process.

- Disable anti-malware software before building the project. Of course, this depends on your security strategy.

Well, I hope that by reading this section, you now have a general understanding of Unity's scripting system, such as the .NET profiles in Unity, the two scripting backends, and some optimization tips for IL2CPP.

In the next section, you will learn how to set up your development environment and use the widely used Visual Studio to develop games in Unity.

# Building Unity games with Visual Studio

Before you start writing any code, it is important to choose suitable development tools. Microsoft's **Visual Studio** is not only a widely used IDE but also the development environment that is installed by default when you install Unity on Windows or macOS:

Figure 1.52 – Visual Studio Installer

While installing Visual Studio, **Visual Studio Tools for Unity** will also be installed. It is a free extension that provides support for writing and debugging C# in Unity.

If you do not install Visual Studio through the Unity Hub, please make sure you installed this extension. You can check it in the **Visual Studio Installer**:

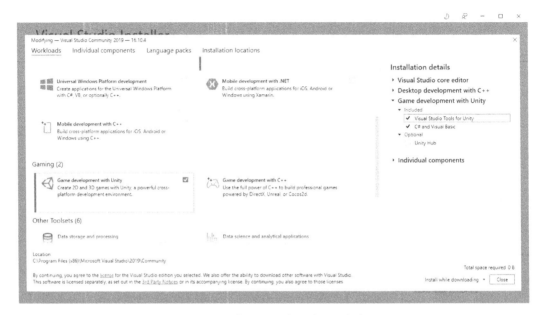

Figure 1.53 – Installing Visual Studio Tools for Unity

After installing the Unity Editor and Visual Studio Community 2019, you can check the **External Script Editor** settings in the **Preferences** window of the Unity Editor:

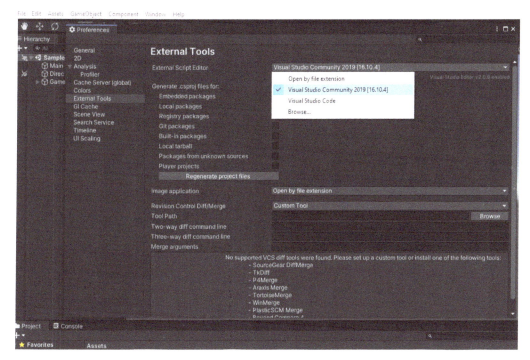

Figure 1.54 – The External Script Editor settings

In addition, you can also select other script editors by modifying this setting, such as **Visual Studio Code** and **JetBrains Rider**.

Then, we can create a new C# script file named `NewBehaviourScript` in the Unity Editor and double-click to open it in Visual Studio:

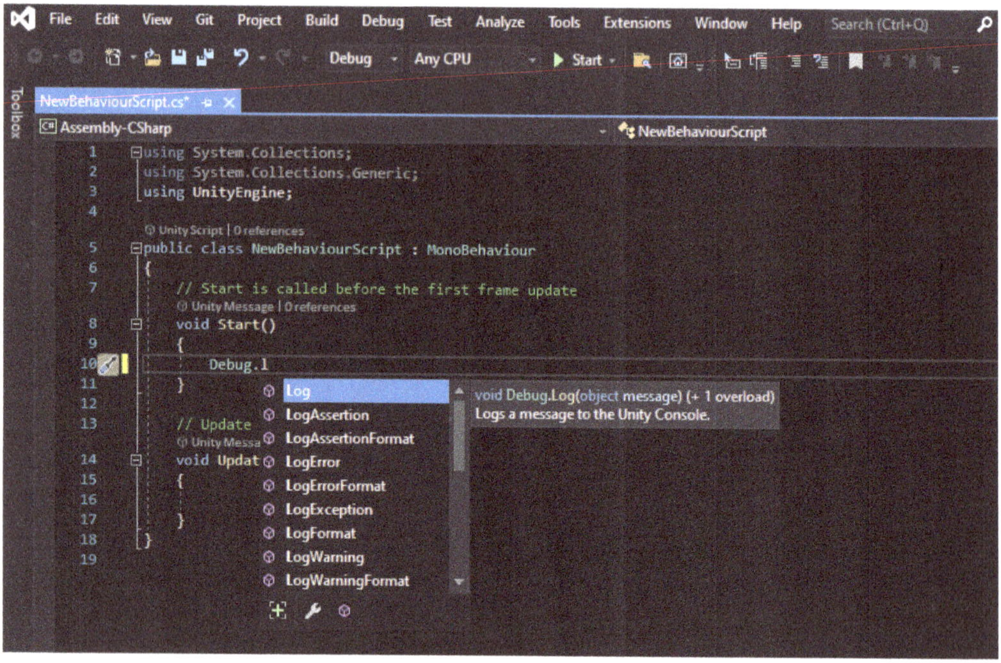

Figure 1.55 – IntelliSense for the Unity APIs

As you can see in the preceding screenshot, there are two built-in methods in the script file by default, namely `Start` and `Update`. Visual Studio supports **IntelliSense** for Unity APIs, so we can write code quickly:

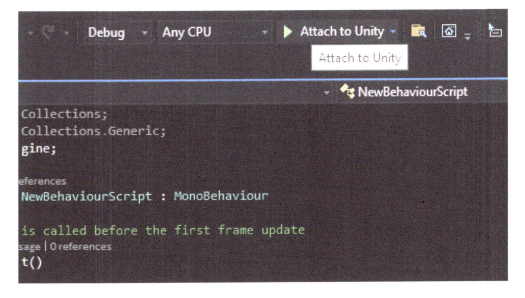

```
NewBehaviourScript.cs
Assembly-CSharp                                              NewBehaviourScript
     1    using System.Collections;
     2    using System.Collections.Generic;
     3    using UnityEngine;
     4
          Unity Script | 0 references
     5    public class NewBehaviourScript : MonoBehaviour
     6    {
     7        // Start is called before the first frame update
          Unity Message | 0 references
          void Start()
          {
    10        Debug.Log("Hello World");
    Location: NewBehaviourScript.cs, line 10 character 9 ('NewBehaviourScript.Start()')
    12
    13        // Update is called once per frame
          Unity Message | 0 references
    14    void Update()
    15    {
    16
    17    }
    18    }
    19
```

Figure 1.56 – Debugging your code

It is also very easy to debug your code in Visual Studio. In the preceding screenshot, I set a breakpoint inside the Start method and clicked the **Attach to Unity** button in Visual Studio:

```
         Debug    Any CPU          ▶ Attach to Unity

                                       Attach to Unity

                                              NewBehaviourScript
Collections;
Collections.Generic;
gine;

eferences
NewBehaviourScript : MonoBehaviour

is called before the first frame update
sage | 0 references
t()
```

Figure 1.57 – Clicking the Attach to Unity button

In order to run this code, I attach this script to a GameObject in the Scene and click on the **Play** button in the Unity Editor to run the game in the **Game** view.

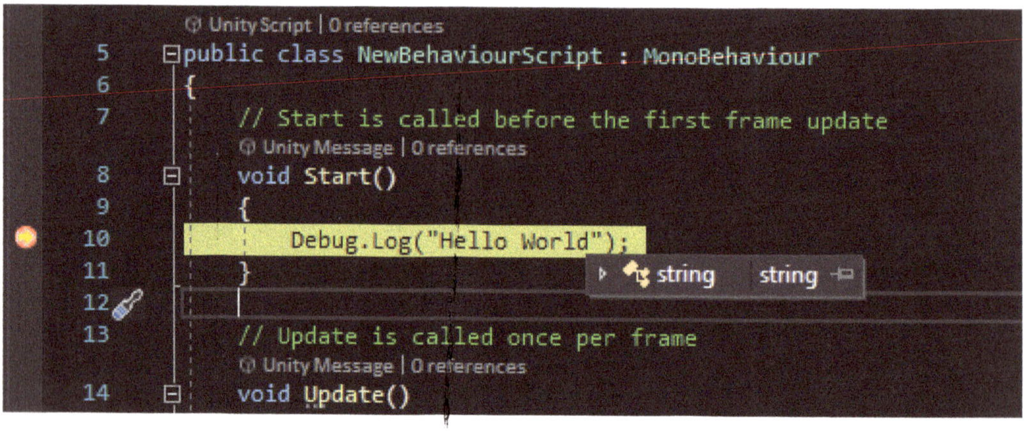

Figure 1.58 – The debugger stopping at the breakpoint

Then, the debugger will stop at the breakpoint, and you can look at the current state of the game.

## Summary

In this chapter, we started by choosing the Unity release and subscription plan that suits your needs. Then, you learned how to install and manage the Unity Editor by using the Unity Hub and explored the five important areas of the Unity Editor – the Toolbar, the **Hierarchy** window, the **Scene** view and the **Game** view, the **Inspector** window, and the **Project** window. You were then introduced to the Unity Editor toolbars and the windows provided by Unity. We also discussed what a game engine is and explored the different features provided by Unity for developers to develop games. We then introduced the .NET profiles in Unity and the scripting backends offered by Unity; you should now know the difference between the Mono scripting backend and the IL2CPP scripting backend. Finally, we demonstrated how to set up Visual Studio for Unity Editor to write code.

In the next chapter, we will start with a detailed introduction to the basic concepts of scripting in Unity, such as GameObjects, components, and some special, important components such as **Transform**. We will also introduce you to the life cycle of a script instance. Then, we will discuss how to create objects from scripts and how to access GameObjects or components through C# code. Some best practices for scripting in Unity will also be introduced. Finally, we will introduce **packages** and the **Package Manager** in Unity.

# 2
# Scripting Concepts in Unity

In the previous chapter, we discussed **scripting** in Unity at a high level. In this chapter, we will introduce this topic in detail. We already know that Unity is internally written in C/C++, but it provides many C# APIs for game developers and allows us to implement game logic in C#. This means that not only can we write our own classes but also many built-in classes are available to us. So, before creating our own C# class, let's learn a bit about Unity's built-in classes first. The life cycle of Unity scripts is another important topic because we need to use different event functions provided by Unity to implement game logic. Then, we will introduce how to create a script in the Unity Editor and use it as a component.

We will cover the following key topics in this chapter:

- Understanding the concepts of scripting in Unity
- The life cycle of a script instance
- Creating a script and using it as a component
- Packages and the Unity Package Manager

# Technical requirements

You can find complete code examples on GitHub under the following repository: `https://github.com/PacktPublishing/Game-Development-with-Unity-for-.NET-Developers`.

Before starting, I want to mention that the following software will be used in this chapter:

- Visual Studio 2019

- Visual Studio Tools for Unity

- Unity 2020.3+

# Understanding the concepts of scripting in Unity

Let's get started with understanding the concepts of scripting in Unity. We know that Unity is not an open source engine; except for enterprise users and users who subscribe to the Pro plan, no one else can access Unity's source code. However, Unity's C# API is open source. Because the C# API is just a wrapper, it does not include the internal logic of the engine. But Unity's open source C# API is also a good reference for us to understand script programming in Unity. You can access it on GitHub: `https://github.com/Unity-Technologies/UnityCsReference`.

## GameObject-components architecture

First of all, I want to let you know that Unity is a **component-based system**. So, the two terms you often hear in Unity game development are **GameObject** and **component**. A GameObject is nothing more than a container for components. It represents an object in the game world, but it does not have any function itself. On the other hand, a component implements the real functionality and can be attached to a GameObject to provide the function for a specific object.

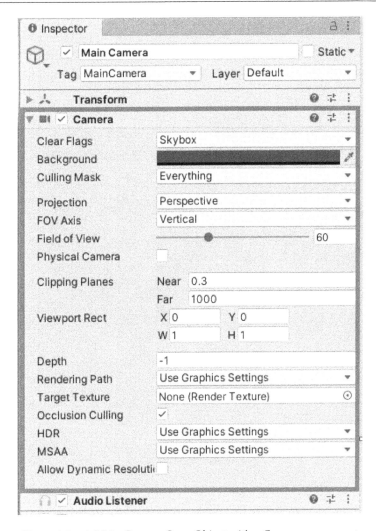

Figure 2.1 – A Main Camera GameObject with a Camera component

For example, you can find a **Main Camera** object in a default Scene in the Unity Editor. It is created by attaching a **Camera** component to a GameObject.

You can enable or disable a set of functions attached to this object by enabling or disabling the GameObject, or enabling or disabling a specific component to enable or disable a specific function.

This way is different from traditional object-oriented programming. It is a bit like LEGO blocks; when an object needs a certain type of function, you only need to add related components to it.

# Common classes in Unity

Unity provides a lot of built-in C# classes, so I will introduce some of the classes we often use in Unity development.

## The MonoBehaviour class

In Unity development, the class you most often encounter is the `MonoBehaviour` class. This is because it is the base class from which every Unity script is derived.

Let's create a new script file in the Unity Editor and name it `ChapterTwo.cs`.

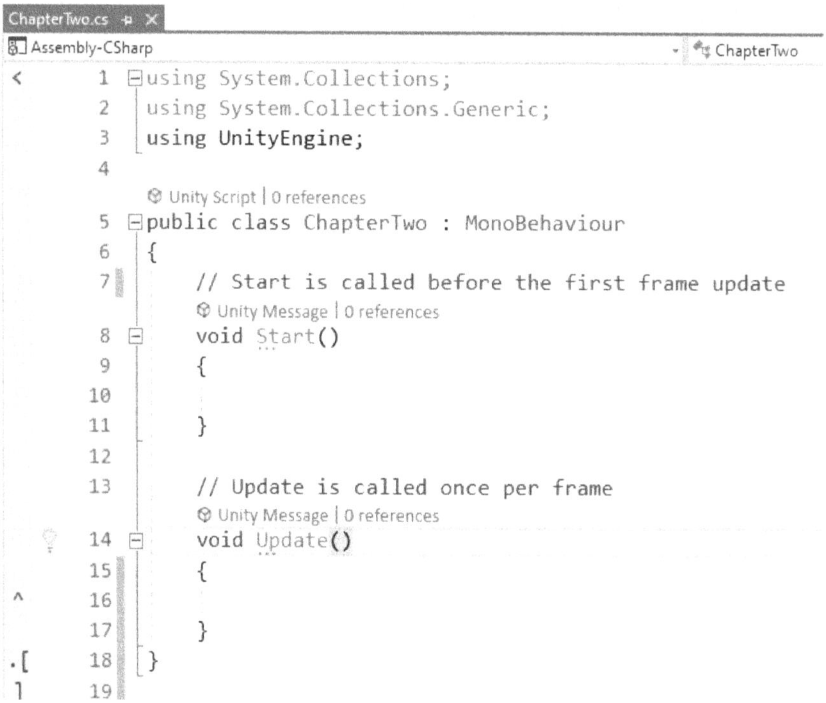
ChapterTwo.cs
Assembly-CSharp
ChapterTwo
using System.Collections;
using System.Collections.Generic;
using UnityEngine;
Unity Script | 0 references
public class ChapterTwo : MonoBehaviour
{
// Start is called before the first frame update
Unity Message | 0 references
void Start()
{
}
// Update is called once per frame
Unity Message | 0 references
void Update()
{
}
}

Figure 2.2 – A default script

Then, we open it in Visual Studio by double-clicking it. You can see our new `ChapterTwo` class derives from `MonoBehaviour`.

So, why is `MonoBehaviour` so important? Because it provides a framework for game developers to interact with the Unity engine. For example, if you want to attach a script to a GameObject in the Scene, the class must inherit from the `MonoBehaviour` class; otherwise, the script cannot be added to the GameObject. When you try to attach a class that does not inherit from `MonoBehaviour` to a GameObject, the Unity Editor will pop up the following error message:

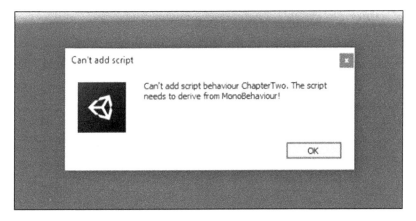

Figure 2.3 – Can't add script error

Without the `MonoBehaviour` class, your code will not be able to access Unity's built-in methods and events, such as the `Start` and `Update` functions that will be created by default in each of your new script files.

`MonoBehaviour` is the most important class in Unity. `Start` and `Update` are the most common built-in functions in Unity. Every time you create a new script file, they will appear in this new file. But if you want to modify the template that creates the script, it is also possible; you only need to modify the script template stored here:

- **Windows**: `%EDITOR_PATH%\Data\Resources\ScriptTemplates`

- **Mac**: `%EDITOR_PATH%/Data/Resources/ScriptTemplates`

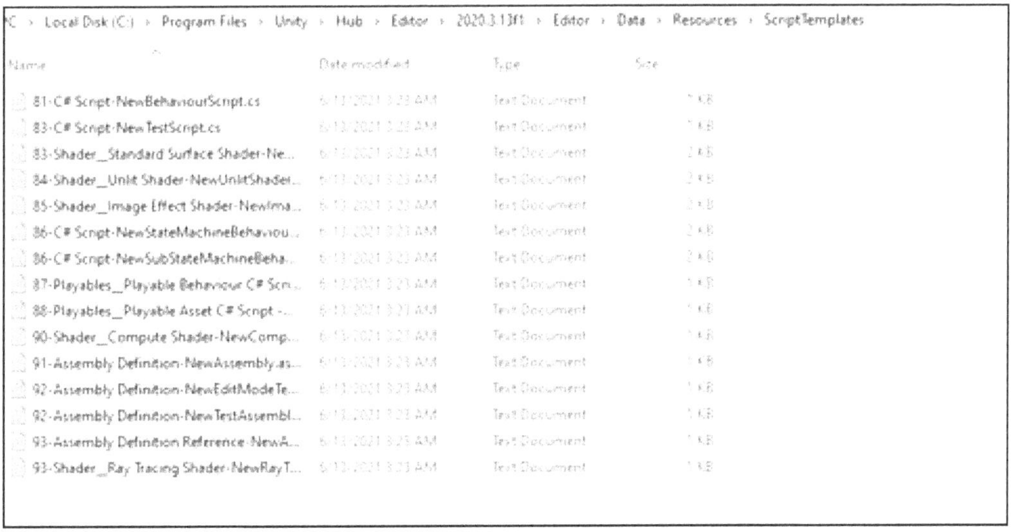

Figure 2.4 – The ScriptTemplates folder

## The GameObject class

We already know that objects in a Scene are called **GameObjects**. In order to access a GameObject from your script, Unity provides the GameObject class to represent it.

When you create a new empty GameObject in a Scene, you'll find this new GameObject contains a name, a tag, a layer, and a **Transform** component.

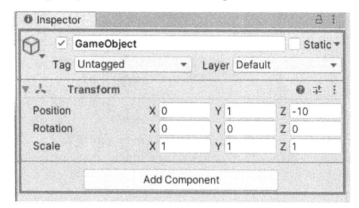

Figure 2.5 – A GameObject

You can modify whether it is a *static* object from the **Inspector** window as well. If the GameObject does not move during runtime, you should check the **Static** property checkbox in the upper-right corner of the **Inspector** window. This is because many systems in Unity can precalculate information about static GameObjects in the Editor to improve performance at runtime.

As we mentioned earlier, a GameObject is a container that can contain various components. Therefore, in scripting, the GameObject class mainly provides a set of methods for managing components, such as the AddComponent method to add a new component to a GameObject and GetComponent to access a component attached to a GameObject.

Let's create a built-in 3D Cube object in the Scene and look at the **Inspector** window of this cube.

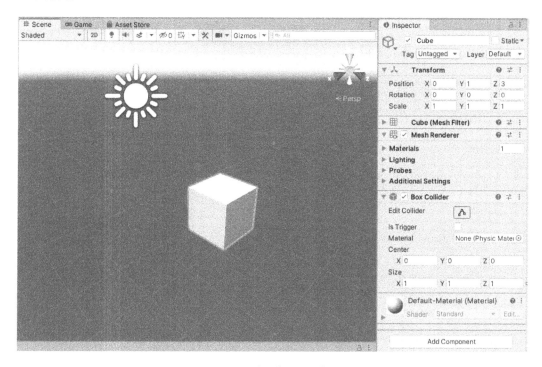

Figure 2.6 – Cube object in the Scene

As you can see in the previous screenshot, this GameObject is called **Cube**, and there are four components that are attached to this Cube object, namely, **Transform**, **Cube (Mesh Filter)**, **Mesh Renderer**, and **Box Collider**. These components provide rendering and physical simulation functions for this object. Therefore, a GameObject is just a container for components, and specific functions come from specific components. You can add a new component by clicking the **Add Component** button in the **Inspector** window, or you can add a component at runtime via code.

In addition to components, the `GameObject` class also offers a collection of methods to find other GameObjects, send messages between GameObjects, or create and destroy a GameObject. For example, you can find GameObjects by using the `GameObject.Find` method, to find a GameObject by name and return it, or the `GameObject.FindWithTag` method, to find a GameObject by tag. You can also use the `Instantiate` method to create a new instance of a `GameObject`, and the `Destroy` method to destroy an instance of a `GameObject`.

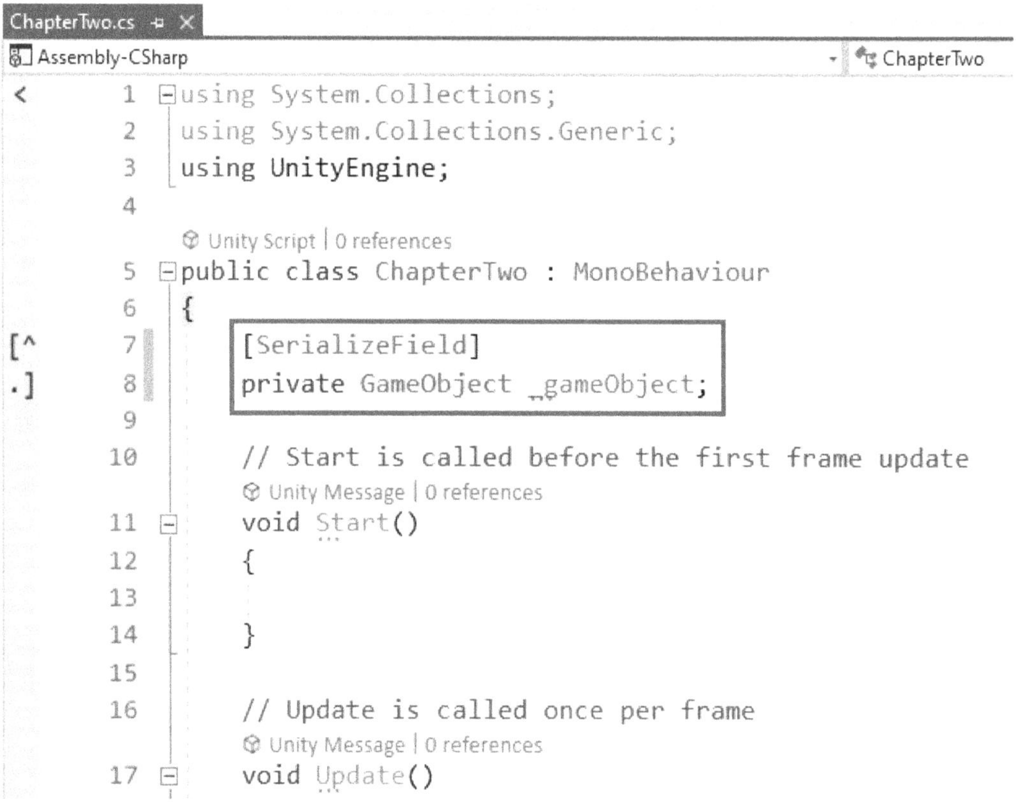
ChapterTwo.cs
Assembly-CSharp
ChapterTwo
using System.Collections;
using System.Collections.Generic;
using UnityEngine;
Unity Script | 0 references
public class ChapterTwo : MonoBehaviour
{
[SerializeField]
private GameObject _gameObject;
// Start is called before the first frame update
Unity Message | 0 references
void Start()
{
}
// Update is called once per frame
Unity Message | 0 references
void Update()

Figure 2.7 – [SerializeField] attribute in a class

It is worth noting that using certain methods to dynamically find a specific GameObject instance at runtime will bring additional overhead, so the easiest way to obtain a reference to another GameObject instance is to declare a public `GameObject` field or use the `[SerializeField]` attribute and declare a private field to maintain the encapsulation of the class. As shown in the previous screenshot, I prefer the second way. We will cover serialization in Unity further in later chapters.

Now, you will find the `GameObject` field is visible in the **Inspector**. You can just drag a GameObject from the Scene or **Hierarchy** panel onto this variable to assign it.

Figure 2.8 – A GameObject variable

## The Transform class

An instance of the `Transform` class will be created automatically when you create a new GameObject in the Scene. This is because every GameObject in the Scene has position, rotation, and scale properties and the `Transform` class is used to store and manipulate the position, rotation, and scale of the GameObject in Unity. So, it is impossible to create a GameObject in Unity without a `Transform` component and you cannot remove it from a GameObject as well.

You can move, rotate, or scale a GameObject by modifying the properties of the `Transform` component in the Unity Editor directly, or you can modify them at runtime by accessing the instance of the `Transform` class.

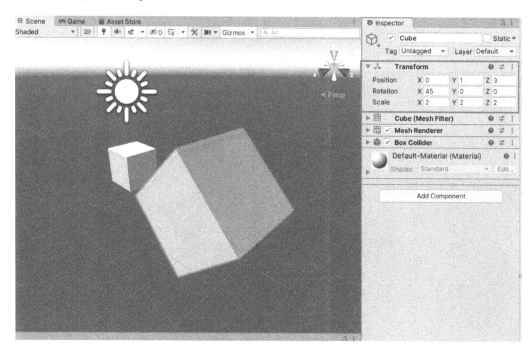

Figure 2.9 – The Transform component

# Prefabs in Unity

**Prefab** is an important concept in Unity. Game developers can use a Prefab to save GameObjects, components, and properties to reuse these resources when developing games with Unity. When instantiating a Prefab, the Prefab acts as a resource template. Next, let's see how to create a new Prefab in Unity.

# How to create a Prefab

First, let's talk about how to create a Prefab. Take a "barbell" as an example. It consists of a **Cube** and two **Sphere** objects.

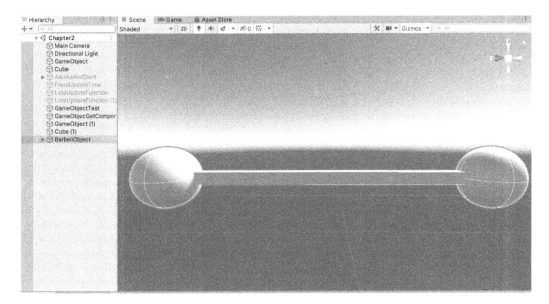

Figure 2.10 – How to create a Prefab

We can create a Prefab of this barbell object by following these steps:

1.  First, find the target GameObject named **BarbellObject** in the **Hierarchy** panel, as shown in *Figure 2.10*.

2.  Drag the target GameObject from the **Hierarchy** panel to the **Project** panel to create a Prefab of it. The newly created Prefab file is shown as a blue cube icon in the Unity Editor.

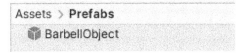

Figure 2.11 – A Prefab file

3.  At this point, if we look at the **Hierarchy** panel again, we can find that the name text of **BarbellObject** and the small cube icon on the left of it have changed from white to blue because it is now a Prefab instance. In this way, we can distinguish whether an object is a Prefab instance on the **Hierarchy** panel.

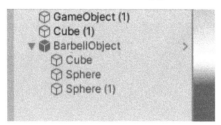

Figure 2.12 – A Prefab instance

As you can see, creating a new Prefab is not complicated. Next, let's explore how to edit an already-created Prefab.

## How to edit a Prefab

Unity provides two ways for developers to edit a Prefab, as follows:

- The first way is to edit a Prefab in Prefab Mode.
- The second way is to edit a Prefab via its instance.

Let's start with Prefab Mode first.

Prefab Mode is a mode specially designed to support editing Prefabs individually. Prefab Mode allows the content of the Prefab to be viewed and edited in a separate Scene. You can enter Prefab Mode in the following ways:

1.  The first way is to click the arrow button of the Prefab instance in the **Hierarchy** view.

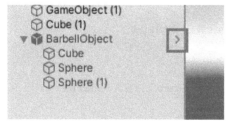

Figure 2.13 – Entering Prefab Mode

2.  The second way is to select the Prefab file in the **Project** panel. A button with the words **Open Prefab** will be displayed in the **Inspector** panel. Click it to enter Prefab Mode.

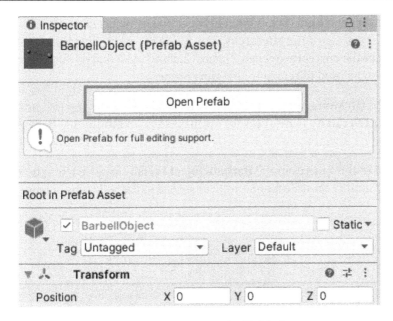

Figure 2.14 – Entering Prefab Mode

3. You can also double-click the Prefab file in the **Project** panel to enter Prefab Mode.

   After entering Prefab Mode, you can modify the Prefab here and you can also find that a navigation bar will be displayed above the Scene view, as shown in the following screenshot:

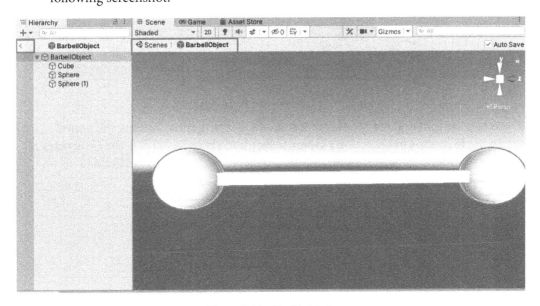

Figure 2.15 – Prefab Mode

4.  Use the navigation buttons to switch between the game Scene and Prefab Mode. In addition, at the top of the **Hierarchy** view, a title bar is also displayed that displays the name of the currently opened Prefab. Clicking the left arrow button in the title bar can also be used to return to the game Scene.

In addition to Prefab Mode, we can also modify a Prefab by modifying the instance of the Prefab in the **Hierarchy** panel. Let's follow these steps to modify the **BarbellObject** Prefab:

1.  Select one of the spheres in the **BarbellObject** Prefab instance from the **Hierarchy** panel and modify its scale from 1 to 2, as shown in the following screenshot:

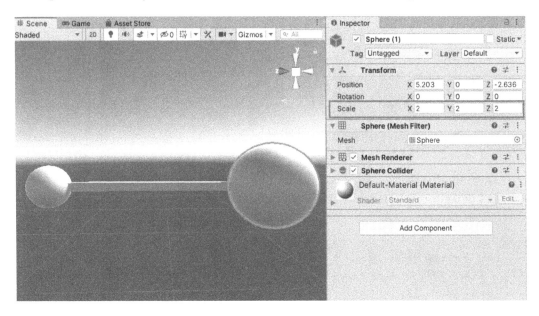

Figure 2.16 – Modifying the Prefab instance

2.  When the root node of the Prefab instance is selected, three buttons will appear in the **Inspector** panel, namely **Open**, **Select**, and **Overrides**. Click the **Overrides** drop-down window to view all modified data items, such as properties and components.

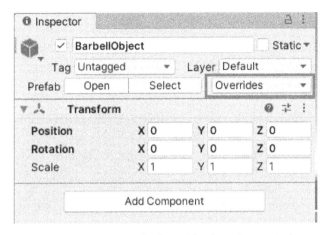

Figure 2.17 – Opening the Overrides drop-down window

3.  In this drop-down window, we can discard or apply all modifications. Here, we should click the **Apply All** button to apply this modification to the Prefab.

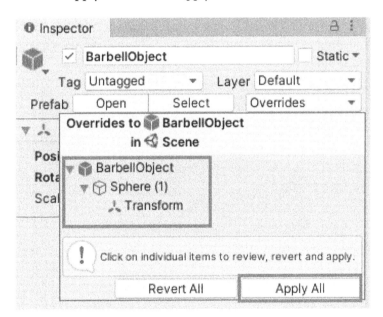

Figure 2.18 – Clicking the Apply All button

Through the two methods described previously, we can easily modify a Prefab in Unity. Next, let's talk about how to instantiate a Prefab at runtime using C# code.

## How to instantiate a Prefab

In Unity development, we can use the `Instantiate` method to create an instance of a Prefab at runtime. There are several variants of the `Instantiate` method. The commonly used Instantiate method variants are shown here:

```
public static Object Instantiate(Object original, Vector3
    position, Quaternion rotation);
public static Object Instantiate(Object original, Vector3
    position, Quaternion rotation, Transform parent);
```

We instantiate a Prefab using these two variants of the `Instantiate` method, both of which can be used to specify the instance's position and orientation, and the latter can also specify the instance's parent.

Let's use the following example to learn how to instantiate a Prefab by calling the `Instantiate` method:

1.  First, let's create a new script called `TestInstantiatePrefab`. In this script, we will assign a reference to a Prefab in the script and call `Instantiate` to create a new instance of this Prefab and assign a parent to the new object:

    ```
    using UnityEngine;
    public class TestInstantiatePrefab : MonoBehaviour
    {
        [SerializeField]
        private GameObject _prefab;
        [SerializeField]
        private Transform _parent;
        private GameObject _instance;

        private void Start()
        {
            var position = new Vector3(0f, 0f, 0f);
            var rotation = Quaternion.identity;

            _instance = Instantiate(_prefab, position,
                rotation, _parent);
        }
    }
    ```

2. Then, we also need to attach this script to a GameObject in the Scene, assign the Prefab to the `_prefab` field of this script, and assign this GameObject as the parent object of the instance of the Prefab that will be created later, as shown in the following screenshot:

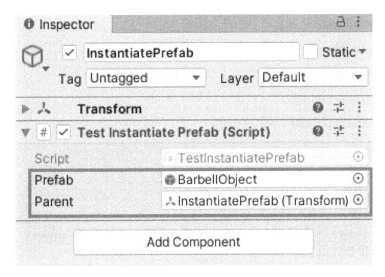

Figure 2.19 – Setting up the component and properties

3. Click the **Play** button of the Unity Editor to run the script. In the following screenshot, you can see that a new instance of the **BarbellObject** Prefab is created and named **BarbellObject(Clone)**, which means that it is an instance of a Prefab, and it is also a child of the `InstantiatePrefab` object:

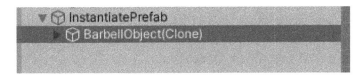

Figure 2.20 – Creating a new instance of a Prefab

In this section, we discussed an important concept in Unity, namely, Prefabs. By reading this section, you should understand what a Prefab is, how to create a Prefab, how to edit a Prefab, and how to instantiate a Prefab at runtime using C# code.

# Special folders in Unity

In addition to the commonly used classes and concepts introduced in the previous sections, there are also some special folders for different purposes in Unity. Some of these folders are related to scripting in Unity. They are as follows:

- The `Assets` folder
- The `Editor` folder
- The `Plugins` folder
- The `Resources` folder
- The `StreamingAssets` folder

Let's look at each one.

## The Assets folder

When a Unity project is created, an `Assets` folder is created to store various resources from models and textures to script files that will be used in this Unity project. This is also the folder you will mainly use when developing a Unity project.

## The Editor folder

The `Editor` folder is used to store script files for the Editor. For example, you can add more functionality to the default Unity Editor by creating some Editor scripts in an Editor folder. Unity compiles the scripts in four independent stages according to the location of the script files. At each stage, Unity will create a separate C# project file (`.csproj`) for this stage. The scripts in the Editor folder will not be available at runtime. If the Editor folder is located in a `Plugins` folder, then a CSharp project file named `Assembly-CSharp-Editor-firstpass` will be created; otherwise, a CSharp project file named `Assembly-CSharp-Editor` will be created.

## The Plugins folder

You should put the plugins or the code that needs to be compiled first in the `Plugins` folder, and Unity will compile the code in this folder first. A CSharp project file named `Assembly-CSharp-firstpass` will be created for the scripts located in this folder. Unity will create a CSharp project file named `Assembly-CSharp` for all other scripts that are in the `Assets` folder but not in the `Plugins` folder and the `Editor` folder.

Figure 2.21 – CSharp project file for different stages

There are some other special folders, such as the `Resources` folder and the `StreamingAssets` folder. We will introduce them in later chapters.

In this section, we discussed the GameObject components architecture of Unity and introduced some of the most common built-in classes in Unity and some special folders related to scripts in Unity. Next, we will learn about another important topic related to scripting in Unity, that is, the life cycle of a script instance.

# The life cycle of a script instance

In the previous section, we introduced basic concepts of scripting in Unity. Now, we will explain another important topic regarding scripting in Unity: the life cycle of a script instance.

We already know that the Unity C# API does not include the internal logic of the engine and the event functions on the script are triggered by the engine's C/C++ code. Therefore, in order to use the Unity engine correctly, it is very important to understand the order of execution for event functions and the life cycle of a C# script in Unity.

We can divide the Unity event functions into the following categories depending on their purpose :

- Initialization
- Update
- Rendering

Let's discuss them next.

## Initialization

If you are familiar with developing .NET applications, you may be surprised by script initialization in Unity because Unity scripts do not use constructors for initialization. Instead, Unity provides some engine event functions to initialize a script instance.

Actually, we have already seen a Unity event function for initialization purposes. Yes, it is the Start() function created by default when creating a new script in Unity.

However, the Start() function is not the first event function that will be triggered when a new instance of the script is created. When a Scene starts, the Awake() event function in each object in the Scene is always called before any Start() functions. Except for the fact that Awake() will be called first, Start() and Awake() work similarly. Both of them are called once during initialization. Now, you may have a question: since we already have the Start function, why do we still need the Awake function?

This is because the Awake function is useful for separating initialization. For example, it is a good idea to use Awake to initialize an object's own references and variables before the game starts. This means that you should not access references to other objects in the Awake function, but should use Start to pass reference information of different objects.

You may be confused, so let me show you some code. Let's consider a case where there are two classes, namely, AwakeAndStartA and AwakeAndStartB. In the first class, there is a List<int> variable and a List<int> property, and the List variable is set up in the Awake function of AwakeAndStartA:

```
public class AwakeAndStartA : MonoBehaviour
{
    private List<int> _listRef;
    public List<int> ListRef => _listRef;


    private void Awake()
    {
        _listRef = new List<int>();
    }


}
```

Now, we get the second class:

```
public class AwakeAndStartB : MonoBehaviour
{
    private void Awake()
    {
        var comp =
        GameObject.Find("A").GetComponent<AwakeAndStartA>();
```

```
        Debug.Log($"comp is null > {comp is null}");
        Debug.Log(comp.ListRef.Count);
    }


}
```

The `AwakeAndStartB` class tries to get the reference of the `AwakeAndStartA` class and also access the `ListRef` property of `AwakeAndStartA` in its Awake function.

If we run the code, we will get the following output; that is, object B can access object A, but not the variables or properties of object A in the Awake function. This is because we should not assume that a reference set up by one GameObject's Awake will be usable in another GameObject's Awake.

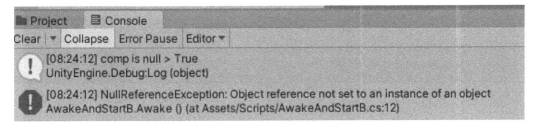

Figure 2.22 – Null reference exceptions

Therefore, in order to use `ListRef` in object B, we can get the reference in the Start function. Let's move the code for printing the number of elements contained in the list from the Awake function to the Start function:

```
public class AwakeAndStartB : MonoBehaviour
{
    private void Start()
    {
        var comp =
          GameObject.Find("A").GetComponent<AwakeAndStartA>();
        Debug.Log($"comp is null > {comp is null}");
        Debug.Log(comp.ListRef.Count);
    }


}
```

This time, the code will print the correct number, as shown in *Figure 2.23*:

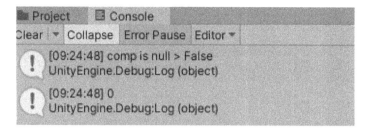

Figure 2.23 – The number of elements contained in the list is 0

Another difference between the Start and Awake functions is that if a script component is not enabled in the Scene, its Start function will not be called but the Awake function will always be called, as you can see in the following screenshot:

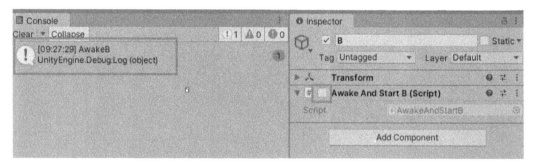

Figure 2.24 – The Awake function is always called

There is a third event function for initialization, which is the OnEnable function. If the script component is enabled in the Scene, then this function will be called after the Awake function and before the Start function. However, there is a big difference between the OnEnable function and the Awake/Start function; that is, the OnEnable function could be called multiple times. This function is called when the component becomes enabled.

# Update

For a game, Update is a very important function because the gameplay logic is driven by Update. Unity offers three different Update functions for different purposes. They are as follows:

- FixedUpdate
- Update
- LateUpdate

`FixedUpdate` is used for physics simulations. So, you should not use this function if your game does not include physics simulations. The `FixedUpdate` function is called at every fixed framerate frame and it could be called multiple times in a single frame. This is because it is very important to ensure a fixed incremental time in a physical simulation. Now, you may be confused again. Let me explain it to you.

By default, the physics simulation needs to be updated every 0.02 seconds. You can change this value in **Project Settings | Time | Fixed Timestep**.

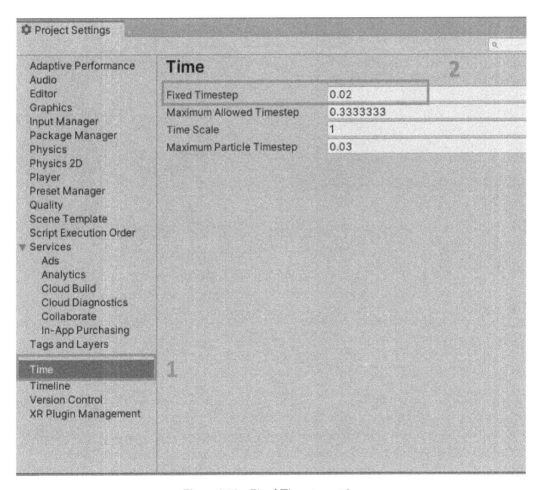

Figure 2.25 – Fixed Timestep setting

Let's consider a case where the framerate of the game itself is low, for example, 25 FPS. This means that the game will take 0.04 seconds to update one frame. Then, the question is, how do we ensure a fixed incremental time for physics simulations?

The answer is not complicated. Unity only needs to call `FixedUpdate` twice in each frame before calling the `Update` function, and in this example, `FixedUpdate` is called every 0.02 seconds. The following screenshot shows the result:

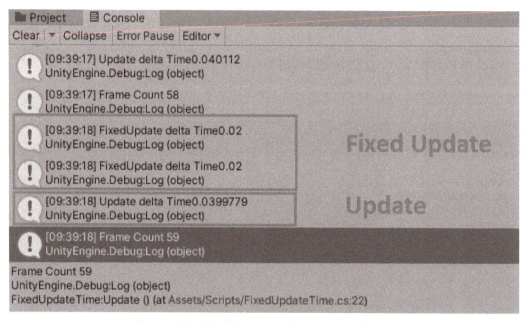

Figure 2.26 – FixedUpdate is called twice in a frame

Therefore, only use the `FixedUpdate` function when using a physics simulation in your project. If your project does not include a physics simulation, then you should not use it.

The `Update` function is another function that will be created by default when you are creating a new script. It is the most commonly used and most important function to implement any type of game logic in Unity. If the script component is enabled in a Scene, then `Update` will be called once per frame.

The third function used to update is the `LateUpdate` function. As its name indicates, `LateUpdate` will be called after the `Update` function. So, we can use it to implement a two-step update in every frame. For example, you have a bunch of GameObjects in the Scene that need to be moved and rotated in the `Update` function, and you will use a camera in the Scene to track the movement of these GameObjects. In order to ensure that all GameObjects have moved completely, you can implement a smooth camera follow in the `LateUpdate` function.

# Rendering

For a game, in addition to the game logic, another important aspect is the game's graphics and rendering. Here, I will introduce three commonly used event functions of rendering. They are as follows:

- OnBecameVisible/OnBecameInvisible
- OnRenderImage
- OnGUI

OnBecameVisible will be called when the renderer is visible to any camera, while OnBecameInvisible is the opposite.

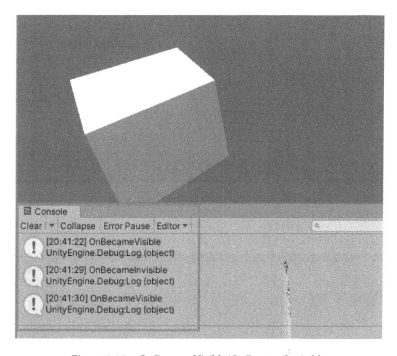

Figure 2.27 – OnBecameVisible/OnBecameInvisible

As you can see in the previous screenshot, when the Cube object moves out of the camera's field of view, OnBecameInvisible will be called, and if it enters the camera's field of view, OnBecameVisible will be called.

If your game logic is very complex, then you can use `OnBecameVisible`/
`OnBecameInvisible` to avoid unnecessary performance overhead. For example, when
a GameObject moves out of view, the functions of the GameObject can be suspended.

`OnRenderImage` is useful for implementing **postprocessing** effects in Unity. This
function will be called after the Scene is completely rendered, and then you can apply
a fullscreen effect to the image, which can greatly improve the appearance of your game.
The following screenshots show the difference between an image with postprocessing
and an image without postprocessing:

Figure 2.28 – Scene with no postprocessing (Unity)

As shown in *Figure 2.29*, applying postprocessing enhances the overall look of the Scene
and delivers stunning effects:

Figure 2.29 – Scene with postprocessing (Unity)

It is worth noting that in order to use `OnRenderImage` correctly, you need to attach the script that implements this function to the GameObject that the Camera component is attached to:

```csharp
public class PostProcessing : MonoBehaviour
{
[SerializeField]
private Material _mat;

    private void OnRenderImage(RenderTexture src,
      RenderTexture dest)
    {
        Graphics.Blit(src, dest, _mat);
    }
}
```

Sometimes, you may need to create some UI to do some prototypes or conduct some tests. Then, `OnGUI` is an ideal option for you. You can create an **Immediate Mode GUI (IMGUI)** in Unity by implementing rendering and handling GUI events in the `OnGUI` function:

```csharp
public class OnGUITest : MonoBehaviour
{
    private void OnGUI()
    {
        if (GUI.Button(new Rect(10, 10, 200, 100),
          "Button"))
        {
            Debug.Log("Hello World!");
        }
    }
}
```

The GUI line is a **Button** control declaration. A Button control with the **Button** header text will be displayed on screen. It is worth noting that the entire Button declaration is placed in an `if` statement. This is because the code in the `if` block needs to be executed when the button is clicked. Specifically, taking the preceding code as an example, when the game is running and the button is clicked, this `if` statement returns `true` and executes the `Debug.Log("Hello World")` line in the `if` block to print out **Hello World** in the **Console** window.

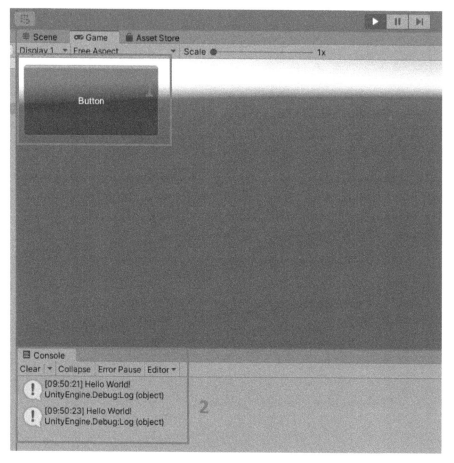

Figure 2.30 – IMGUI

The preceding screenshot shows an IMGUI button and the message printed in the Console window by clicking this button.

In this section, we explained the life cycle of a script instance and some commonly used event functions offered by the Unity engine. In the next section, we will explore how to create a script file that will interact with the engine and add it as a component to a GameObject in the Scene.

# Creating a script and using it as a component

In addition to Unity's built-in components, we can also create script components. When you create a script and attach it to a GameObject, you can see the component you created in the GameObject's **Inspector** window, just like Unity's built-in components.

## How to create a new script in Unity

It is very easy to create a new C# script in Unity. I will introduce two different ways to do it.

Firstly, you can right-click the **Project** panel in the Unity Editor, and then a menu will pop up. You only need to select **Create | C# Script**, then the Unity Editor will create a C# file in the folder identified in the **Project** panel.

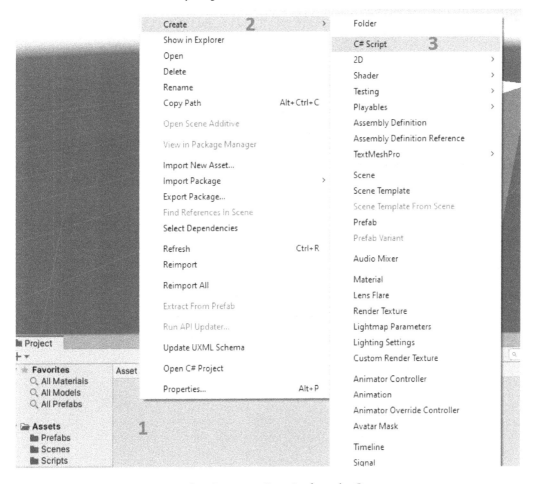

Figure 2.31 – Creating a new C# script from the Create menu

The filename of the new script is `NewBehaviourScript.cs` by default. You can change the name when creating it.

Figure 2.32 – Changing the name of the script when creating it

For example, in the previous screenshot, the new C# file will be created in the `Assets/ Chapter 2/Scripts` folder. In this way, the newly created script will not be automatically attached to a GameObject in the Scene. You need to manually add it to a GameObject later.

On the other hand, you can also create a script and attach this script to a GameObject directly. What you need to do is to select a GameObject in the Scene and click **Add Component | New script** in the **Inspector** window to create a new script. This script will be automatically attached to the GameObject, and you can find the script file you just created in the `Assets` folder of your project.

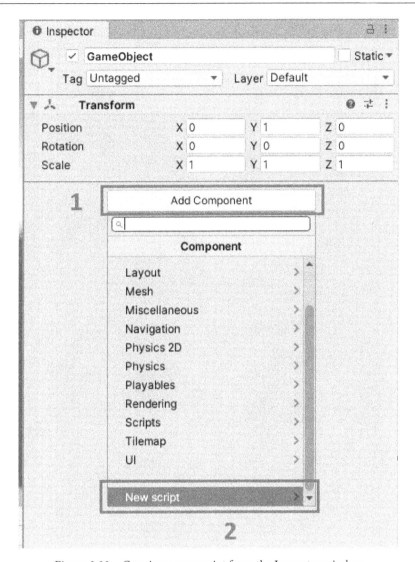

Figure 2.33 – Creating a new script from the Inspector window

Similar to creating a new script in the **Project** window, you also need to type the name of this script because the default name of the script is NewBehaviourScript.cs.

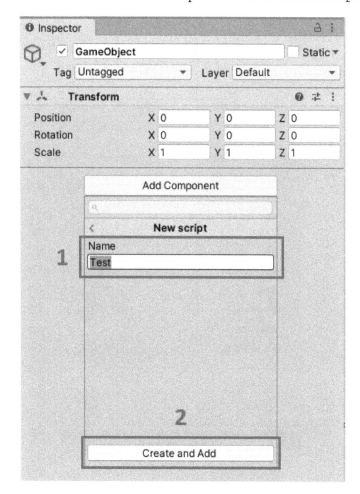

Figure 2.34 – Changing the name of the script when creating it

If you want to open the script in the IDE, we have set Visual Studio 2019 as the IDE for the Unity project; you can double-click the script file to open it in Visual Studio 2019. You will find the name of the C# class is the same as the name of the script file.

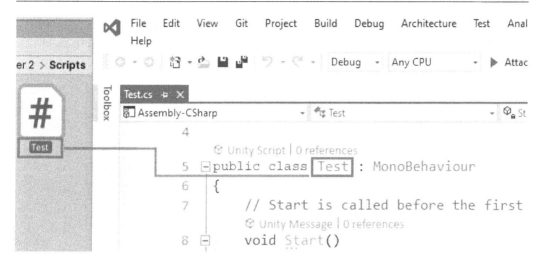

Figure 2.35 – C# class name and the filename of the script

# Adding a script as a component to a GameObject in the Scene

In the previous section, we introduced how to create a new script and attach it to a GameObject automatically. But we still need to learn how to add a script to a GameObject manually in the Editor and add a script component to a GameObject through C# code at runtime.

## Adding a script component to a GameObject in the Editor

The easiest way to add a script as a component to a GameObject in the Unity Editor is to drag the script file to the GameObject.

However, the following two situations may cause the script to not be added to the GameObject:

- The filename and the class name are different. This is why the name of the script is the same as the class name when the script is created. However, you may change one of them by mistake. So, if you cannot add a script to a GameObject, check the filename and the class name first.

- The second reason is relatively obvious: there are compile errors in the script. In this situation, the **Console** window will print out the compile errors. You need to fix all of these errors so that you can add them to a GameObject.

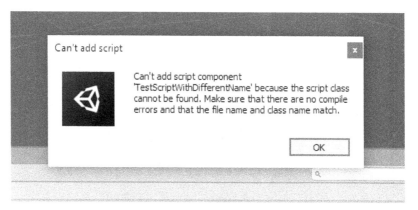

Figure 2.36 – Can't add script message

You can also add a script component to the GameObject from the **Inspector** window. The following steps demonstrate how to do it:

1. Select the GameObject that you want to attach the script to.

2. Click the **Add Component** button in the **Inspector** window. Not only will the scripts we created be added, but also lots of built-in components can be added to the GameObject.

3. In order to quickly find the script that needs to be added, we can enter the name of the script in the search box.

4. Finally, select the target script in the drop-down box.

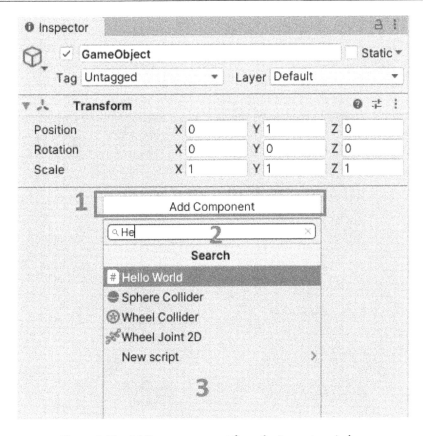

Figure 2.37 – Adding a component from the Inspector window

## Adding a script component to a GameObject at runtime

In addition to manually adding script components to a GameObject in the Editor, we can also add components to a GameObject through C# code at runtime.

Let's open the Test.cs file we just created in Visual Studio 2019 and add a new field, as shown here:

```
[SerializeField]
private HelloWorld _helloWorld;
```

> **Note**
> A field is a variable of any type that is declared directly in a class or struct.

Here, you can see that the name of the private field of the `HelloWorld` type is `_helloWorld`, and you will also find that a `[SerializeField]` attribute is placed on the declaration of `_helloWorld`. This is to allow Unity to serialize this private field. We will discuss the serialization system in Unity in later chapters, but you should understand that when Unity serializes a script, it only serializes public fields by default. If a variable can be serialized by Unity, then it can be displayed and modified in the Unity Editor. So, you can use a public field here. However, generally, it is a good idea to use fields only for variables that have private or protected accessibility. This is why Unity provides developers with the `[SerializeField]` attribute, which will force Unity to serialize private fields.

Then, we add the Test script component to a GameObject in the Scene by dragging it to the GameObject.

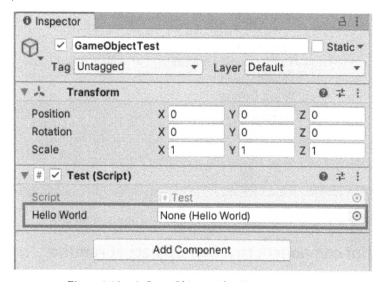

Figure 2.38 – A GameObject with a Test component

You can see that there is the serialized field of the Test script component attached to the GameObject in the previous screenshot. The value of this field is **None**, which means we need to assign a value to it. Let's add more code to the `Test.cs` script to attach the `HelloWorld` script component to the same GameObject and assign a reference to this new `HelloWorld` component to this field.

Because we only want the code to run once, we can modify the `Start` function, as follows:

```
void Start()
{

    _helloWorld =
        gameObject.AddComponent<HelloWorld>();

}
```

Here, we are calling the `AddComponent<T>` method, which is a generic method to add the `HelloWorld` component to this GameObject, and it will return the reference to the attached component, so we can assign this value to the `_helloWorld` field.

> **Note**
>
> A generic method is a method that is declared with type parameters. The preceding code shows how to call the `AddComponent<T>` method by using `HelloWorld` for the type argument.

It is worth noting that in addition to the generic method, there is also a version of `AddComponent`, which is `AddComponent(string className)`, a method with a string argument. It has been deprecated, so you should no longer use this method, but instead use the generic version.

Play the game by clicking the **Play** button in the Unity Editor.

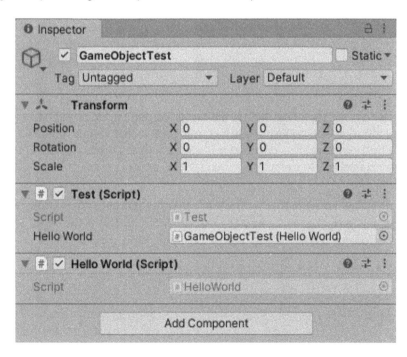

Figure 2.39 – Attaching the HelloWorld component at runtime

Looking at the **Inspector** window again, you can see in the preceding screenshot that there is a `HelloWorld` component attached to the GameObject, and the reference to this component is assigned to the field of the Test component.

Well done. Now we have learned how to add components to the GameObjects in the Scene. Next, let's explore how to access components on the same GameObject or different GameObjects through C# code.

## Accessing a component attached to a GameObject

When we develop a Unity project, we often need to access other components, because we can reuse the functions defined by different components.

Here, let's add some code to the `HelloWorld.cs` script to print a **Hello World!** message to the Console window in the Editor:

```
public void SayHi()
{
    Debug.Log("Hello World!");
}
```

> **Note**
>
> The `Debug.Log` line in the `SayHi` method is a commonly used method to print messages that can help you debug your game to the Console window. The `Debug` class also offers many other methods, such as `LogError`, `LogWarning`, and `Assert`.

We can think of this as a feature that we want to reuse in different scripts. Then, we also need to create a new script called `TestGetComponent.cs`. This is the script where we will place the code to access the `HelloWorld` component at runtime:

```
public class TestGetComponent : MonoBehaviour
{
    void Update()
    {
        var helloWorld =
            gameObject.GetComponent<HelloWorld>();
        if (helloWorld == null)
```

```
    {
        return;
    }
    helloWorld.SayHi();
    }
}
```

As we already know that the Update function runs at every frame of the game, in order to demonstrate how to access a component, we can put the code in the Update function, as shown in the code of the TestGetComponent class.

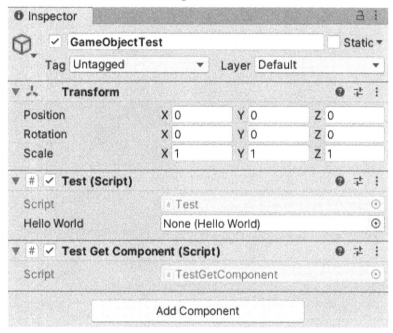

Figure 2.40 – A GameObject with the TestGetComponent component

Then, we attach the `TestGetComponent` script to the same GameObject as a component, play the game, and look at the Console window. The **Hello World!** message appears there.

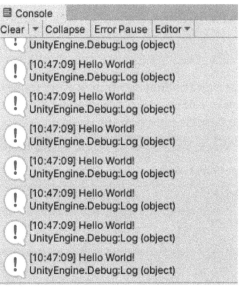

Figure 2.41 – Hello World! appears in the Console window

**Note**

For performance reasons, it is recommended to not use this function in every frame.

In this case, we accessed other components attached to the same GameObject. Additionally, we can also access other components on different GameObjects.

Firstly, we need to get the reference of the target GameObject. Here, we can either assign the referenced object to this script in the Editor or use the `GameObject.Find` method to find the target object at runtime. From the perspective of game performance, don't call the `GameObject.Find` method to find the target object in a method such as `Update` that is called at every frame. If you can't assign a reference to your script in the Editor, for example, the referenced object is dynamically created at runtime, then you can use this method to find the target object and cache the target object instead of finding the target object at every frame. In this example, we can find the target object and cache it in the `Start` method, as shown:

```
private GameObject _targetGameObject;

    private void Start()
```

```
    {
        // Using Find method to find game objects is not
            recommended,
        // this is just to demonstrate how to call this
            method to find
        // the target object at runtime.
        _targetGameObject =
            GameObject.Find("GameObjectTest");
    }
```

Then, let's change the Update function of the TestGetComponent class, as follows:

```
    void Update()
    {

        var helloWorld =
            _targetGameObject.GetComponent<HelloWorld>();
        if (helloWorld == null)
        {
            return;
        }
        helloWorld.SayHi();
    }
```

Here, we are using the GameObject.Find(string name) function to find a GameObject by name and return it. The name of the target GameObject is GameObjectTest.

There are other functions that can be used to look for a GameObject at runtime, such as GameObject.FindWithTag(string tag), which returns one active GameObject tagged tag. However, in order to use this function correctly, the tag must first be declared in the tag manager. You can manage these tags from **Project Settings** | **Tags and Layers**.

However, as we mentioned earlier, the Find method and its variants are not recommended to find GameObjects. This example is just to demonstrate how to call the method at runtime to find the target object if you need to find dynamically created objects at runtime.

Next, we create a new GameObject and attach the `TestGetComponent` script to it. At the same time, remove the `TestGetComponent` script from the target GameObject named `GameObjectTest`.

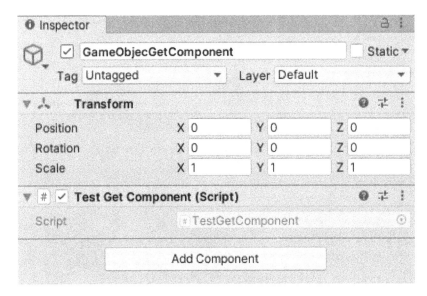

Figure 2.42 – A GameObject with the TestGetComponent component

Play the game and look at the Console window. The same **Hello World!** message appears there again.

In this section, we learned how to create a new script in Unity and how to attach a script as a component to a GameObject, and also discussed how to access a component through code at runtime to reuse functions. Next, let's explore the Unity Package Manager and packages in Unity.

# Packages and the Unity Package Manager

If you are a .NET developer, then I believe that you must know the **NuGet** package manager. The Package Manager in Unity is very similar to NuGet, which enables game developers to share and consume useful code. But they are different. In Unity, you can reuse not only useful code but also digital assets, Shaders, plugins, and icons. A package in Unity is a container that includes the contents mentioned earlier.

In this section, I will introduce packages and Package Manager in Unity so that you can understand the package mechanism in Unity and how to use the Unity Package Manager to manage packages.

# Unity Package Manager

Unity provides game developers with a tool called the Unity Package Manager to manage the packages in a project and add new packages to the project. We can open the Package Manager window by clicking **Window | Package Manager**.

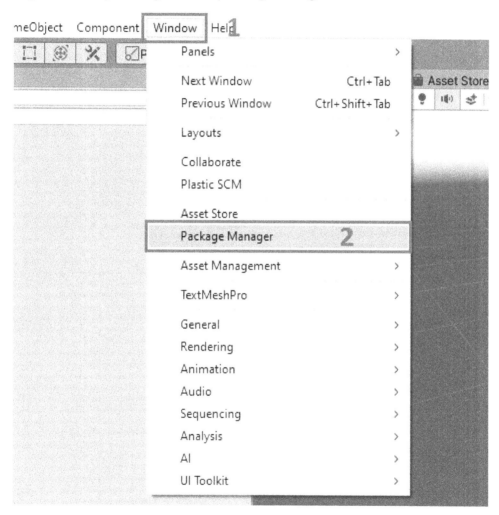

Figure 2.43 – Opening the Package Manager window from the Window menu

By default, this window shows the installed packages in your project and the version of each package. If a new version of a package is available, there will be an **upgrade** icon beside the version number. You can also sort these packages, for example, in ascending order by name or descending order by release date.

Figure 2.44 – Unity Package Manager

On the right side of the window, detailed information on the currently selected package will be displayed, such as the package name, publisher, release date, version number, document link, and description. You can also remove a package from your project by clicking the **Remove** button in the lower-right corner.

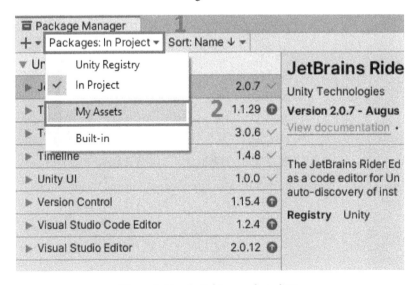

Figure 2.45 – Switching package lists

This window can also display different lists. For example, you can view, download, and import assets purchased from the Unity Asset Store (`https://assetstore.unity.com/`) by selecting the **My Assets** option from the drop-down menu.

The assets purchased from the Asset Store may be free or paid. The Asset Store provides a variety of assets, covering everything from textures, models, and animations to entire project examples.

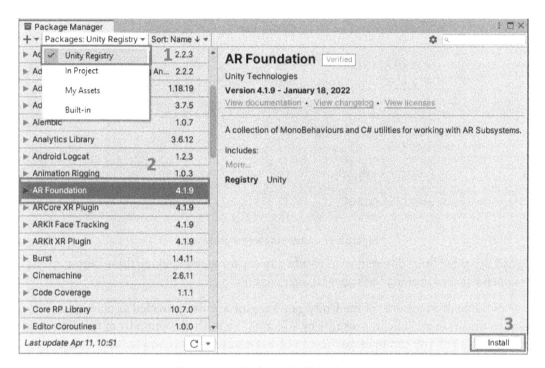

Figure 2.46 – Packages in Unity Registry

You can also install a package from Unity Registry. By selecting the **Unity Registry** option from the drop-down menu, you can browse all packages registered in Unity Registry. If you want to install a package, you need to select it and click the **Install** button in the lower-right corner.

In addition to installing a package from Unity Registry, the Unity Package Manager also provides other ways to install a package, that is, installing a new package from a local folder, installing a new package from a local tarball file, and installing a new package using a Git URL.

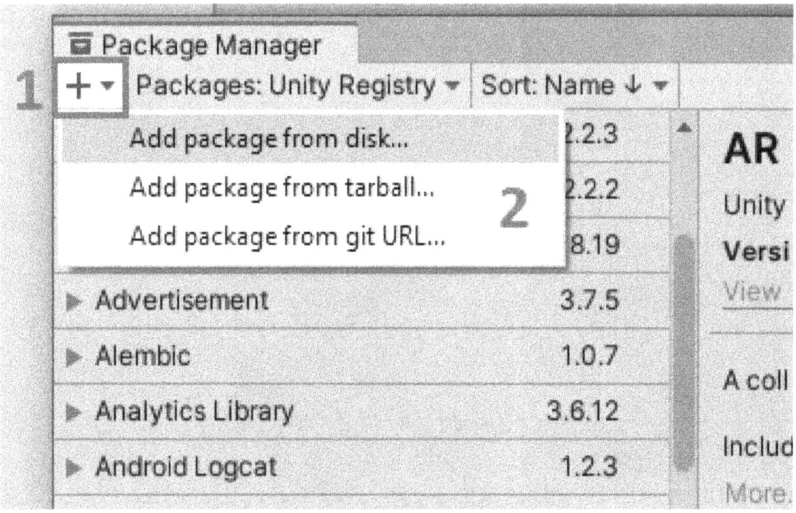

Figure 2.47 – Installing a new package

You can use these three different ways to add a new package by clicking the + button in the upper-left corner of the Package Manager window.

Some of the built-in features of the Unity game engine are also provided as packages. You can view the list of all built-in packages by selecting the **Built-in** option from the drop-down menu. Here, you can manage these built-in features. You can reduce the runtime build size of your game by disabling packages that you do not need. For example, if you develop a game without VR or AR functionality, you can disable XR-related packages by clicking the **Disable** button in the lower-right corner of the **Package Manager** window.

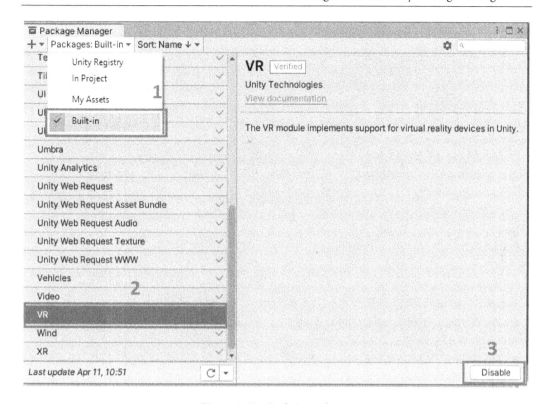

Figure 2.48 – Built-in packages

# Package

A package is a container that contains features to meet various needs of a project. You can add a new feature to your game by adding a package. For example, the **AR Foundation** package will provide AR functionality. You can also remove a package to reduce the size of your game. Therefore, the use of packages makes Unity game development more flexible and decoupled.

However, if you are not careful, using a package may also make your game full of bugs. This is because different packages may be in different states.

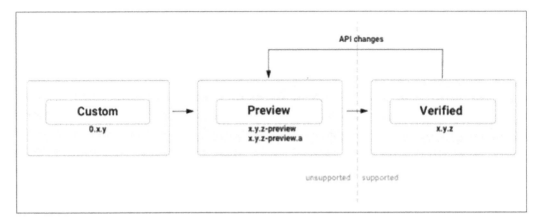

Figure 2.49 – Package life cycle with Unity Package Manager (Unity)

A package developed and maintained by Unity may be in one of the following two states:

- Preview packages
- Verified packages

A package in preview means that it is currently ready for testing, and it may go through many changes in later versions. Unity cannot guarantee future support for preview packages, so you should not use them in production.

By default, you cannot find packages in the preview state in the Package Manager window. If you really need to use the preview packages, for example, to test new features for future projects, you can follow these steps to allow the Package Manager window to display the packages in the preview state:

1. Open the **Project Settings** window for the Package Manager by clicking the gear icon and then clicking the **Advanced Project Settings** item.

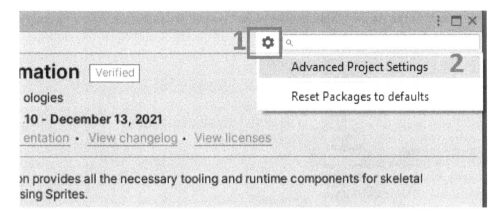

Figure 2.50 – Advanced Project Settings

2.  Check the **Enable Preview Packages** option.

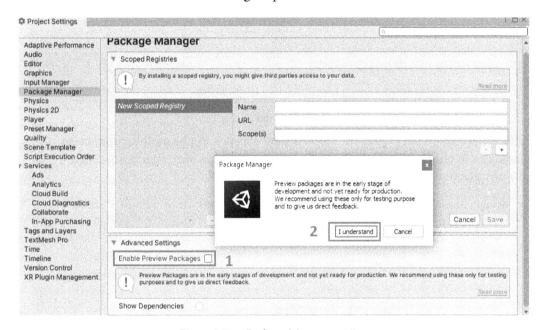

Figure 2.51 – Package Manager settings

3.  Then, look at the Package Manager window. You will see that the preview packages appear in the package list. In the package list, all packages in the preview state are marked with **Preview**.

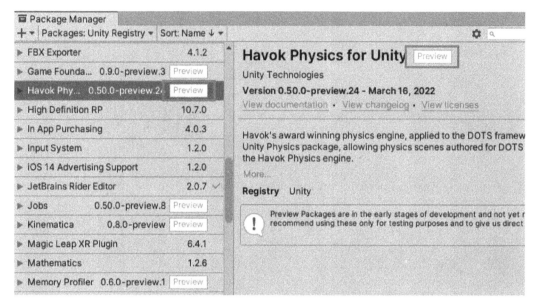

Figure 2.52 – Preview packages

On the other hand, a package in the verified state means it can be used in production. A package will only be considered a verified package if it has been rigorously tested and Unity guarantees supporting that verified package.

The Package Manager window displays the verified packages list by default. Packages in the verified state are marked with **Verified**.

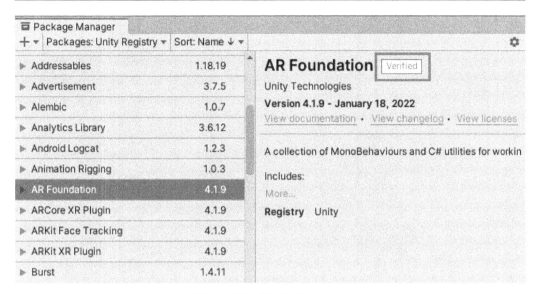

Figure 2.53 – Verified packages

# Summary

In this chapter, we started by introducing some of the most commonly used classes in Unity script programming, and then explained the life cycle and important event functions of a script instance, as well as discussing how Unity initializes a script and how the game logic is updated in a script.

We also discussed how to create a new script in Unity and how to attach a script as a component to a GameObject. In addition to manually adding components in the Editor, we can also use C# code to dynamically add a component or access a component at runtime.

Finally, we demonstrated how to add or remove a package through the Unity Package Manager to provide a feature or reduce the size of the game. At the same time, we also explained the difference between preview packages and verified packages.

In the next chapter, we will learn about the UI system in Unity and, at the same time, we will also introduce how to optimize UI performance in Unity.

# Part 2: Using C# Scripts to Work with Unity's Built-In Modules

After gaining a general understanding of the Unity game engine and knowing how to write scripts in Unity, we can start to learn the main modules in the Unity engine one by one, such as creating a UI in Unity and applying physics in a game.

This part includes the following chapters:

- *Chapter 3, Developing UI with the Unity UI System*
- *Chapter 4, Creating Animations with the Unity Animation System*
- *Chapter 5, Working with the Unity Physics System*
- *Chapter 6, Integrating Audio and Video in a Unity Project*

# 3

# Developing UI with the Unity UI System

The UI is very important for a game, and Unity offers three different UI solutions for game developers. They are the **Immediate Mode Graphical User Interface (IMGUI)**, the **Unity UI (uGUI)** package, and the **UI Toolkit**. IMGUI is a relatively old UI solution in Unity and it is not recommended for building a runtime UI. The UI Toolkit is the latest UI solution; however, it is still missing some features you can find in the uGUI package and IMGUI. The uGUI package is a mature UI solution in Unity that is widely used in the game industry. Therefore, this chapter will introduce how to use uGUI to develop the UI of your game.

We will cover the following key topics in this chapter:

- C# scripts and common UI elements in Unity

- C# scripts and the UI Event System in Unity

- The **Model-View-ViewModel (MVVM)** pattern and the UI

- Performance tips to increase performance of the UI

Let's get started!

# C# scripts and common UI components in Unity

uGUI has been provided as a built-in package in the Unity Editor since Unity 2019; therefore, we can see the content of the uGUI package directly in the **Project** window, which also includes the C# source code.

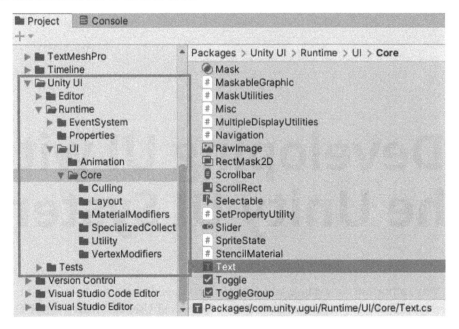

Figure 3.1 – The uGUI package

As we mentioned in the previous chapter, the Unity development workflow is primarily built around the structure of components. uGUI is no exception. It is a **component-based** UI system that uses different components to provide different UI functions. For example, every button, text, or image you see in the UI is actually a `GameObject` with a set of components.

As shown in *Figure 3.1*, we can find the C# source code of many commonly used UI elements, such as **Text**, **Slider**, and **Toggle**. However, some UI components are implemented using C++ code inside the engine, such as **Canvas**, and the code of such components cannot be viewed from within the Unity Editor.

In this section, we will introduce the commonly used UI components in Unity. We can divide these components into the following four categories, according to their functions:

- `Canvas`
- `Image and Raw Image`
- `Text`
- `Selectable UI components`

# Canvas

**Canvas** is the most basic and important UI component of uGUI. To understand how to use uGUI correctly and efficiently, it is essential to understand **Canvas** first.

**Canvas** is the component used to render UI elements in uGUI. All UI elements should be located inside the area of a canvas, which is very simple to create in a scene.

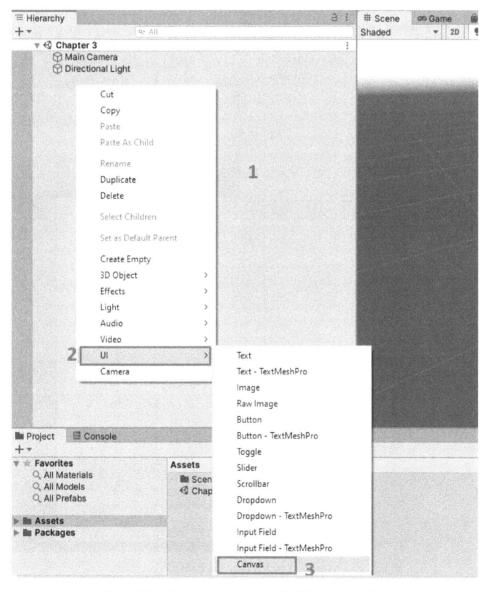

Figure 3.2 – Creating a canvas from the Hierarchy window

As shown in *Figure 3.2*, you can create a new canvas as follows:

1. Right-click in the **Hierarchy** window to open the menu.
2. Select **UI | Canvas**.

In addition to creating a new **Canvas** object from the **Hierarchy** window, we can also create a new **Canvas** object by clicking **GameObject | UI | Canvas**.

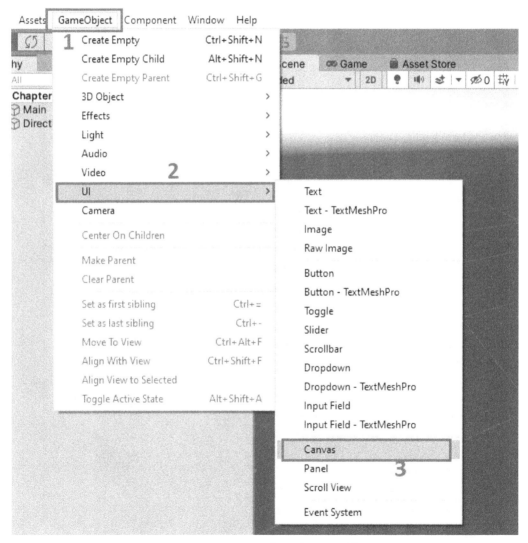

Figure 3.3 – Creating a canvas from the GameObject menu

As you can see in *Figure 3.2* and *Figure 3.3*, we can also create other different UI elements from these menus, such as **Text**, **Button**, **Image**, and **Slider**. Since all UI elements are the children of **Canvas**, if you want to create a new UI element directly and there is no canvas, a new **Canvas** object will be created automatically. The new UI element will be a child object of the **Canvas** object parent.

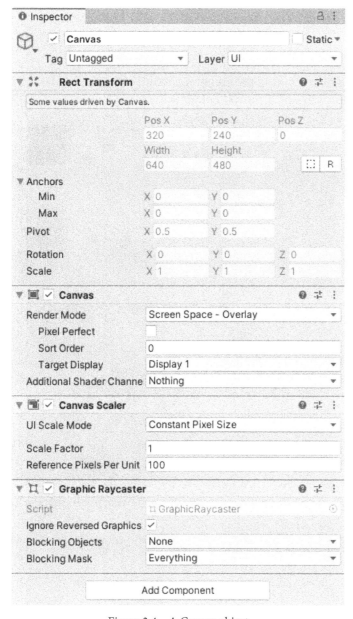

Figure 3.4 – A Canvas object

Once a **Canvas** object is created, we can see that there is not only a **Canvas** component attached to this GameObject but also **Rect Transform**, **Canvas Scaler**, and **Graphic Raycaster** components. As mentioned previously, **Canvas** is the component used to render UI elements, so all UI components must be children of Canvas; otherwise, they will not be rendered by Unity.

We will explore them separately in order.

## The Canvas component

If you select the **Canvas** object in the scene, you may be surprised to find that its position is strange. By default, it is not in the field of view of **Main Camera**.

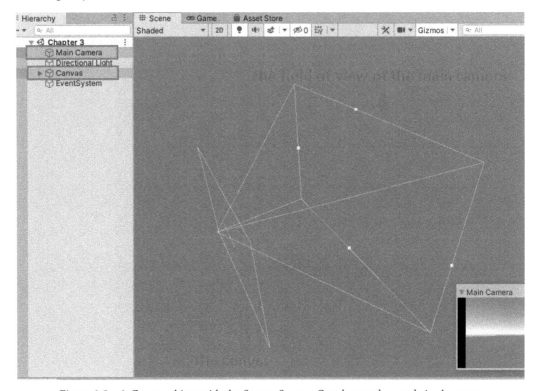

Figure 3.5 – A Canvas object with the Screen Space - Overlay render mode in the scene

This is because the **Canvas** component attached to this GameObject provides three different **render modes**, as follows:

- **Screen Space - Overlay**

- **Screen Space - Camera**

- **World Space**

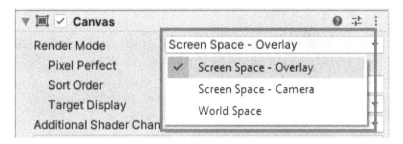

Figure 3.6 – Render modes

The **Screen Space - Overlay** render mode places UI elements on the screen that are rendered on top of the scene. Therefore, the cameras located in the scene used to render the game scene will not affect the rendering of the UI. This is the default render mode provided by the **Canvas** component.

As the name implies, the **Screen Space - Camera** render mode is somewhat similar to the previous one. However, as can be seen from the name, the second render mode will be affected by the camera.

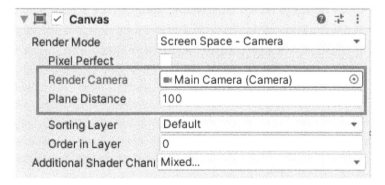

Figure 3.7 – The Screen Space - Camera render mode

As you can see in *Figure 3.7*, if the **Screen Space - Camera** render mode is selected, we need to specify a camera for this canvas and set a distance between them. Furthermore, if we still select this canvas in the scene, we will find that it has been moved into the field of view of this particular camera.

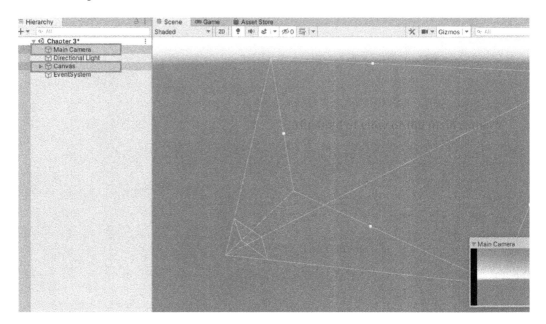

Figure 3.8 – A Canvas object with the Screen Space - Camera render mode in the scene

In this case, the UI elements are rendered by this camera, which means that the camera settings affect the appearance of the UI. This is different from the **Screen Space - Overlay** render mode.

*Figure 3.9* shows that when the **Field of View** value of this camera is changed from 100 to 30, the game scene and the UI have changed:

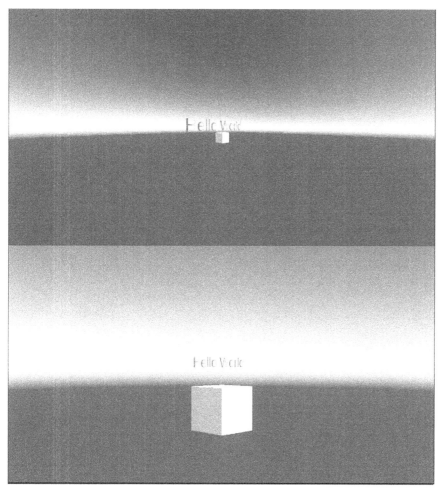

Figure 3.9 – The field of view (FoV) of the camera is 100 in the upper half and 30 in the lower half

The last render mode is **World Space**. In this mode, the canvas will work like any other GameObject in the scene. The biggest difference between this mode and the **Screen Space - Camera** render mode is that we can manually adjust the size, position, and even rotation angle of the canvas, just like a normal GameObject.

As shown in *Figure 3.10*, we can use the **Rect Transform** component of this **Canvas** object to adjust its **Width** and **Rotation** values:

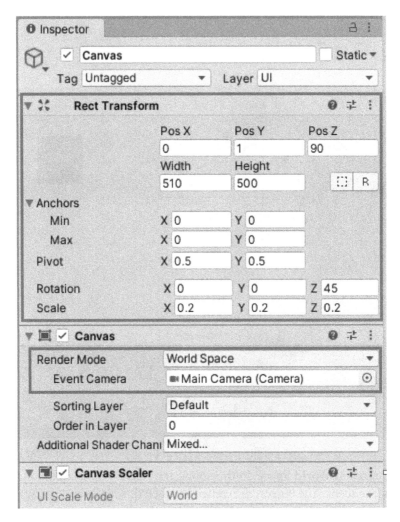

Figure 3.10 – The World Space render mode

*Figure 3.11* shows the **Canvas** object in the scene after manually setting the **Width** and **Rotation** values:

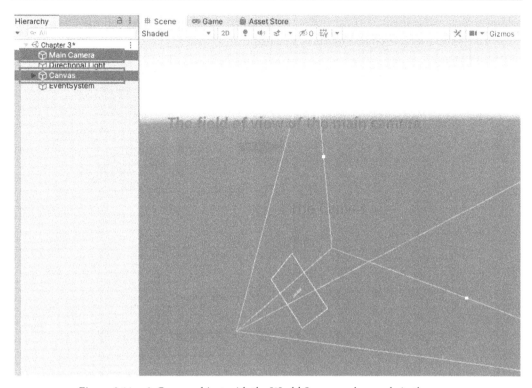

Figure 3.11 – A Canvas object with the World Space render mode in the scene

Here, we use the **RectTransform** component to set the size of the canvas. Every UI object will contain a **RectTransform** component, just like every normal GameObject will contain a Transform component. Next, we will explore the **RectTransform** component.

## The Rect Transform component

The **Rect Transform** component is similar to the regular **Transform** component. The biggest difference is that the former is used for UI elements instead of regular GameObjects. When a new UI element object is created, the **Rect Transform** component will be automatically attached to it.

Looking at this component, you can see some properties that can be seen on the **Transform** component, such as **Position**, **Rotation**, and **Scale**. There are also some unique properties.

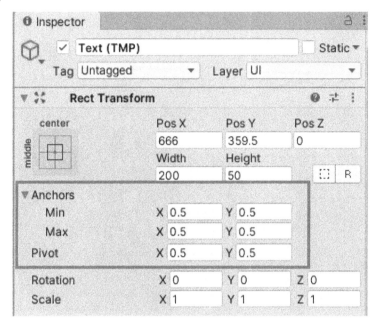

Figure 3.12 – A Rect Transform component

These unique ones are **Anchor** and **Pivot**. We will discuss these in turn.

## Anchors

The anchors are numerical values indicating the position of the four corners of the area as seen from the **Rect Transform** parent. The lower left is represented by AnchorMin.x and AnchorMin.y, and the upper right is represented by AnchorMax.x and AnchorMax.y. By default, the lower left is 0.5 and 0.5, and the upper right is also 0.5 and 0.5, centered relative to the parent, as shown in *Figure 3.12*.

We can directly modify the value of anchors – for example, we can change the lower-left corner from 0.5 and 0.5 to 0 and 0, so that the lower-left corner of the parent and child are the same. Then, we change the upper-right corner from 0.5 and 0.5 to 0.5 and 1, which means that the position of the upper-right corner of the child is half of the *x* axis position of the upper-right corner of the parent. The result is shown in *Figure 3.13*:

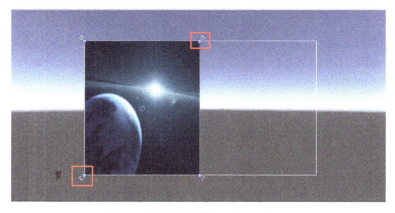

Figure 3.13 – Modifying the anchors

Anchors are very useful when developing the UI in Unity. For example, if you want to display the UI at the top of the screen, such as a title, you need to specify the distance from the top of the parent. If you want to display the UI at the bottom of the screen, such as a footer, you need to specify the distance from the bottom of the parent.

In order to make it easier for developers to use anchors, Unity provides some anchor presets, as shown in *Figure 3.14*:

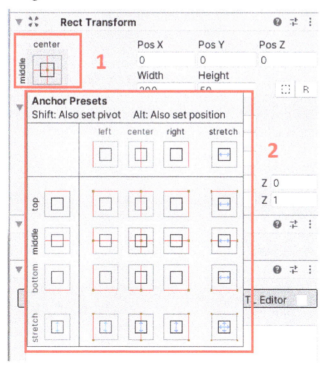

Figure 3.14 – Anchor Presets

## Pivot

The **Pivot** point is the origin of this rectangle area. The value of the **Pivot** point is specified in normalized values between 0 and 1. When the UI element is scaled or rotated, it will scale or rotate around that point:

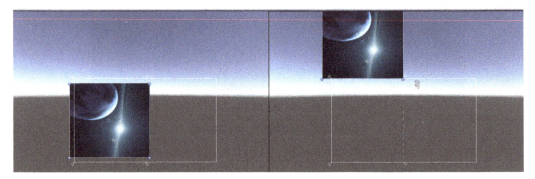

Figure 3.15 – Rotate 45 degrees along the *z* axis around the center and

45 degrees along the *z* axis around the upper-right corner

*Figure 3.15* shows a 45-degree rotation along the *z* axis around the center, which has a **Pivot** point value of 0.5 and 0.5, and a 45-degree rotation along the *z* axis around the upper-right corner, which has a **Pivot** point value of 1 and 1.

## The Canvas Scaler component

Along with the **Canvas** component, a **Canvas Scaler** component is also created automatically. The **Canvas Scaler** component is used to control the overall scale and pixel density of UI elements inside a canvas. By using **Canvas Scaler**, we can implement a resolution-independent UI layout:

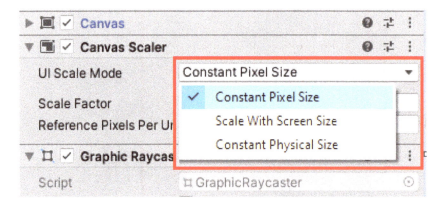

Figure 3.16 – The Canvas Scaler component

There are three **UI Scale Mode** types provided by a **Canvas Scaler** component:

- **Constant Pixel Size**
- **Scale With Screen Size**
- **Constant Physical Size**

If the canvas render mode is **ScreenSpace - Overlay** or **ScreenSpace - Camera**, then we can set the UI Scale Mode. On the other hand, if the canvas render mode is **World Space**, the UI Scale Mode cannot be modified. Next, we will introduce these three different modes.

**Constant Pixel Size** is the default UI Scale Mode. In this mode, the size of the UI elements will retain the same size in pixels regardless of screen size.

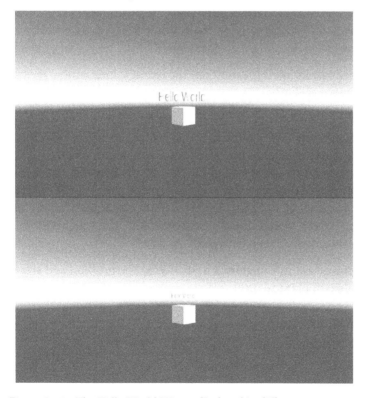

Figure 3.17 – The Hello World UI text displayed in different screen sizes
(1920 x 1080 in the upper half and 3840 x 2160 in the lower half)

As shown in *Figure 3.17*, a **Hello World** UI text will retain its own size in pixels. When the screen resolution is relatively low (*1920 x 1080*), the text will be displayed larger. When at a higher screen resolution (*3840 x 2160*), the text will be displayed smaller.

If you want to keep the UI elements displayed consistently under different screen resolutions, the **Scale With Screen Size** mode is an ideal option.

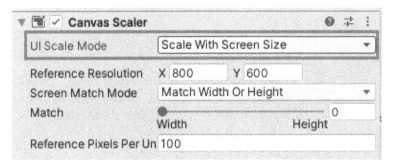

Figure 3.18 – The Scale With Screen Size mode

If **UI Scale Mode** is set to **Scale With Screen Size**, the position and size of the UI elements will be specified according to the value of pixels in the **Reference Resolution** properties, as shown in *Figure 3.18*.

If the current screen resolution is greater than the reference resolution, the canvas will be scaled to fit the screen resolution. Conversely, if the current screen resolution is less than the reference resolution, the canvas will shrink to fit the screen resolution.

If the screen resolution ratio is the same as the reference resolution ratio, it is very easy to scale and shrink the UI elements. But when the screen resolution ratio is different from the reference resolution ratio, scaling the canvas will distort it. In order to avoid this situation, the resolution of the canvas will also depend on the setting of **Screen Match Mode**, which you can also see in *Figure 3.18*. By default, the **Screen Match Mode** setting is **Match Width or Height**, which allows you to scale the canvas area with the width or height as the reference, or a value in between.

When **UI Scale Mode** is set to **Constant Physical Size**, the position and size of UI elements are specified in physical units such as *millimeters* and *inches*.

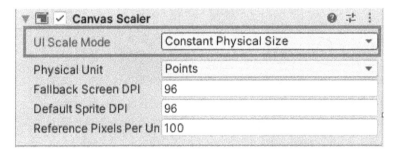

Figure 3.19 – The Constant Physical Size mode

In addition to the **Canvas Scaler** component, another component is also automatically created, which we will take a look at next.

## The Graphic Raycaster component

As the name suggests, the **Graphic Raycaster** component is used to perform raycasting against a list of UI elements within a canvas to determine which of the UI elements has been hit. So it can translate the player's input into UI events. It should be noted that there needs to be an **Event System** component in the scene for **Graphic Raycaster** to work properly. About the **Event System** component, we will introduce it later in the section *"C# scripts and the UI Event System in Unity"*.

This is useful when you need to determine whether the cursor is over UI elements in the scene, such as UI text or UI images. For example, say you want the player to be able to drag and drop a UI image into your game to change its position, then you have to know whether the player's cursor is over the UI image and get data about the cursor movement when the drag occurs. In this case, you need to create a script that implements the `IPointerDownHandler` and `IDragHandler` interfaces defined in the `UnityEngine.EventSystems` namespace, meaning that you can get events when the player clicks and drags the image, as shown here:

```csharp
using UnityEngine;
using UnityEngine.EventSystems;

public class DragAndDropExample : MonoBehaviour,
    IPointerDownHandler, IDragHandler
{
    private RectTransform _rectTransform;

    public void OnPointerDown(PointerEventData eventData)
    {
        Debug.Log("This UI image is clicked!!!");
        _rectTransform = GetComponent<RectTransform>();
    }

    public void OnDrag(PointerEventData eventData)
    {
        Debug.Log("This UI image is being dragged!!!");
```

```
        if (RectTransformUtility
        .ScreenPointToWorldPointInRectangle
        (_rectTransform, eventData.position,
        eventData.pressEventCamera,
        out var cursorPos))
        {
                _rectTransform.position = cursorPos;
        }
    }
}
```

Let's break down the code as follows:

- We add the `UnityEngine.EventSystems` namespace with the `using` keyword to get events related to clicking and dragging UI elements.

- The `DragAndDropExample` class implements the two interfaces, namely, `IPointerDownHandler` and `IDragHandler`.

  - Specifically, we implement the `OnPointerDown` method in the `IPointerDownHandler` interface, which will be called when the UI element is clicked.

  - And we implemented the `OnDrag` method in the `IDragHandler` interface. When a drag occurs, this method will be called every time the cursor is moved.

- In the implementation of the `OnPointerDown` method, which takes `PointerEventData` as an argument, gets an instance of the `RectTransform` component, and assigns it to the `_rectTransform` field.

- In the implementation of the `OnDrag` method, which also takes `PointerEventData` as an argument, gets the cursor position, and modifies the `position` property of the `_rectTransform` field to move the UI element.

In order for this script to work, you need to attach the script to the UI element in the scene that you want to drag and drop.

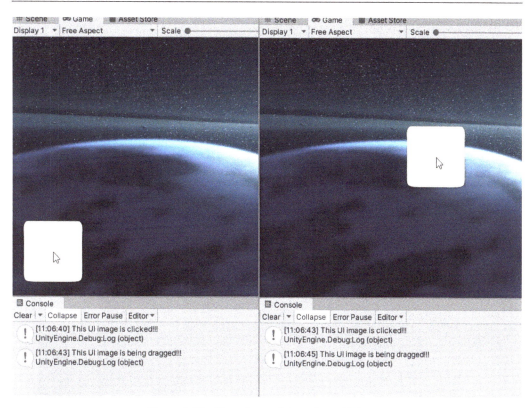

Figure 3.20 – Dragging and dropping a UI image

*Figure 3.20* shows the UI image drag and drop interaction based on the **Graphic Raycaster** component.

The components described previously are automatically created when a **Canvas** object is created. Next, we will introduce other UI elements.

# Image

Displaying images is an important function of the UI. There are two types of components provided by uGUI that display images – the **Image** component and the **Raw Image** component.

We will now explain these features and how to use them properly.

## The Image component

You can use the **Image** component to display an image on your UI.

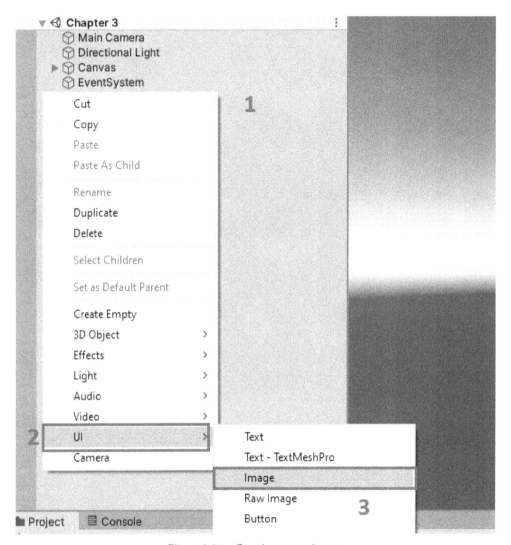

Figure 3.21 – Creating a new image

As shown in *Figure 3.21*, you can create a new image as follows:

1.  Right-click in the **Hierarchy** window to open the menu.

2.  Select **UI** > **Image**.

If you want to create a background image for your game UI, you can also select
**UI** > **Panel**. The panel is nothing but an image.

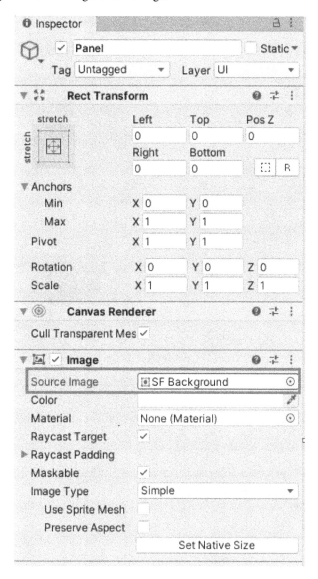

Figure 3.22 – The Image component

In this case, we create a panel as the background. As you can see in *Figure 3.22*, here we specify a texture called **SF Background** as the source image of this **Image** component. It should be noted that the texture used by the **Image** component must be set to the Sprite type when imported into Unity.

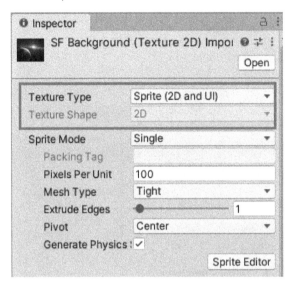

Figure 3.23 – Texture Import Settings

**Texture Type** can be set in the texture's **Import Settings** panel, as shown in *Figure 3.23*.

> **Note**
>
> Sprites are 2D graphic objects used for the UI and other elements of 2D gameplay.

The advantage of using sprites as an image source is that the corners will not be stretched or distorted when resizing the sprites.

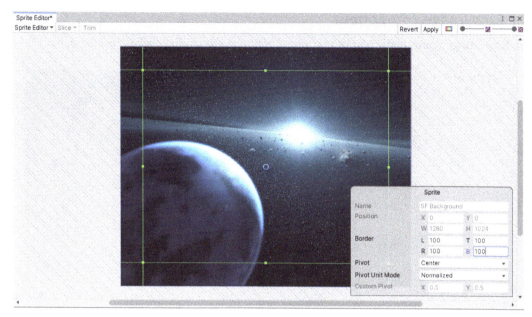

Figure 3.24 – The Sprite Editor

This is because **Sprite Editor** in Unity provides the option of **9-slicing** the image, which divides the image into nine regions. As shown in *Figure 3.24*, in this case, when the image is resized, the corners of the image will remain the same.

Note

9-slicing is a common technique in UI implementation. The main advantage of using 9-slicing is that it can handle the stretching of the image very well. Once an image is stretched, there will be problems such as distortion and blurring, but some parts of the image can be stretched. For example, a UI background frame, the middle part of which is usually a solid color, can be stretched, but the four corners of the image may have some special patterns that cannot be stretched. At this time, we can use the 9-slicing technique to divide the whole image into nine grids, and each of the four corners is in a grid. Then, we can only stretch and enlarge the middle part of the image and keep the four corners as they are.

Therefore, in most cases, using the **Image** component to display UI images is the preferred choice.

# The Raw Image component

The **Raw Image** component is another component used to display images on the game UI.

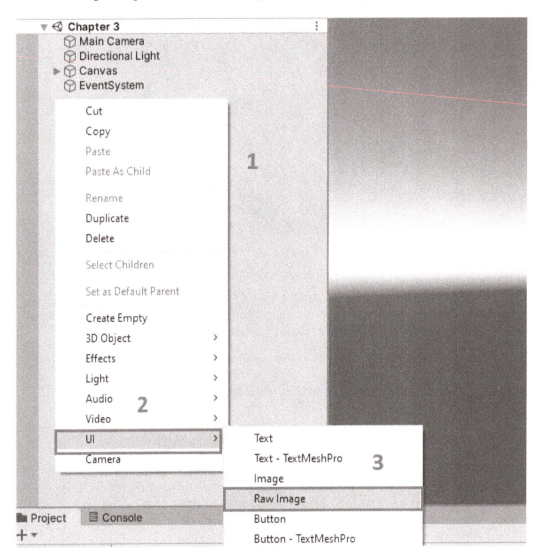

Figure 3.25 – Creating a new raw image

As shown in *Figure 3.25*, you can create a new image, as follows:

1.  Right-click in the **Hierarchy** window to open the menu.

2.  Select **UI > Raw Image**.

The difference between the **Raw Image** component and the **Image** component is that the source of an **Image** component must be a **Sprite** type. Conversely, **Raw Image** accepts any texture. Also, the function of the **Raw Image** component is simpler than an **Image** component, as shown in the following screenshot:

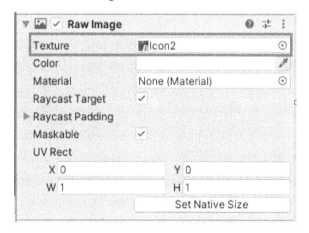

Figure 3.26 – A Raw Image component

The following code snippet shows how to modify the image displayed by the **Image** and **Raw Image** components:

```
using UnityEngine;
using UnityEngine.UI;

public class ImageAndRawImage : MonoBehaviour
{
[SerializeField]
private Image _image;
[SerializeField]
private Sprite _sprite;
[SerializeField]
private RawImage _rawImage;
```

```
[SerializeField]
private Texture _texture;

    void Start()
    {
        _image.sprite = _sprite;
        _rawImage.texture = _texture;
    }
}
```

It should be noted that in order to be able to access UI-related classes in the code, we need to use the `UnityEngine.UI` namespace.

Another important part of the UI is **text**. Next, let's explore the two components provided by uGUI to display text.

# Text

The simplest way to display characters in uGUI is to use the **Text** component. However, it is also troublesome to adjust the spacing between characters and express decorations with **Text** alone. **TextMeshPro** is another option, which provides gorgeous character expression. In this section, we will explore the **Text** and **TextMeshPro** components in turn.

## The Text component

The **Text** component is a component commonly used to display UI text since the early days of uGUI. Creating text for the game UI is very simple; just follow these step:

1.  Right-click in the **Hierarchy** window to open the menu.
2.  Select **UI** > **Text**.

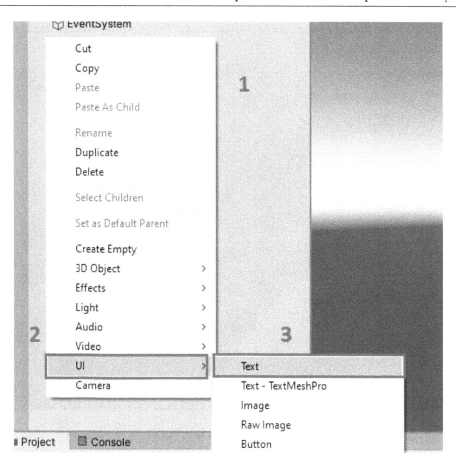

Figure 3.27 – Creating text

A **Text** object will be created in the canvas; we can find it in the **Scene** view of the Unity Editor, as shown in *Figure 3.28*:

Figure 3.28 – Text in the Scene view

You can see that the text content is in a white frame, which represents the **Rect Transform** component attached to this **Text** object and identifies its size. If changing the font size causes the text content to exceed this white frame, the text content cannot be displayed. Therefore, remember to consider the **Rect Transform** component of **Text** when changing the font size.

In addition to changing the font size, you can also change the font used or enable
**Rich Text**.

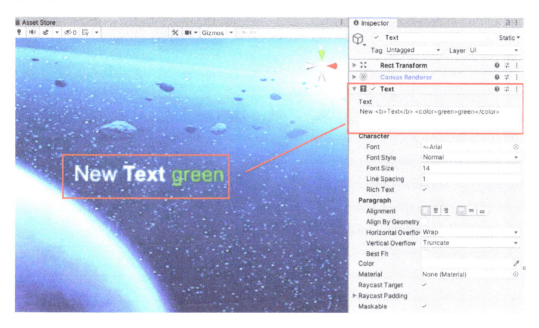

Figure 3.29 – The Text component

As you can see in *Figure 3.29*, if the **Rich Text** checkbox is ticked, then we can use markup
tags, such as `<b></b>`, `<i></i>`, and `<color></color>`, within the text to provide
style changes to the text.

However, the function provided by the **Text** component is relatively simple. When the
**Text** component changes, the polygon used to display the text needs to be recalculated,
resulting in graphic reconstruction, which can cause potential performance problems, and
when displayed in high resolution, the text rendered by this component looks very blurry.
Therefore, after the original **Text** component, Unity also provides another text solution for
the UI. Next, we will introduce the **TextMesh Pro** component.

## The TextMeshPro component

**TextMeshPro** (**TMP**) is the ultimate text solution for the UI provided by Unity. It is a powerful mechanism for text rendering that can be used to replace the **Text** component. **TextMesh Pro** has been designed to take advantage of **Signed Distance Field** (**SDF**) rendering, allowing it to render text beautifully at any resolution. You can also create custom shaders for **TextMesh Pro** to get effects such as outlines and soft shadows.

It should be noted that it is not included in the default Unity UI package, but is included in the TextMeshPro package. So if you can't find **TextMesh Pro** when creating UI text, then you should first check whether this package has been added to your project.

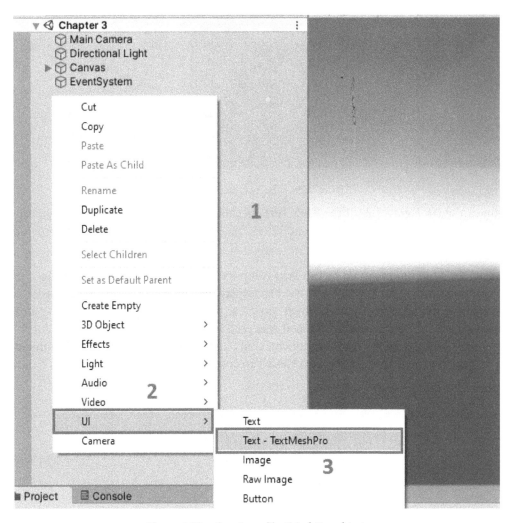

Figure 3.30 – Creating a TextMeshPro object

Creating **TextMeshPro** text for the game UI is very simple; just follow these steps:

1.  Right-click in the **Hierarchy** window to open the menu.
2.  Select **UI > Text > TextMeshPro**.

Figure 3.31 – The TextMeshPro component

As shown in *Figure 3.31*, the text rendered by **TextMeshPro** is sharper than that rendered by the **Text** component.

In addition to rendering the text sharper, **TextMeshPro** also provides improved control over text format and layout. As shown in *Figure 3.32*, you can directly change the style of the text through the editor. There are several common styles to choose from, such as *bold* and italics. Similarly, you can also use tags to modify the text style, just like the **Text** component, and features such as **Spacing Options**, **Alignment**, and **Wrapping**, can be used to control the text layout.

In addition, you can also achieve more rendering effects, such as clicking the outline option of the shader to add outline effects to the text.

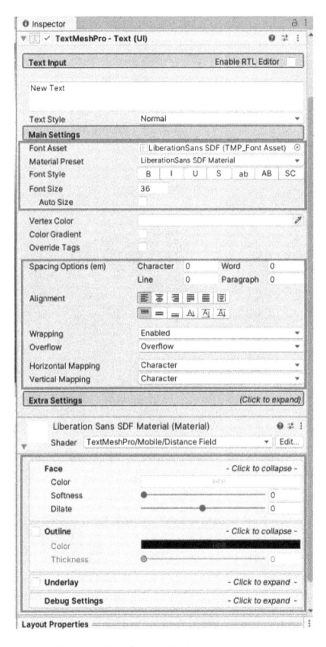

Figure 3.32 – The TextMeshPro component

Using TextMesh Pro to implement your UI text is a good choice.

# Selectable UI components

You can use selectable components in uGUI to handle interactions. These components include **Button**, **Toggle**, **Slider**, **Dropdown**, **Input Field**, and **Scrollbar**. In this section, we will mainly discuss the most commonly used component, namely, the **Button** component.

## Button

Creating a **Button** element for the game UI is very simple; just follow these steps:

1.  Right-click in the **Hierarchy** window to open the menu.

2.  Select **UI > Button - TextMeshPro**.

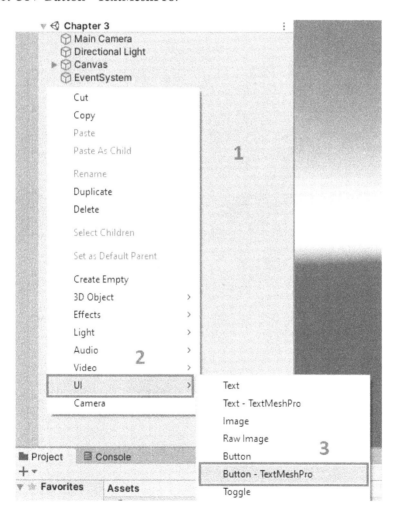

Figure 3.33 – Creating a Button object

As shown in *Figure 3.33*, there are two options to create a button in the menu, namely, **Button** and **Button -TextMeshPro**. Here, we select **Button -TextMeshPro** so that the text content on the button is rendered by **TextMeshPro**.

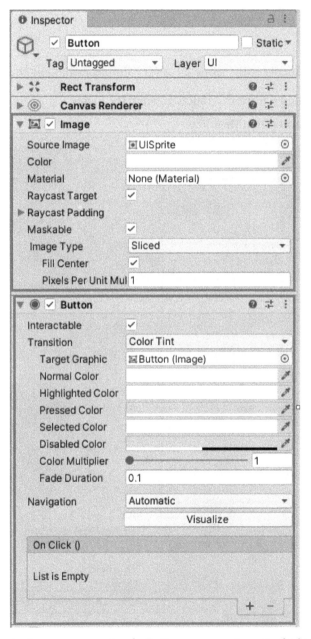

Figure 3.34 – An Image component and a Button component are attached to the button

Once a default button object is created, this object includes not only a **Button** component but also an **Image** component. This is because the **Button** component only provides the function of interacting with the user; it does not provide the function of graphic display. Therefore, the image of the button needs an **Image** component to display.

## Selected states

The **Button** component has five **selected states** inherited from the `Selectable` class, namely, `Normal`, `Highlighted`, `Pressed`, `Selected`, and `Disabled`, which are defined by an enumeration named `Selectable.SelectionState`. Therefore, as shown in *Figure 3.34*, there are five different colors in the **Transition** section corresponding to these five different selected states, which means that when the user interacts with this button, this button will provide different feedback according to the different states.

## onClick

The important role of a button is to receive user clicks and trigger corresponding events. In Unity, it is very easy to set up button `onClick` events. You can either manually set up button `onClick` events in the editor or set button `onClick` events programmatically.

In order to set up a new event to the button in the editor, we can click the + button at the bottom of the **On Click ()** section, as shown in *Figure 3.35*. This will create a new action.

Figure 3.35 – Setting up a new onClick event in the editor

We can also programmatically set the button `onClick` event; the following code shows how to do this:

```
using UnityEngine;
using UnityEngine.UI;

public class ButtonClickExample : MonoBehaviour
{
    // Start is called before the first frame update
    void Start()
    {
        var button = GetComponent<Button>();

        button.onClick.AddListener(() =>
```

```
        {
            Debug.Log(You have clicked the button!);
        });
    }
}
```

In this section, we learned about commonly used UI components and got an understanding of uGUI, the UI solution provided by Unity. Next, we will explore the UI Event System in Unity. If there is no event system in the scene, UI elements such as buttons cannot interact with players, so it's an important topic.

# C# scripts and the UI Event System in Unity

**EventSystem** is a mechanism for sending events to objects in a game that supports keyboards, mice, screen touches, and so on. EventSystem consists of multiple modules for sending events. If there is no **EventSystem** object in the scene, then, when creating a canvas, an **EventSystem** object will be automatically created along with it.

Figure 3.36 – EventSystem

As shown in *Figure 3.36*, the **Inspector** window of the **EventSystem** object exposes very few functionalities. This is because EventSystem is designed as a manager for cooperation between various **input modules**.

It should be noted that there can be, at most, one **EventSystem** object in a scene. If there are multiple **EventSystem** objects in the scene, a warning message will be displayed, as shown in *Figure 3.37*:

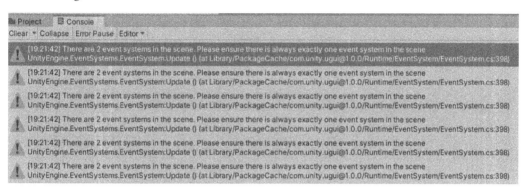

Figure 3.37 – A warning message when there are multiple EventSystem objects

When the game is running, **EventSystem** will look for the **InputModule** component attached to the same GameObject. This is because **InputModule** is the class responsible for the main logic of **EventSystem**. We can also find the Input Module used in this case, as shown in *Figure 3.36*, namely, **Standalone Input Module**. Next, we will introduce Input Modules.

# Input Modules

Unity provides two built-in Input Modules, namely, the **Standalone Input Module** and the **Touch Input Module**. In the past, the Standalone Input Module was used for keyboards, mice, and game controllers, and the Touch Input Module was for touch panels such as smartphones. Nowadays, the Standalone Input Module is compatible with all platforms and the Touch Input Module has been deprecated, so you can treat the Input Module as the Standalone Input Module.

The purpose of the Input Module is to map hardware-specific inputs (such as touches, joysticks, mice, and game controllers) to events sent through the messaging system.

# The new Input System package

In addition to this default built-in Input Module, Unity also provides a new, more powerful, flexible, and configurable **Input System** package.

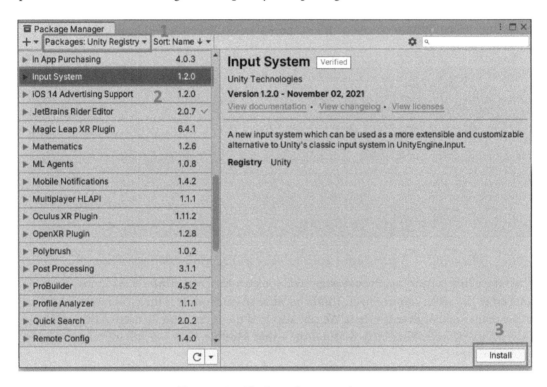

Figure 3.38 – The Input System package

If you want to use the new input system, then you need to install the package from the Package Manager window, as shown in *Figure 3.38*. Moreover, a newly created **EventSystem** component will still use the legacy **Standalone Input Module** component by default, so you need to manually replace it with the new **InputSystemUIInputModule** component, as shown in *Figure 3.39*:

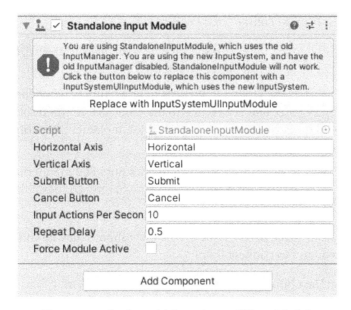

Figure 3.39 – Replace with InputSystemUIInputModule

By reading this section, we learned that in order to ensure that the game UI can correctly respond to player input, an **EventSystem** component and an Input Module are necessary. Next, let's move on to discussing how to create UI in Unity using the Model-View-ViewModel (MVVM) pattern.

# The Model-View-ViewModel (MVVM) pattern and the UI

A common challenge in Unity development is to find elegant ways to decouple components from each other, especially when developing the UI because it involves UI logic and UI rendering. **Model–View–ViewModel** (**MVVM**) is a software architectural pattern that helps developers separate the **ViewModel**, which is the UI logic, from the **View**, which is the UI graphics. In this section, we will explore how to implement an MVVM pattern in Unity.

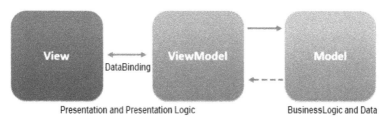

Figure 3.40 – MVVM

As its name suggests, MVVM consists of three parts:

- **Model**: This refers to the data access layer, which can be `Database`, or `PlayerPrefs`, which stores player preferences in Unity, and so on.

- **View**: This represents the Unity UI. It needs to be a Unity component that inherits from `MonoBehaviour` and is attached to the UI object. Its main role is to manage UI elements and trigger UI events, but it does not implement any concrete UI logic itself.

- **ViewModel**: This can be a pure C# class and does not need to inherit from `MonoBehaviour`. It does not need to consider what the UI looks like; it only needs to implement concrete logic.

We can see that there are three parts in MVVM, so how should they be connected? Generally, we use two ways to connect them:

- **Data binding**: Data binding is the key technology of MVVM. It is used to bind and connect the properties of `ViewModel` and `View`. Elements bound to data will automatically reflect every data change. By using data binding, a `ViewModel` can modify the value of the UI control in the View.

- **Event-driven programming**: This method is used to raise events from the View triggered by user actions, which are then processed by the ViewModel.

There are some mature MVVM framework implementations for Unity, such as the **Loxodon Framework**, which is a lightweight MVVM framework built specifically to target Unity. You can find its repository on GitHub (`https://github.com/vovgou/loxodon-framework`) or add it to your project via Unity Asset Store directly.

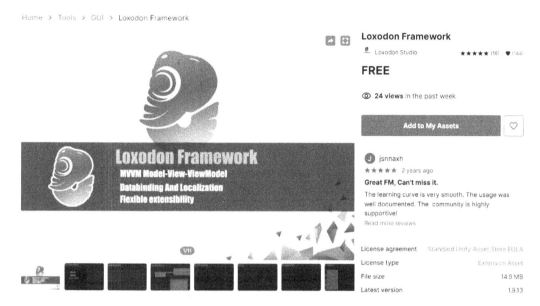

Figure 3.41 – Loxodon Framework

Since our next example will use this framework, I recommend that you import this framework into your project first. After importing this framework, you should find it in the `Assets` folder of your project.

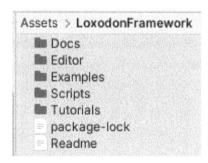

Figure 3.42 – The LoxodonFramework folder

Now, let's perform the following steps to implement a sample MVVM UI via **LoxodonFramework** in Unity:

1.  First, let's set up `LoxodonFramework` in our game scene. We need to create a new canvas and add the **GlobalWindowManager** component to this canvas, as shown in *Figure 3.43*. A **GlobalWindowManager** component is a container that is used to manage views.

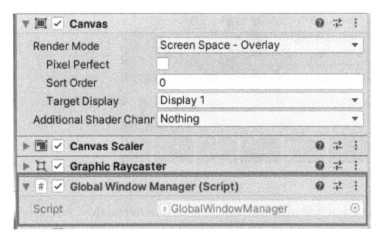

Figure 3.43 – The GlobalWindowManager component

2.  Next, we need to define a view. As we mentioned earlier, a view represents UI elements in Unity. As you can see from the following code, this view is relatively simple, containing only a button UI element and a text UI element, and this `SampleView` class inherits from the `Window` class in the Loxodon Framework. In the following code, you can also find the `BindingSet` class, which is used to bind and connect properties of `ViewModel` and `View`:

```
using UnityEngine;
using UnityEngine.UI;
using Loxodon.Framework.Views;
using Loxodon.Framework.Binding;
using Loxodon.Framework.Binding.Builder;
using Loxodon.Framework.ViewModels;
using TMPro;

public class SampleView : Window
{
    [SerializeField]
```

```
    private Button _submitButton;

    [SerializeField]
    private TextMeshProUGUI _message;

    private SampleViewModel _viewModel;

    protected override void OnCreate(IBundle bundle)
    {
        _viewModel = new SampleViewModel();
        BindingSet<SampleView, SampleViewModel>
          bindingSet =
          this.CreateBindingSet(_viewModel);

        bindingSet.Bind(_message).For(v =>
          v.text).To(vm => vm.Message).OneWay();
        bindingSet.Bind(_submitButton).For(v =>
          v.onClick).To(vm => vm.Submit);
        bindingSet.Build();
    }
}
```

Let's break down this example:

- The two _submitButton and _message fields of this SampleView class refer to a Button component and a TextMeshProUGUI component, respectively.

- In the OnCreate method, we first create a BindingSet instance to bind SampleView to its corresponding ViewModel class – that is, SampleViewModel. We will introduce how to create the SampleViewModel class later.

- Then, we bind the text property of the _message field in SampleView to the Message property in SampleViewModel by calling the Bind method of BindingSet. You can see in the code that we use OneWay binding here, which means that only the view model can modify the value of the UI element in the view.

- We also bind the onClick event of the _submitButton field in SampleView to the Submit method in SampleViewModel. Finally, we call the Build method of BindingSet to build the binding.

3.   At the same time, we also need to create these required UI elements in the Unity scene, as shown in the following figure. Let's call it **SampleUI**.

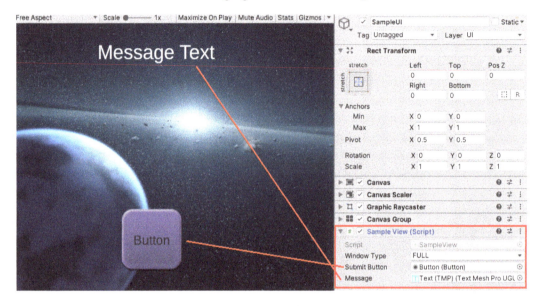

Figure 3.44 – Setting up the UI elements

4.   Then, let's create a new folder called `Resources` and create a prefab for this sample UI by dragging it from the **Hierarchy** window to the **Resources** folder, as shown in the following screenshot. So far, we have created UI elements and a **View** component that represent UI elements in the MVVM architecture. **SampleUI** can be removed from the scene because we will load its prefab and create the UI at runtime.

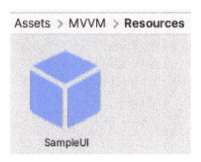

Figure 3.45 – The SampleUI prefab

5.  We also need a `SampleViewModel` class, which implements concrete logic. The `SampleViewModel` class inherits from the `ViewModelBase` class in the Loxodon framework, and the logic is implemented in the `Submit` method, which modifies the `Message` property. In the view we created earlier, we bound the button's `onClick` event to the `Submit` method in the `SampleViewModel` class, and we also bound the view's `text` property of the Text UI element to the `Message` property of `SampleViewModel`. Therefore, after the `Submit` method modifies the `Message` property, the modified message content will be displayed on the UI:

```
using Loxodon.Framework.ViewModels;

public class SampleViewModel : ViewModelBase
{
    private string _message;
    private int _count;

    public SampleViewModel() { }

    public string Message
    {
        get { return _message; }
        set => Set<string>(ref _message, value,
            Message);
    }

    public void Submit()
    {
        _count++;
        Message = $The number of times the button is
            clicked: {_count};
    }
}
```

6.  Finally, start up code is needed to register services and create the UI. The following
    start up code supports loading the prefab of **SampleUI** and setting up the view. You
    can find the `ApplicationContext` class in the following code; we use it to store
    data and services that can be accessed by other classes in the Loxodon Framework.
    Then, the code registers the `IUIViewLocator` service to load the UI prefab and
    create the UI elements:

```
public class Startup : MonoBehaviour
{
    private ApplicationContext _context;

    private void Awake()
    {
        _context = Context.GetApplicationContext();

        // Register services
        IServiceContainer container =
          _context.GetContainer();
        container.Register<IUIViewLocator>(new
          ResourcesViewLocator ());

        var bundle = new
          BindingServiceBundle
          (_context.GetContainer());
        bundle.Start();
    }

    private IEnumerator Start()
    {
        // Create a window container
        var winContainer =
          WindowContainer.Create(MAIN);

        yield return null;

        IUIViewLocator locator =
          _context.GetService<IUIViewLocator>();
```

```
   var sampleView =
     locator.LoadWindow<SampleView>(winContainer,
     SampleUI);
   sampleView.Create();
   ITransition transition =
     sampleView.Show().OnStateChanged((w, state)
     =>
     {
     });

   yield return transition.WaitForDone();
  }
 }
```

7. Let's run the game. As you can see in the following screenshot, we create a view that displays the message text at the top and a **Submit** button at the bottom. Once we click the **Submit** button, an event will be triggered and processed by the `SampleViewModel` class to update the message information, and the view will also update the UI text to display the latest information through data binding.

Figure 3.46 – The sample UI with MVVM

This way, the UI graphics and UI logic are separated. UI designers and programmers can work at the same time without relying on each other, thereby improving the efficiency of UI development in Unity.

In this section, we discussed how to use MVVM to implement the UI in Unity. Next, we will learn what we must pay attention to when implementing the UI in Unity – that is, optimizing UI performance.

# Performance tips to increase performance of the UI

The UI is an important part of a game, so if you do not implement it properly, it may cause potential performance issues. In this section, we will discuss the best practices for implementing the game UI in Unity to optimize the performance problems caused by the UI.

## The Unity Profiler

The first best practice tip is to be good at using the Unity **Profiler**. The Profiler is a tool that you can use to get performance data about your game, including **CPU Usage**, **GPU Usage**, **Rendering**, **Memory**, **UI**, and **UI Details**. In order to view performance data about the UI, perform the following steps:

1.  Click **Window** > **Analysis** > **Profiler** to open the **Profiler** window.
2.  Click the **UI** or **UI Details** module area in the **Profiler** window to view performance data related to the UI, such as the CPU time consumed by **Layout** and **Render**.

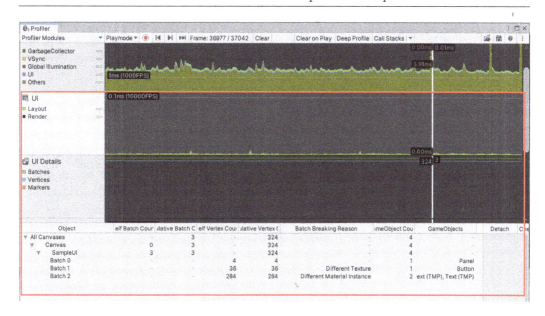

Figure 3.47 – The UI area in the Profiler window

In addition to the **UI** and **UI Details** areas, the **CPU Usage** area in the **Profiler** window also provides performance information related to the UI. In the **CPU Usage** area, you can see the CPU time consumed by a specific marker, such as **UGUI.Rendering. RenderOverlays**, as shown in the following screenshot:

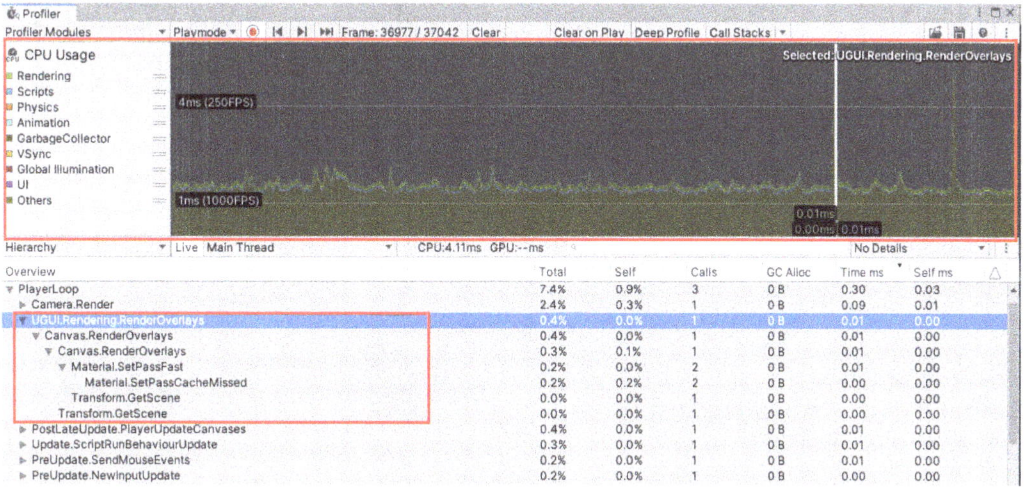

Figure 3.48 – The CPU Usage area in the Profiler window

This was just a brief introduction to the Profiler tool. In the following chapters, we will discuss the Unity Profiler in detail.

# Multiple canvases

The second-best practice tip is a very important aspect that needs to be considered when implementing the UI in Unity, especially when your game UI is very complex. If necessary, you may need to create multiple canvases to manage and display different UI elements. As we have mentioned before, a canvas generates meshes representing the UI elements placed on it and regenerates the meshes when the UI elements change.

Suppose that you build the UI of the entire game in a single canvas with thousands of UI elements, and when one or more UI elements on the canvas change, all the meshes used to display the UI regenerate. This may be expensive, and you may experience CPU spikes that take a few milliseconds.

Therefore, it is a good idea to create multiple different canvases to manage them, based on the update frequency of UI elements. For example, frequently updated dynamic UI elements such as progress bars and timers can be in one canvas, and infrequently updated static UI elements such as UI panels and background images can be in another. Of course, there is no magic bullet; you need to manage the canvas on a project-by-project basis.

# Use Sprite Atlas

As we introduced when discussing UI images, sprites are 2D graphic objects used for the UI and other elements of 2D gameplay. When importing a new texture into the Unity Editor, we can set the texture type of this texture to a sprite. So, your game project may contain a lot of sprite files. If so, many sprites are treated as separate individuals, and rendering performance may decrease. This is because Unity will issue a **draw call** for each sprite in the scene, and multiple draw calls may consume a lot of resources and negatively affect your game performance.

> **Note**
> A draw call is a call to the graphics API to draw objects (for example, to draw a triangle).

As shown in the following screenshot, there are two draw calls to render **Button1** and **Button2** because these two buttons use two different textures:

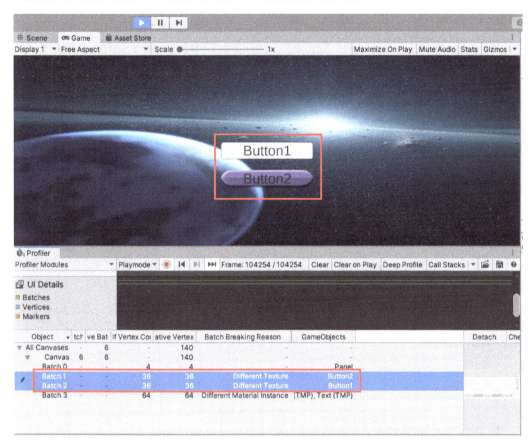

Figure 3.49 – Multiple draw calls

So, it is a good idea to combine several textures or sprites into a combined texture.

We can perform the following steps to use the **Sprite Atlas** provided by Unity to combine textures:

1.  If the **Sprite Atlas** packing is disabled, enable it in **Edit** > **Project Settings** > **Editor** > **Sprite Packer** > **Mode**.

2.  Click **Assets** > **Create** > **2D** > **Sprite Atlas** to create a Sprite Atlas asset.

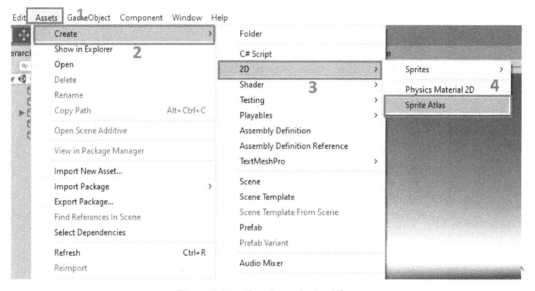

Figure 3.50 – Creating a Sprite Atlas

3.  Under the **Objects for Packing** drop-down menu of the Sprite Atlas asset, select the + symbol to add textures or folders to the Sprite Atlas.

However, we still need to be aware that although Sprite Atlas can effectively reduce the count of draw calls, improper use can easily lead to a waste of memory. When a sprite is active in an atlas, Unity loads all the sprites in the atlas to which the sprite belongs. If there are many sprites in an atlas, even if only one sprite is referenced in the scene, the whole atlas will be loaded, which will cause large memory consumption. In order to solve this problem, the sprites can be packaged into multiple smaller atlases according to their purpose. For example, the sprites used in the login panel can be packaged as a login panel atlas, while the sprites used in the game character panel are packaged as a character panel atlas.

# Summary

In this chapter, we started by introducing some of the most commonly used UI component classes of the uGUI solution, such as the **Canvas**, **Rect Transform**, and **Image** components. We then explained the Event System in Unity, the legacy Input Module, and the new more powerful Input System package provided by Unity.

We also discussed how to decouple components from each other when developing the UI in Unity by using the MVVM architectural pattern.

Finally, we explored some best practices for implementing the game UI in Unity to optimize the performance problems caused by the UI.

In the next chapter, we will learn about the animation system in Unity and, at the same time, we will also introduce how to optimize animation performance in Unity.

# 4

# Creating Animations with the Unity Animation System

Whether for 2D games or 3D games, if you want a game to be lively and interesting, good animation is essential. As a very popular game engine, Unity provides easy-to-use and powerful animation development tools. In this chapter, we will explore the animation system in Unity, sometimes referred to as **Mecanim**, to make Scenes and characters in your game not static, but dynamic. Then, we will demonstrate how to implement 3D and 2D animation in Unity with two examples. Finally, we'll cover how to improve the performance of the animation system in Unity.

We will cover the following key topics in this chapter:

- Exploring the Unity animation system's concepts
- Implementing 3D animation in Unity
- Implementing 2D animation in Unity
- Improving the performance of Unity's animation system

By the end of this chapter, you will be able to create 3D and 2D animations in Unity, as well as knowing how to control animations through C# code and how to optimize animation performance.

# Technical requirements

Before starting, I recommend you first download **Unity-Chan! Model** from Unity Asset Store: `https://assetstore.unity.com/packages/3d/characters/unity-chan-model-18705`.

This cute 3D girl model asset is produced by Unity Technologies Japan, and it's available for all developers to download and make games with it.

The following content is included in this asset:

- 3D models with beautiful textures
- "Unity-Chan!" original shaders
- 31 animations
- 31 still poses
- 12 emotions made from blend shapes
- A sample locomotion scene and other sample Scenes

Now, let's get started!

# Exploring the Unity animation system's concepts

Animation is an important aspect of game development. In this section, we will first learn the basic concepts of the Unity animation system. Specifically, we will introduce the following concepts:

- What Animation Clips are and how to create an Animation Clip in Unity
- How to create an Animator Controller to manage a set of animations for characters
- How to use the Avatar system to work with animation rigging
- What the Animator component is and how to use it to assign animation to a GameObject

Let's move on!

# Animation Clips

Animation in Unity can range from simple cube rotation to complex character movement and actions, and they are all based on **Animation Clips**, which are used to store keyframe-based animations in Unity.

We can manually create an Animation Clip file in the Unity Editor to implement some simple traditional keyframe animation effects via the **Animation** window, such as simple movement, rotation, and so on.

The following steps show how to animate a GameObject in a Scene:

1. Right-click in the **Hierarchy** window (on the right-hand side) and select **3D Object | Cube** from the pop-up menu to create a new **Cube** object in the Scene.

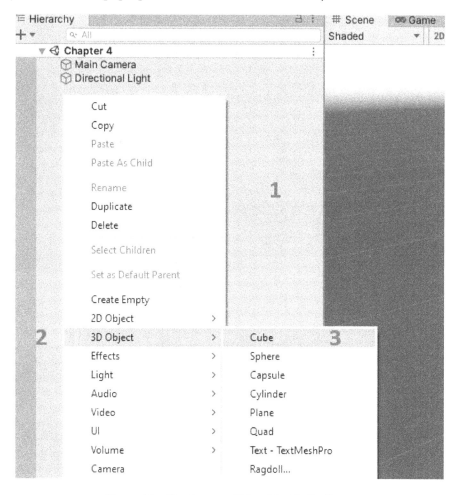

Figure 4.1 – Create a new Cube object in the Scene

2.  Select the **Cube** object in the Scene view. Then navigate to **Window | Animation | Animation** to open the **Animation** window. In addition to opening this window from the menu, we can also use the *Ctrl + 6* shortcut to open it:

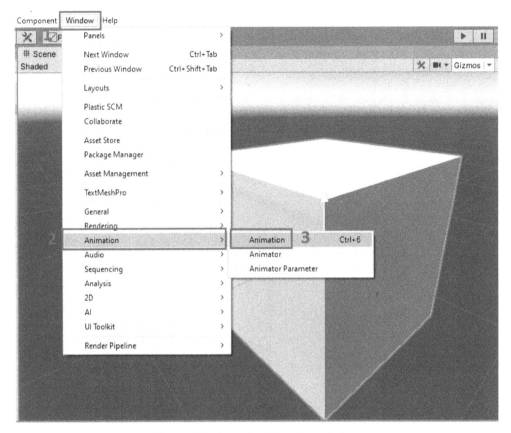

Figure 4.2 – Open the Animation window

3.  Click on the **Create** button in the **Animation** window to create a new Animation Clip:

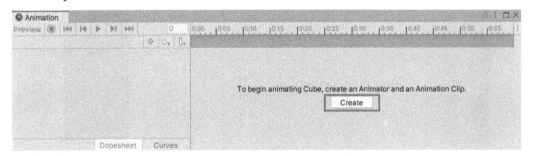

Figure 4.3 – The Animation window

4.  Click on the **Add Property** button to display a list of available properties that can be animated. As shown in *Figure 4.4*, we can not only modify **Position** and **Rotation** but also modify the properties of other components. Here we can add **Scale** as the property that will be animated by clicking the + button next to it:

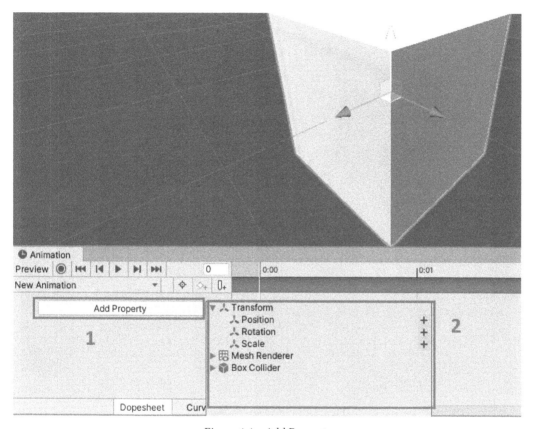

Figure 4.4 – Add Property

5.  When a property is added, two keyframes are created by default: the first keyframe and the second keyframe are at *0:00* and *1:00* on the timeline respectively. So, we need to create a third keyframe to change the **Scale** property of this Cube:

I.   Place your cursor at *0:10* on the timeline.

II.  Right-click in the **Animation** window and click on **Add Key** from the pop-up menu to add a new keyframe.

III.    Set **Scale.x** to 0.5, as shown in *Figure 4.5*:

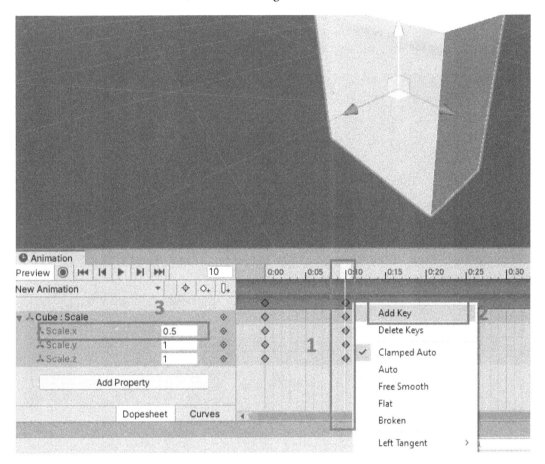

Figure 4.5 – Add keyframes

6.    In order to preview the animation, click on the **Play** button to play the Animation Clip. You will see that the volume of the Cube shrinks rapidly and then slowly enlarges.

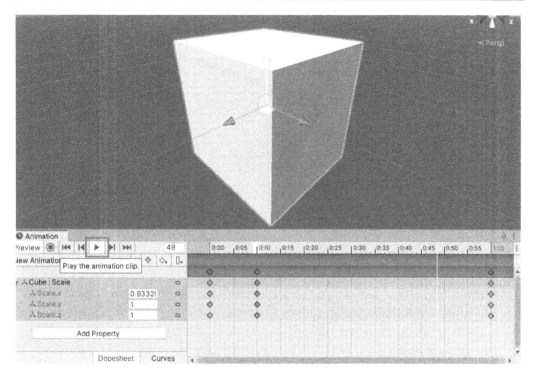

Figure 4.6 – Play the Animation Clip

We can also use **recording mode** to create an Animation Clip in Unity, as demonstrated in these steps:

1.  The steps to create a new GameObject and open the **Animation** window are the same as before. So, let's start directly with how to use recording mode to create an Animation Clip for a Sphere object in the Scene. We can click the record button to enable keyframe recording mode, as shown in *Figure 4.7*:

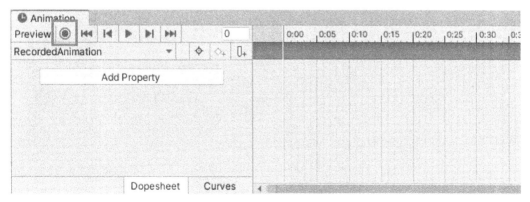

Figure 4.7 – Enable keyframe recording mode

2.  After clicking the record button, it will enter recording mode. Now, we can modify the point in time that we want it to be at by dragging on the timeline:

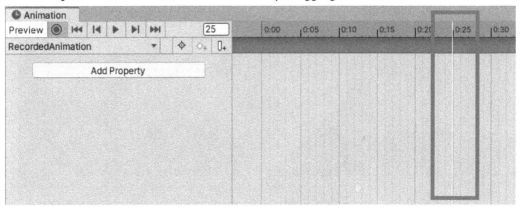

Figure 4.8 – Drag on the timeline

In recording mode, whether you move, rotate, or scale the target GameObject in the Scene, Unity will automatically add the keyframe of the current time point to the Animation Clip. Here we can move the GameObject from its original position (0, 0, 0) to a new position, let's say, (1, 0, 0). And you can see in the following figure that Unity created keyframes for the Sphere object:

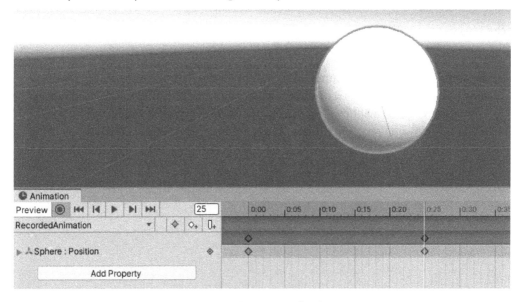

Figure 4.9 – Unity creates keyframes

3.  Finally, click on the record button again to exit recording mode and click the **Play** button to play the Animation Clip we just created:

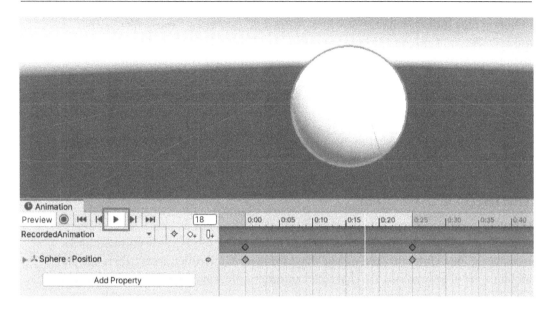

Figure 4.10 – Play the Animation Clip

In addition, importing external animation assets into the Unity Editor can also automatically create Animation Clip files.

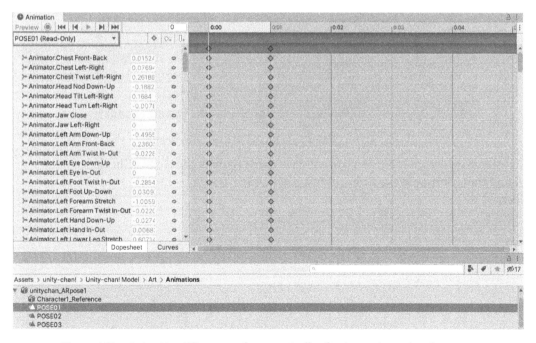

Figure 4.11 – Animation Clips created automatically after importing animation assets

Animation files such as generic **FBX** files, **Autodesk® 3ds Max® (.max)** files, native **Autodesk® Maya® (.mb or .ma)** files, and **Blender™ (.blend)** files need to be imported into our Unity project first before they can be used by Unity. After animation files are imported, Unity will generate Animation Clip files. We can open the **Animation** window to view an Animation Clip by double-clicking the Animation Clip file in the Unity Editor, as shown in *Figure 4.11*.

Now that you understand what an Animation Clip is and how to create a new one in Unity, let's move on to the next concept: the Animator Controller.

## Animator Controller

Imagine that our game character has multiple animations. For example, say a character can both run and attack – it is very important to manage both of these animations for the character. In a Unity project, we use the **Animator Controller** asset to arrange and maintain a set of animations for characters or other animated GameObjects.

An **Animator Controller** will reference the Animation Clips it uses and use a so-called **state machine** to manage various animation states and transitions between them.

We can import the **Unity-Chan! Model** asset that we downloaded earlier.

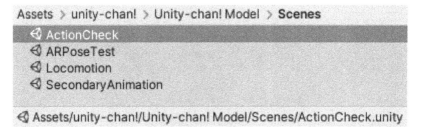

Figure 4.12 – Unity-Chan! Model ActionCheck Scene

This asset provides multiple demo Scenes; we chose to open the **ActionCheck** Scene. You can find these Scenes in the `Assets/unity-chan!/Unity-chan! Model/Scenes` folder.

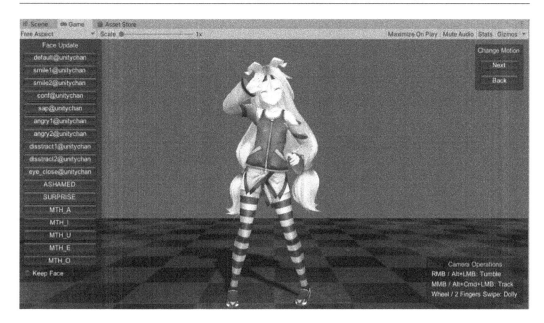

Figure 4.13 – Unity-Chan! Model

As *Figure 4.13* shows, the Unity-Chan! model has been set up in the Scene. If we open the Animator Controller file used by this model, we can see all the Animation Clips used by this model and the transitions between Animation Clips in the state machine displayed in the **Animator** window, as shown in *Figure 4.14*.

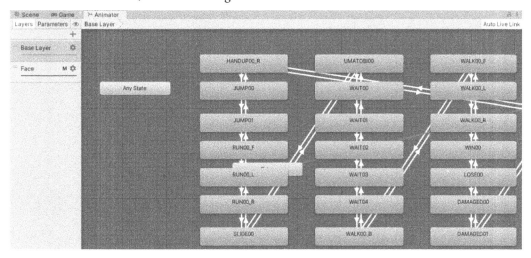

Figure 4.14 – Animator Controller

We can also follow these steps to manually create an **Animator Controller** asset in the Unity Editor:

1. Select the **Project** view and right-click to open the menu.
2. Select **Create | Animator Controller** to create a new **Animator Controller** asset:

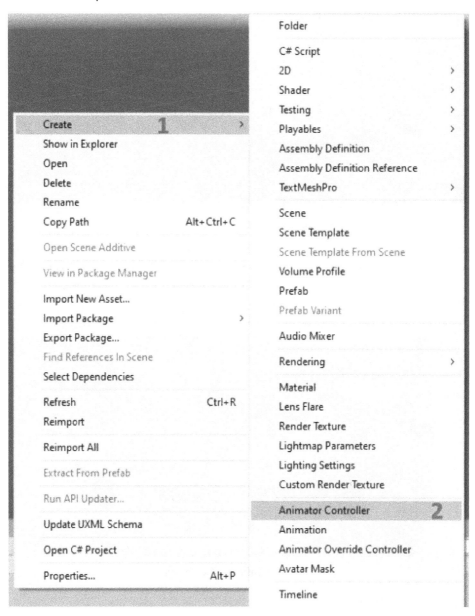

Figure 4.15 – Create a new Animator Controller

3.  Double-click the **Animator Controller** asset we just created to open the **Animator** window.

4.  Here, drag the **POSE01** Animation Clip into the **Animator** window directly to create a new state. In the state machine, a state is represented by a box because the **POSE01** Animation Clip is the first animation we dragged, so we can see this animation is connected to the entry point of the Animator Controller, indicating that this animation will be the default animation.

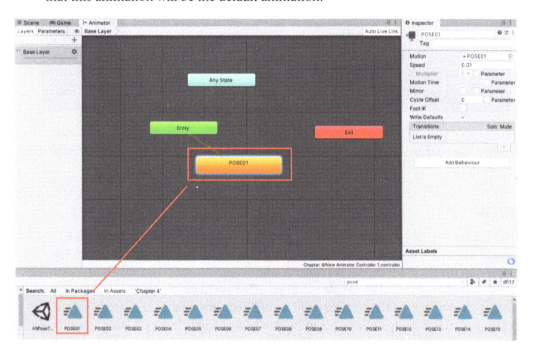

Figure 4.16 – Create a new state

5.  Create the second state by dragging the **POSE02** Animation Clip into the **Animator** window.

6.  Select the **POSE01** state and right-click to open a menu, then select **Make Transition** to make a transition between **POSE01** and **POSE02**.

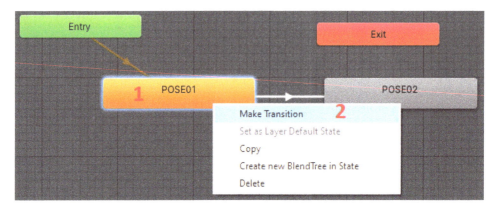

Figure 4.17 – Make Transition

Now, we've created an Animator Controller asset and added some Animation Clips to the state machine.

## Avatar

Unlike the animation we created for Unity's built-in Cube model earlier, the model imported from the external tools into the Unity Editor may be more complicated. For example, the Unity-Chan! model is a human-like model. A model in Unity is represented by a mesh of **triangles**, and a triangle is composed of **vertices**. When the model is animated, the position of the vertices will be modified. Obviously, when many vertices make up a model, moving each vertex individually is an inefficient operation. Therefore, a common technique in computer animation is not to move each triangle individually during the animation but to skin the model before it is animated. This technique is called **skeletal animation** or **rigging.**

Unity uses a system called **Avatar** to identify whether the animation model is a humanoid layout and which parts of the model correspond to the head, body, arms, legs, and so on.

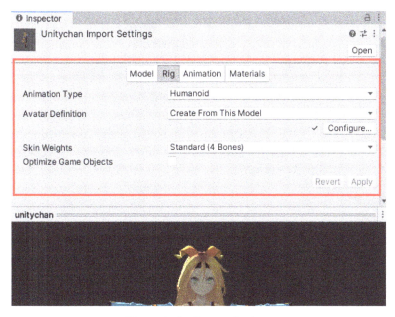

Figure 4.18 – Import Settings

We can open the **Import Settings** window for Unity-Chan! by clicking the model in the Unity Editor. As *Figure 4.18* shows, we can specify the kind of rig it is in the **Rig** tab of the window, and in this case, **Animation Type** for this model is **Humanoid**. The animation system will try to match the model's existing bone structure with the **Avatar** bone structure. If the bone structure can be successfully mapped, then an **Avatar** asset will be created automatically as shown in *Figure 4.19*.

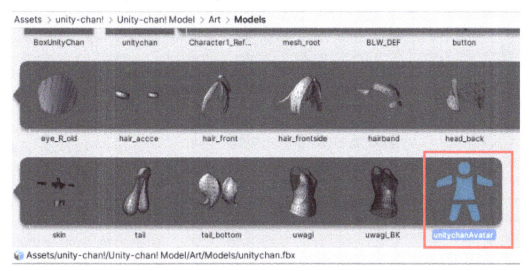

Figure 4.19 – unitychanAvatar

> **Note**
>
> **Bones** are a hierarchical set of interconnected parts of **skeletal animation**.
> **Skinning** makes each vertex of the triangle depend on the bone.

On the other hand, if the animation system cannot automatically match the model's existing bone structure with the **Avatar** bone structure, we need to configure the **Avatar** manually. In addition, even if the bone structure can be successfully mapped, sometimes we want to manually adjust things to achieve better results. At this time, we can also modify it by configuring the **Avatar** asset.

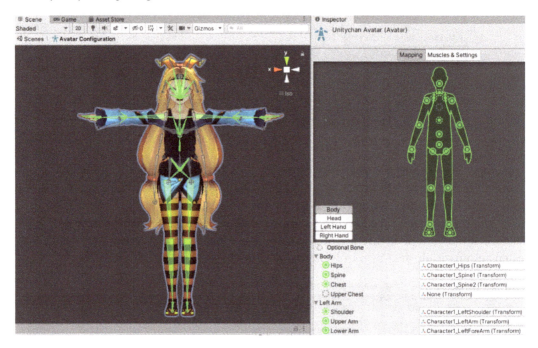

Figure 4.20 – Configure Avatar

Follow these steps to configure it:

1. Click on the model to open the **Import Settings** window for it.

2. Click the **Configure** button in the **Rig** tab of the window to open the **Avatar Inspector** window.

3. Configure the bones in the **Avatar Inspector** window as shown in *Figure 4.20*.

## Avatar Mask

After we create a mapping between the bones of the model and the bone structure of Unity's Avatar system, we can play the animation of this character. However, sometimes we may not want to animate all the bones of the character. A common example is that the walking animation may involve a character swinging their arms, but if they pick up a phone to make a call, their arms should hold the phone instead of swinging as they walk. In this case, we want to restrict an animation to specific body parts, and the **Avatar Mask** asset provided by Unity can help us achieve this goal.

We can create a new **Avatar Mask** asset by selecting **Assets | Create | Avatar Mask** as shown in *Figure 4.21*.

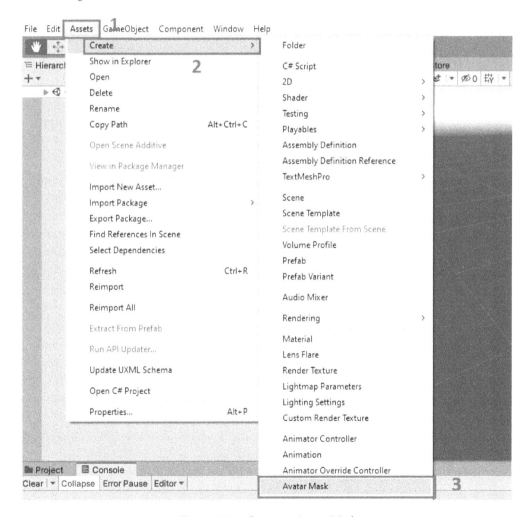

Figure 4.21 – Create an Avatar Mask

After creating a new **Avatar Mask** asset, we can configure it to define which parts of the animation should be masked. As *Figure 4.22* shows, the **Avatar Mask Inspector** window allows us to click on a diagram of a humanoid body to select or deselect certain parts to mask. Here we mask an arm of Unity-Chan!, which means some animation will not affect this arm at runtime.

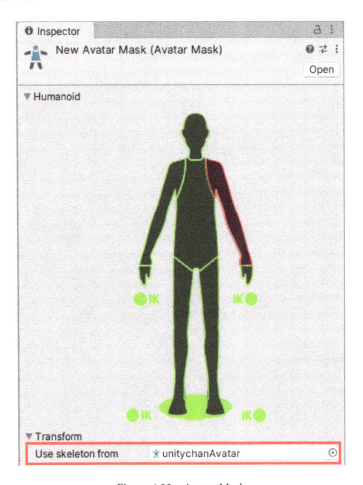

Figure 4.22 – Avatar Mask

In order for this Avatar Mask asset to take effect, we need to apply it to an Animator Controller.

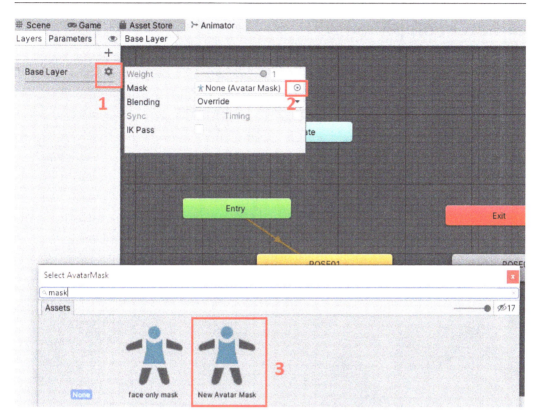

Figure 4.23 – Apply the Avatar Mask asset

In this case, we will apply this Avatar Mask asset to the Animator Controller we created earlier as follows:

1.  Double-click the **New Animator Controller** file to open the **Animator** window.

2.  Click on the gear icon of the **Base Layer** item to open the **Layer settings** panel.

3.  Then, click the radio button next to the **Mask** field and select the New **Avatar Mask** asset to apply from the **Select AvatarMask** window that pops up, as shown in *Figure 4.23*.

In this way, we can limit the animation to specific body parts.

Next, we will explore another important concept in Unity's Animation development solution, namely the **Animator** component. By using Animator, we can use this Animator Controller asset in our game.

# Animator component

In the previous sections, we explored Animation Clips, Animator Controllers, and Avatar in Unity. However, just creating Animation Clips, Animator Controllers, and Avatar assets is not enough to animate the characters in a game Scene. We still need the Animator component to assign animation to the GameObject in the Scene.

> **Note**
>
> Animator Controllers and Animator components have similar names but different functions. An Animator component uses an associated Animator Controller to apply animations to a GameObject.

If you see an Animator component on a GameObject, you will find that it will bring together all the various assets we discussed before. It is the root of the binding system in Unity's animation solution, so it is very important.

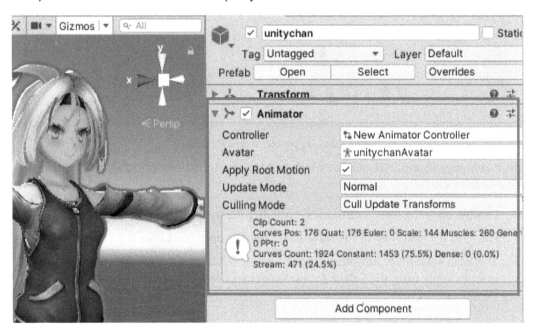

Figure 4.24 – Animator component

Here we can drag the Unity-Chan! model into the Scene to create a new character GameObject, and add an Animator component to the GameObject, as *Figure 4.24* shows. This is how the Animator component was configured:

- The Animator component needs to reference an Animator Controller, which defines the Animation Clips to be used. We can assign the Animator Controller that we created in the *Animator Controller* section to it, with the name **New Animator Controller**.

- Since the Unity-Chan! model is a humanoid model, provide the corresponding Avatar asset to this Animator component.

- The **Apply Root Motion** setting of the Animator component determines whether or not any change to the position or rotation of the root node will be applied.

- The **Update Mode** setting of the Animator component determines the update mode of the Animator component. There are three different options, namely **Normal**, **Animate Physics**, and **Unscaled Time**.

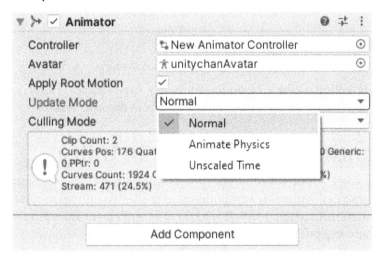

Figure 4.25 – The Update Mode setting

- The last setting is **Culling Mode**, which determines whether the animations of the Animator component should play off-screen. There are three different options, namely **Always Animate**, **Cull Update Transforms**, and **Cull Completely**.

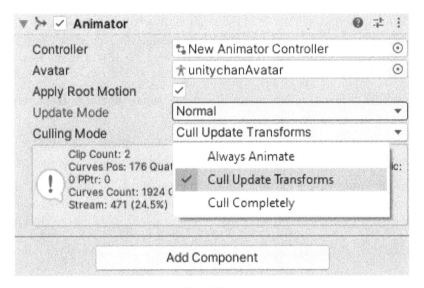

Figure 4.26 – The Culling Mode setting

After reading this section, we have an understanding of the concepts of Unity's animation system. We will use this system to create 3D animations in the next section!

# Implementing 3D animation in Unity

We have covered some important concepts, such as **Animation Clips**, **Animator Controllers**, **Avatar**, and **Animator components**, in the Unity animation system in the previous sections. In this section, you will learn how to implement animation for 3D models with these concepts.

# Importing animation assets

First, we need to know how to import animation assets into Unity from **digital content creation (DCC)** software. As a demonstration, we still use the Unity-Chan! model asset as an example. We can find all the animation assets in the `/Assets/unity-chan!/Unity-chan! Model/Art/Animations` folder as shown in the following screenshot:

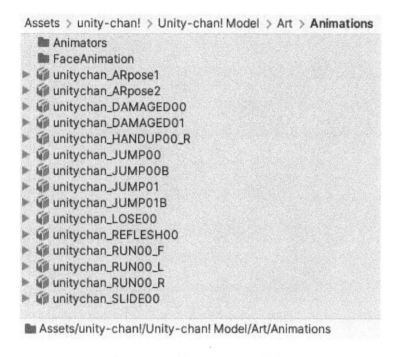

Figure 4.27 – The Animations folder

Here we can select one animation asset in the **Project** window to open its **Import Settings** window.

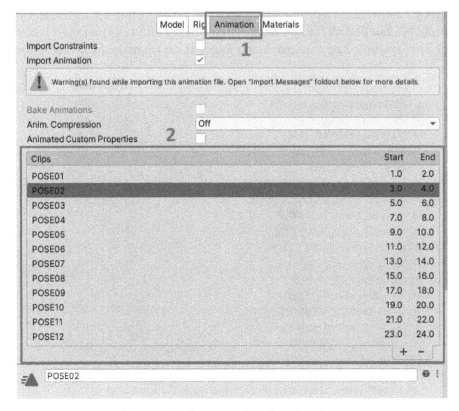

Figure 4.28 – Import settings for animations

As shown in *Figure 4.28*, click **Animation** in the **Inspector** window to switch to the **Animation** tab, and you can see all the Animation Clips contained in the animation asset.

## Animation compression

In addition, in the **Animation** tab, we can also find animation-related import settings.

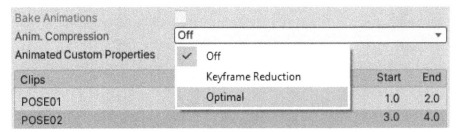

Figure 4.29 – The Anim. Compression setting

As shown in *Figure 4.29*, there is a setting called **Anim. Compression**, whose value is **Off** by default, which means that Unity doesn't reduce the keyframe count on import. In this case, Unity will keep the highest precision animation, but at the expense of a large animation size. If reducing the size of the animation, whether on our hard disk or in memory, is important, we can consider the two other **Anim. Compression** options, which are **Keyframe Reduction** and **Optimal**.

| Anim. Compression | Optimal |
|---|---|
| Rotation Error | 0.5 |
| Position Error | 0.5 |
| Scale Error | 0.5 |

Rotation error is defined as maximum angle deviation allowed in degrees, for others it is defined as maximum

Figure 4.30 – Optimal

If the **Keyframe Reduction** option is selected, the **Animation Compression Error** options will be displayed, as shown in the preceding figure. The values mean how much to reduce the precision of the Animation Clip to. The default value is 0.5; the smaller the value, the higher the precision.

If the **Optimal** option is selected, Unity will decide how to compress the Animation Clip.

## Animation Events

We can also modify the properties of a single Animation Clip. After selecting an Animation Clip in the list, we can scroll down to see the settings for this particular Animation Clip.

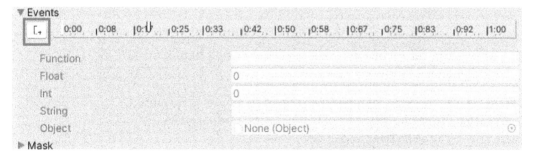

Figure 4.31 – Settings for the Animation Clip

As the preceding figure shows, there is an option for the Animation Clip to add an Animation Event, which allows us to call functions in a script at a specified point in the timeline.

In order to create a new Animation Event, first, we need to position the point in the timeline where we want to add the event, then click the **Add Event** button in the upper-left corner. A small white marker on the timeline will be created, which indicates the new event.

After creating a new event, we also need to configure it by following these steps:

1.  As we can see in *Figure 4.32*, there are multiple fields to fill in, and we entered the name `PrintStringFromAnimationEvent` in the **Function** field, which means this event is set up to call the `PrintStringFromAnimationEvent` function in a script attached to the GameObject. Several other fields can pass in different types of parameters for this function, such as `Float`, `Int`, and `String`.

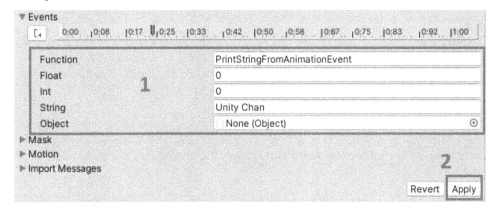

Figure 4.32 – Add an Animation Event

2.  After setting the event, remember to click the **Apply** button to make the configuration of the event take effect.

At the same time, we need to implement a function whose name must exactly match the name already filled in the function field, namely `PrintStringFromAnimationEvent`:

```
public void PrintStringFromAnimationEvent(string
    stringValue)
{
    Debug.Log("PrintStringFromAnimationEvent is called
        with a value of " + stringValue);
}
```

This function will accept a string type parameter; once this Animation Event is triggered, this function will be called, and the string value will be printed in the **Console** window, as shown in the following figure:

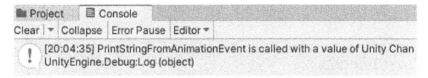

Figure 4.33 – Print the string value in the Console window

Now that we know more about importing animation assets into Unity and how to set up Animation Events, let's turn our attention to setting up the Animator Controller!

# Configuring the Animator Controller

After importing animation assets, we need to set up an Animator Controller to reference these Animation Clips that will be used in our game. In fact, we created an Animator Controller when we introduced it earlier and referenced two Animation Clips that will be used. However, we did not configure this Animator Controller; for instance, we did not configure how to switch between the two animations. In this section, we will explore how to configure an Animator Controller and use C# code to switch between different animations.

## Adjusting the animation speed

We can view the settings of a specific state in the Animator Controller by selecting the state in the **Animator** window.

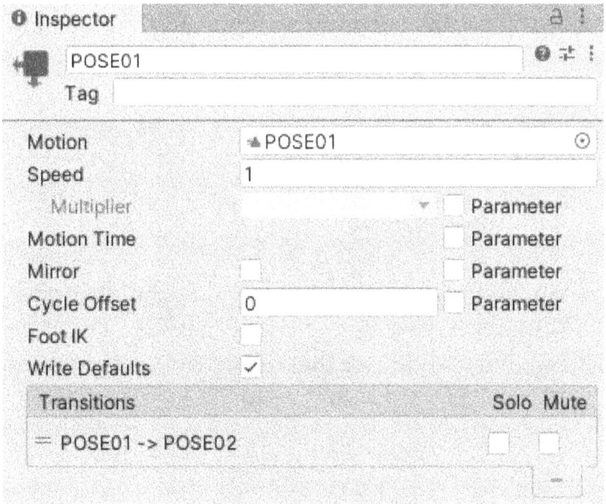

Figure 4.34 – Settings of an animation's state

There are multiple settings, such as **Speed**, **Multiplier**, and **Motion Time**. First, let's review the **Speed** setting and adjust the animation speed. The default value of **Speed** is 1. If the **Speed** value is 0.5, the play speed of **Motion Time** will be halved, so it needs twice the play time. Similarly, a speed of 2 will make the play speed of **Motion Time** twice the normal speed and halve the play time.

## Animator parameters

As we can see in *Figure 4.35*, there are other settings that require parameters to be used. These parameters are called **animation parameters** and they are variables defined in the Animator Controller that can be accessed and assigned values from a C# script. Therefore, they are an important part of using C# code to control animation. In order to add new parameters and edit existing parameters, we should switch to the **Parameters** section of the **Animator** window by clicking the **Parameters** button in the top-right corner.

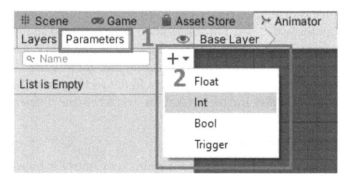

Figure 4.35 – The Parameters section

As shown in the preceding figure, the parameters can be one of the following four types:

- **Float**
- **Int**
- **Bool**
- **Trigger**

As a demonstration, we can add a new parameter called **SpeedMultiplier**. Then, we open the settings of the **POSE01** animation state again and check the **Parameter** checkbox after the **Multiplier** setting, and you can see that the newly created **SpeedMultiplier** parameter appears.

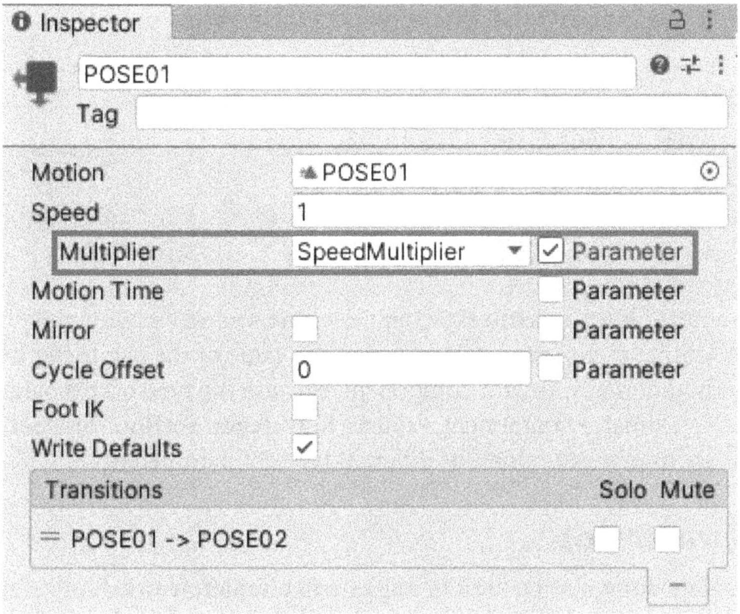

Figure 4.36 – The SpeedMultiplier parameter

As we mentioned earlier, these parameters can be accessed and assigned values using C# code. Therefore, we can create a new script to access and set a value of the SpeedMultiplier parameter as shown in the following code snippet:

```
using UnityEngine;

public class AnimationParametersTest : MonoBehaviour
{
[SerializeField]
private Animator _animator;
[SerializeField]
private float _speedMultiplier;

    // Start is called before the first frame update
    void Start()
    {
        if(_animator == null)
        {
```

```
            _animator = GetComponent<Animator>();
        }


        _animator.SetFloat("SpeedMultiplier",
          _speedMultiplier);
    }
}
```

Here, we create a new C# script named `AnimationParametersTest` and obtain a reference to the Animator component, then we set the value of the parameter by calling the **SetFloat** method of the Animator component, because the type of this parameter is float. Similarly, the Animator component also has, **SetInteger**, **SetBool**, and **SetTrigger** methods, which are used to set values for different types of parameters.

## Configuring transitions

Animation parameters can also be used to implement animation switching. We can use animation transitions to connect two animation states and switch between them. However, by default, animation transitions will automatically switch between two connected animation states, but we obviously prefer to be able to control the switching of animations when developing games.

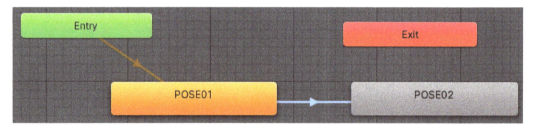

Figure 4.37 – Animation transitions

We can set a transition to occur only when certain conditions are true, and animation parameters can be used to determine whether these conditions are met, so we can use them here to control the switching of the animation.

The following steps demonstrate how to add a new parameter, set up a condition, and control the switching of different animations from C# code:

1.  Switch to the **Parameters** section of the **Animator** window and create a new `bool` variable parameter called **Run**; its default value is **false**, as shown in the following screenshot.

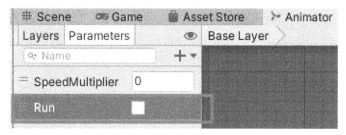

Figure 4.38 – A new parameter

2.  Select the transition we want to apply the condition to; here we choose the transition from **POSE01** to **POSE02**.

3.  In the **Inspector** window of the **POSE01 -> POSE02** transition, we can give a name such as Go To Run to this transition, as shown in *Figure 4.39*.

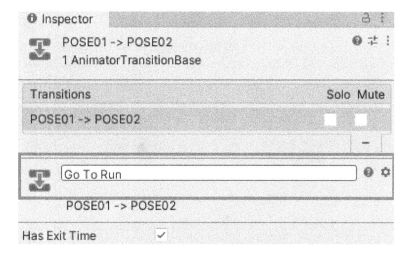

Figure 4.39 – Name the transition

4.  At the bottom of the same window, all the conditions for this transition are listed. By clicking the + button, add a new condition for this **Go To Run** transition, as shown here:

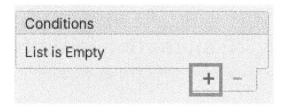

Figure 4.40 – Add a new condition

5.  After adding a new condition, we also need to select a parameter, the value of which is considered as the condition. The parameter here is **Run**. When the value of the **Run** parameter is **true**, it can be considered that the condition is met.

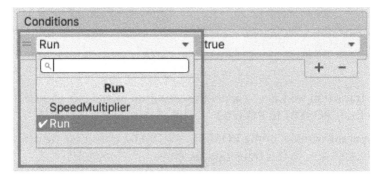

Figure 4.41 – Select a parameter as the condition

6.  However, the default value of the **Run** parameter is **false**. Therefore, in order to switch from **POSE1** to **POSE2**, create a C# script to set the value as follows:

```csharp
⚙ Unity Message | 0 references
private void Update()
{
    bool canRun = Input.GetKey(KeyCode.R);
    _animator.SetBool("Run", canRun);
}
```

Figure 4.42 – C# code snippets

The Input.GetKey method will return true when the user holds down the key identified by KeyCode; otherwise, it will return false, and then we use this value to set the value of the **Run** parameter. Therefore, we can control the switching of the animation by pressing the key.

After reading this section, we have learned how to implement animation for 3D models and how to control animation through C# code in Unity. Next, we will discuss how to implement animation for 2D assets.

# Implementing 2D animation in Unity

In this section, we will use the tools we explored earlier to implement 2D animation in Unity.

The implementation of 2D animation is different from the implementation of 3D animation. A common implementation technique for 2D animation is to use **Sprite Animations**, which are Animation Clips that are created for 2D assets.

There are many ways to create Sprite Animations; we can create them directly in the **Animation** window of the Unity Editor or create them in external tools, such as Aseprite, a popular animation sprite editor, and Piskel, a free online sprite editor.

Here, we use the sprite animation created by an external tool. You can download this asset from Unity Asset Store here: `https://assetstore.unity.com/packages/2d/characters/free-pixel-mob-113577`.

Figure 4.43 – A Sprite Sheet

After downloading the assets, we can find that this image contains many different Sprites, as shown in *Figure 4.43*. We call this a **Sprite Sheet**, which is an image containing sequential Sprites commonly used for animation for 2D assets.

> **Note**
>
> If an image contains a set of non-sequential Sprite images, we call it a **Sprite Atlas**, which is often used to implement UI.

Then, we should import this image file into the Unity Editor by performing the following steps:

1. As shown in *Figure 4.44*, since this image contains a series of Sprite images, we set **Sprite Mode** to **Multiple** in the **Import Settings** window:

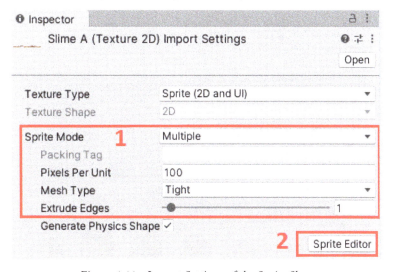

Figure 4.44 – Import Settings of the Sprite Sheet

2.  Click the **Sprite Editor** button to open **Sprite Editor** in Unity. **Sprite Editor** provides tools that allow us to modify Sprite Sheets, such as slicing Sprite Sheets into individual Sprites.

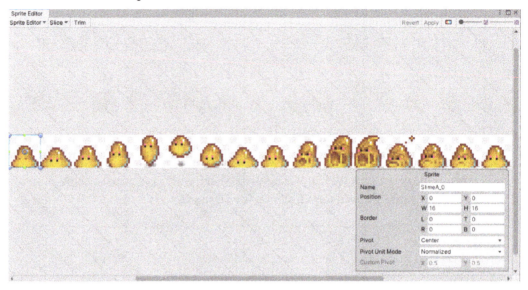

Figure 4.45 – Sprite Editor in Unity

3.  By clicking the **Slice** button, we can open the drop-down menu where we can set **Type** to **Grid By Cell Count**. Since there are **16** individual Sprite images here, in the **Grid By Cell Count** option, change the value of **Column** to 16, change the value of **Row** to 1, and then click the **Slice** button at the bottom of the drop-down menu and close **Sprite Editor**, as shown in the following screenshot:

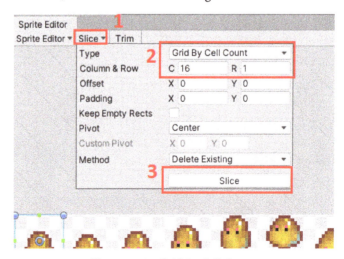

Figure 4.46 – Grid By Cell Count

4.   Then we can select this **Sprite Sheet** asset in the **Project** window to expand it, and you can see all the individual Sprites in it:

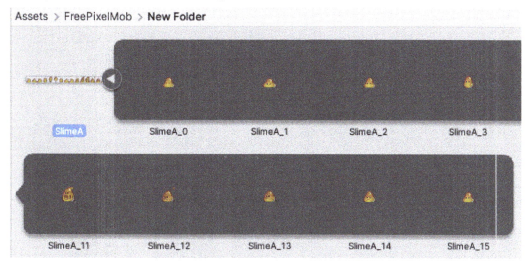

Figure 4.47 – Sprites in the Sprite Sheet

At this point, we have imported the assets and created these sprites. The next question is, how do we use these Sprites to create Animation Clips in Unity? The answer is not complicated. We only need to select the Sprites that make up the Animation Clip we want to create and drag them into the scene. The Unity Editor will automatically create the Animation Clip and ask us to select the folder where the Animation Clip file will be stored, as shown in the following screenshot:

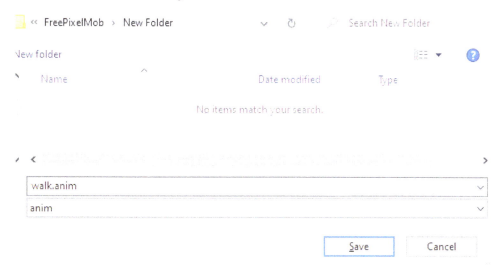

Figure 4.48 – Create an Animation Clip file

In this case, we select the first eight Sprites from the Sprite Sheet and drag them to the Scene view of the Unity Editor. Then, we rename the Animation Clip file to `walk` and save it.

Unity will create the Animation Clip file, as I mentioned earlier, and a new Animator Controller asset as well. A new GameObject with an Animator component attached will also be created in the Scene, which references the Animator Controller, as shown in *Figure 4.49*.

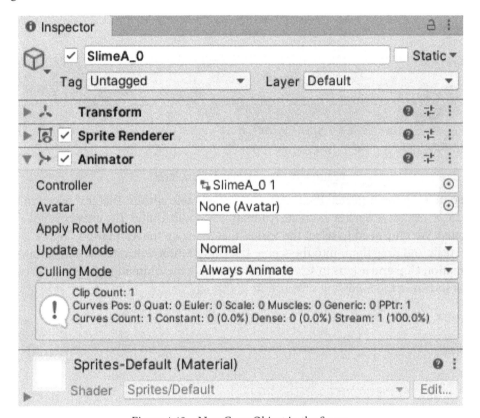

Figure 4.49 – New GameObject in the Scene

Now we can run the game to play the animation by clicking the **Play** button in the Unity Editor, and we can see the **walk** animation is playing!

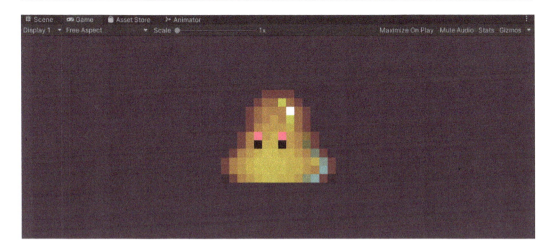

Figure 4.50 – Play the walk animation

After reading this section, you have learned how to implement animation for 2D assets; next, we will share some tips to improve animation performance.

# Improving the performance of Unity's animation system

In Unity, the implementation of animation may cause excessive memory usage and CPU overhead. In this section, we will talk about how to avoid performance problems caused by animation. Specifically, we'll first introduce the **Unity Profiler** tool and how to use it to view animation-related performance metrics, and then we'll look at how to reduce the CPU overhead and memory footprint of animations.

## The Unity Profiler

First, we should learn how to use tools to view and locate performance bottlenecks rather than relying on subjective guesses and experience. Of course, it's not that experience is not important, but using tools will help you locate problems more quickly.

The Unity Editor provides developers with a Profiler tool, which we can use to view the detailed memory usage of the game and real-time CPU overhead.

In order to view performance data about the CPU overhead of animation, we should follow these steps:

1.   Click **Window | Analysis | Profiler** to open the **Profiler** window.

2.   Click the **CPU Usage** module area in the **Profiler** window to view the performance data on the CPU overhead, such as the CPU time consumed by **Animator.Update**, as shown in *Figure 4.51*.

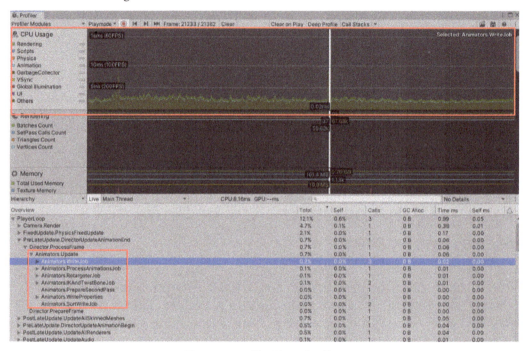

Figure 4.51 – The Unity Profiler

3.   The Unity Profiler also allows us to switch from the **Hierarchy** view to the **Timeline** view, which is more intuitive in some cases.

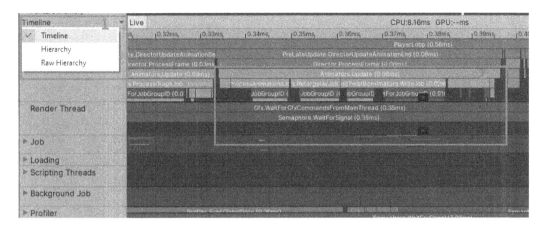

Figure 4.52 – The Timeline view in the Profiler window

In addition to the **CPU Usage** module, we can also view the detailed memory consumption of the game in the **Memory** module.

4.  Click on the **Memory** module area in the **Profiler** window to view the performance data of memory consumption. The default display mode is **Simple** mode, and the memory consumption is counted by types in the **Profiler** window. For example, the memory usage of **Textures** is about 106.3 MB, and the memory usage of **Meshes** is about 4.5 MB, as shown here:

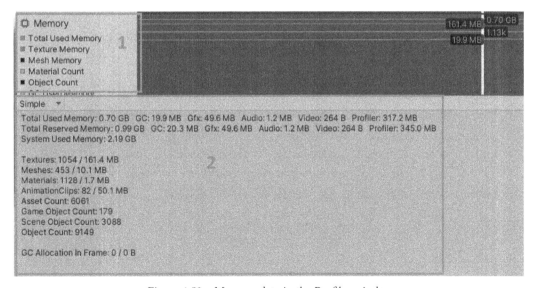

Figure 4.53 – Memory data in the Profiler window

5.  Compared with **Simple** mode, **Detailed** mode is more powerful. We can switch from **Simple** mode to **Detailed** mode by selecting **Detailed** from the drop-down menu in the upper-left corner.

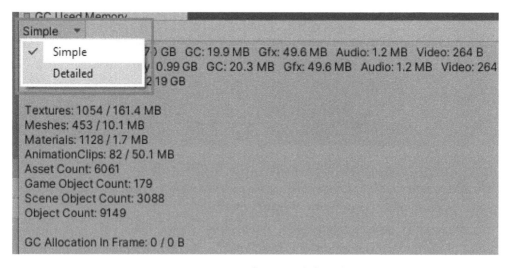

Figure 4.54 – Switch to Detailed mode

**Detailed** mode does not display memory consumption data in real time like **Simple** mode. Instead, we need to manually click the **Take Sample Playmode** button to sample the game memory at the current time.

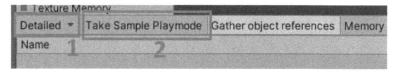

Figure 4.55 – Take a memory sample

Depending on the number of objects created in the game or how much memory is consumed, the sampling time will be different. But once the sampling is complete, we will see the detailed memory overhead; for example, in the following screenshot, there are 82 Animation Clips taking up 50.1 MB of memory.

| Name | Memory | Ref cour |
|------|--------|----------|
| Detailed ▼  Take Sample Playmode  Gather object references  Memory usage in the Editor is not the same as it would be in a Player | | |
| ▶ Other (490) | 0.65 GB | |
| ▶ Not Saved (4187) | 178.4 MB | |
| ▼ Assets (4493) | 138.2 MB | |
| ▼ AnimationClip (82) | 50.1 MB | |
|   WAIT02 | 5.7 MB | 1 |
|   WAIT01 | 4.4 MB | 1 |
|   WAIT04 | 3.7 MB | 1 |
|   WAIT03 | 3.7 MB | 1 |
|   WALK00_B | 3.2 MB | 1 |
|   WIN00 | 2.1 MB | 1 |
|   REFLESH00 | 1.9 MB | 1 |
|   WAIT00 | 1.9 MB | 1 |
|   DAMAGED01 | 1.9 MB | 1 |
|   LOSE00 | 1.7 MB | 1 |
|   JUMP01B | 1.5 MB | 1 |
|   JUMP00 | 1.4 MB | 1 |
|   JUMP00B | 1.4 MB | 1 |
|   SLIDE00 | 1.1 MB | 1 |
|   WALK00_L | 1.0 MB | 1 |
|   WALK00_R | 1.0 MB | 1 |
|   WALK00_F | 1.0 MB | 1 |

Figure 4.56 – Detailed memory data in the Profiler window

From the preceding introduction, we can see that the optimization of animation should mainly focus on CPU overhead and memory consumption. Therefore, the following two best practices need to be considered when using Unity's animation system to implement animation.

# Animator's Culling Mode

In order to reduce the CPU overhead of animation, we should set the **Animator** window's **Culling Mode** property to **Cull Update Transforms** or **Cull Completely**.

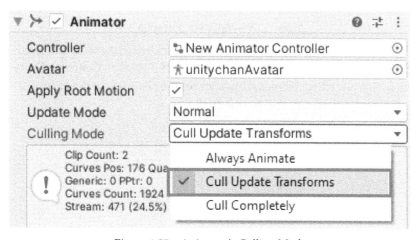

Figure 4.57 – Animator's Culling Mode

By setting it to **Cull Update Transforms**, Unity will disable some features of the animation system such as Retarget, Inverse Kinematics (IK) Transforms when the Animator is not visible on screen. If it is set to **Cull Completely**, Unity will completely disable the animation when the Animator is not visible. Therefore, the goal of reducing CPU overhead can be achieved.

# Anim. Compression

Another best practice is to set **Anim. Compression** in the animation import settings window to save memory.

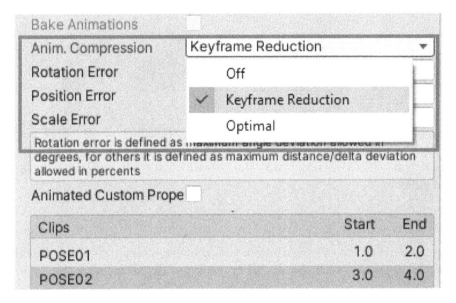

Figure 4.58 – Anim. Compression

By setting it to **Keyframe Reduction**, Unity will reduce keyframes on import and compress keyframes when storing animations in files. If it is set to **Optimal**, Unity will decide how to compress, either by reducing keyframes or by using a dense format.

# Summary

In this chapter, we started by introducing some of the most important concepts of the Unity animation system, such as Animation Clips, Animator Controllers, Avatar, and the Animator component. Then, we demonstrated how to implement 3D animation in Unity, including how to import animation assets into the Unity Editor, how to create an Animation Event on an Animation Clip, how to set up animation parameters to control an animation via C# code, and so on.

We also discussed how to implement 2D animation in Unity. The implementation of 2D animation is different from the implementation of 3D animation. A common implementation technique for 2D animation is to use Sprite Animations, which are Animation Clips that are created for 2D assets.

Finally, we explored some best practices for implementing animation in Unity to optimize the performance problems caused by the animation system.

In the next chapter, we will learn about the Physics system in Unity, and at the same time, we will also introduce how to optimize Physics performance in Unity.

# 5
# Working with the Unity Physics System

A physics simulation in a game is not only an indispensable function for implementing the realism in the game. Adding a physics simulation to your game can usually improve the fun and playability of the game. Generally speaking, it determines how objects move and how they collide with one another, such as the collision between a player and a wall and the effect of gravity. As a popular game engine, Unity provides developers with a variety of tools, allowing developers to integrate physics simulation functions in their games.

We will cover the following key topics in this chapter:

- Concepts in the Unity Physics system
- Scripting with the Physics system
- Creating a simple game based on the Physics system
- Increasing the performance of the Physics system

By the end of this chapter, you will be able to apply the physics simulation correctly and efficiently in Unity to add more realism or fun to your game.

Now, let's get started!

# Technical requirements

You can find complete code examples on GitHub under the following repository:
`https://github.com/PacktPublishing/Game-Development-with-Unity-for-.NET-Developers`.

# Concepts in the Unity Physics system

A  simulation is a useful function of a game. Unity provides different tools for different purposes. For example, if we want to develop a 3D game, then we can use the built-in 3D physics integrated with the **Nvidia PhysX engine**. If we want to add a physics simulation to a 2D game, then we can choose the built-in 2D physics integrated with the **Box2D engine**.

> **Note**
>
> **PhysX** is an open source, real-time physics engine middleware SDK developed by Nvidia as a part of the Nvidia GameWorks software suite. **Box2D** is a free, open source 2D physics simulator engine.

In addition to these built-in Physics solutions, Unity also provides Physics engine packages. These are the `Unity Physics` package and the `Havok Physics for Unity` package. They are different from the built-in Physics systems. They need to be installed separately using `Unity's Package Manager`, and they are used in projects with Unity's **Data-Oriented Technology Stack (DOTS)**. We will introduce DOTS in later chapters.

> **Note**
>
> **Havok Physics** is designed primarily for video games and allows for the real-time collision and dynamics of Rigidbodies in 3D.

In this chapter, we will focus on the built-in physics and will first learn the basic concepts of the Unity Physics system.

# Collider

Similar to the rendering function, a physics engine also needs to understand the shape of `GameObjects` in a game scene in order to perform physics simulation correctly. When developing a Unity project, we can use the **Collider** component to define the shape of a GameObject for physical collision calculations.

It should be noted that the shape defined by a collider does not have to be exactly the same as the shape of the model. We can even create a collider without a model display. For example, we can create a new cube in the scene, and a collider component will be created and attached to this cube automatically. Then, the shape of the collider can be modified from the Inspector window, as shown in the following image; its shape is different from the shape of the model.

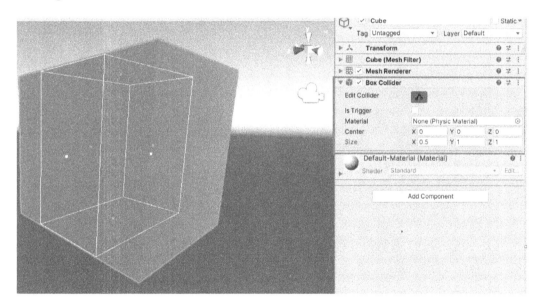

Figure 5.1 – Modifying the shape of the Collider (green frame)

In order to reduce the complexity of physics simulation and improve the performance of the game, we often use some rough shapes, such as the **Box Collider** and the **Sphere Collider**. Next, we will explore one of the most commonly used colliders, namely, the Box Collider.

## Primitive colliders

Unity provides a set of primitive colliders for game developers, including the Sphere Collider and the Box Collider. The Box Collider is one of the most commonly used colliders in Unity. It will be automatically created and assigned to the Cube object in a scene, as we see in *Figure 5.1*. We can also add a new Box Collider to a GameObject manually, as follows:

1.  Create a new GameObject in the scene by clicking the **Create Empty** button.

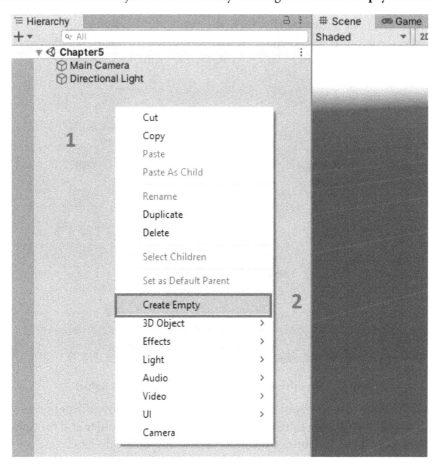

Figure 5.2 – Creating a new GameObject

2.  Select this newly created GameObject and click the **Add Component** button in the Inspector window.

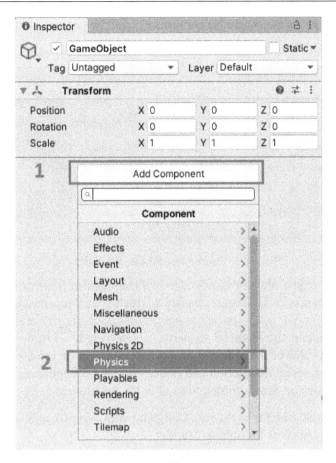

Figure 5.3 – Add Component

3.  Here, we can select the **Physics > Box Collider** button or enter `Box Collider` in the search box to add the Box Collider component to this GameObject.

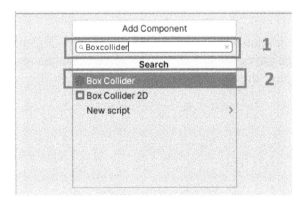

Figure 5.4 – Adding the Box Collider

Now we have added a new **Box Collider** component, and the properties of this Box Collider are shown in the following screenshot:

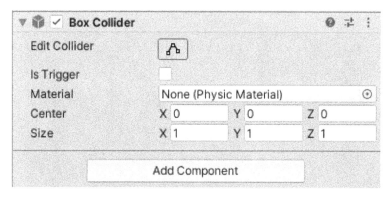

Figure 5.5 – Properties of the Box Collider

The **Edit Collider** button at the top allows us to edit the shape of this box in the scene. Below this button, there is an **Is Trigger** checkbox which, if enabled, means this collider will be used as a trigger. We will introduce more details about triggers later. The third property of this collider is the **Material** property, for referring to a **Physics Material** instance. The default value of the **Material** property is null, and we can assign an instance of Physics Material to adjust the friction and bouncing effects of colliding objects. The last two properties, **Center** and **Size**, are used to modify the position and size of this box.

As we mentioned earlier, similar to the Box Collider, Unity also provides other colliders with primitive shapes, such as the Sphere Collider.

We use them in cases where the accuracy of physical collision simulations is not high, but if the game requires accurate physical collision simulations, then we can also use another collider, namely, **Mesh Collider**.

## Mesh Collider

Sometimes, we need to develop some game projects that require high physical simulation accuracy. In this case, the physical shape of the GameObject is often required to be consistent with the shape of the model mesh of the GameObject. This is why we need a Mesh Collider.

There are different ways to create and add a Mesh Collider to a GameObject. Because the Mesh Collider needs the information of the mesh, so, the first way to create a Mesh Collider is by importing the model into the Unity Editor. You can check the **Generate Colliders** checkbox to import the mesh that automatically attaches mesh colliders, as shown in *Figure 5.6*:

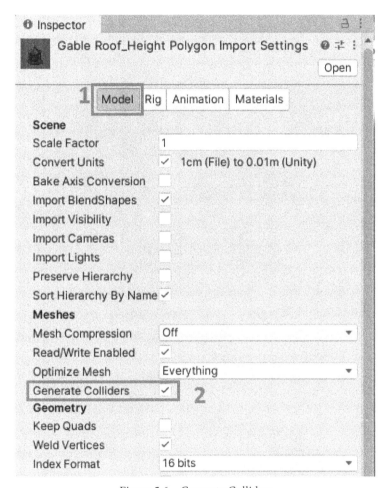

Figure 5.6 – Generate Colliders

Unity also allows us to add a **Mesh Collider** component to a GameObject manually. The steps for adding a Mesh Collider are similar to the steps for adding a Box Collider in the previous section. After selecting the target GameObject, click the **Add Component** button, and then select **Physics > Mesh Collider** to add it to the GameObject, as shown in *Figure 5.7*:

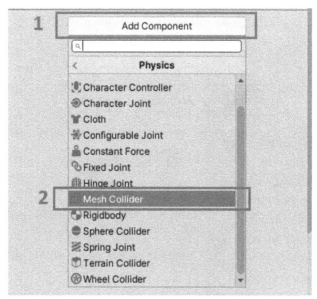

Figure 5.7 – Adding a Mesh Collider to a GameObject

Since the mesh of a model may consist of many vertices and triangles, and the Mesh Collider will be generated based on the mesh, the computational cost of a Mesh Collider is much larger than that of the colliders introduced before. Even by default, Unity does not calculate the collision between mesh colliders, but only calculates the collision between a Mesh Collider and primitive colliders, such as a Box Collider and a Sphere Collider.

In order to enable collision detection between mesh colliders, we need to reduce their complexity by checking the **Convex** checkbox of the Mesh Collider component, as you can see in the following screenshot:

Figure 5.8 – Properties of a Mesh Collider

By enabling this checkbox, **Mesh Collider** is limited to 255 triangles. If we look at the GameObject in the scene at the same time, we can see that the Mesh Collider is only roughly consistent with the model's mesh, and that the complexity has been greatly reduced.

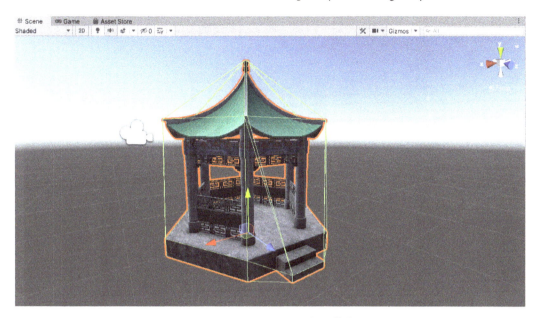

Figure 5.9 – A convex Mesh Collider

However, if we run the game now, we will find that no physical effects are applied to the game; for example, objects will not fall due to gravity. This is because our game still lacks an important component. Let's explore this next!

# Rigidbody

The **Rigidbody** component is an indispensable component for applying physical effects in Unity. By adding Rigidbody to a GameObject, physics will control the GameObject, such as applying gravity to it. Rigidbodies are usually used with colliders; if two Rigidbodies collide with one another, unless the two GameObjects have colliders attached, they will not have a collision effect between them but will pass through each other.

Now, let's add a **Rigidbody** component to a GameObject in the scene:

1. Create a new cube in the scene by clicking the **3D Object > Cube** button.

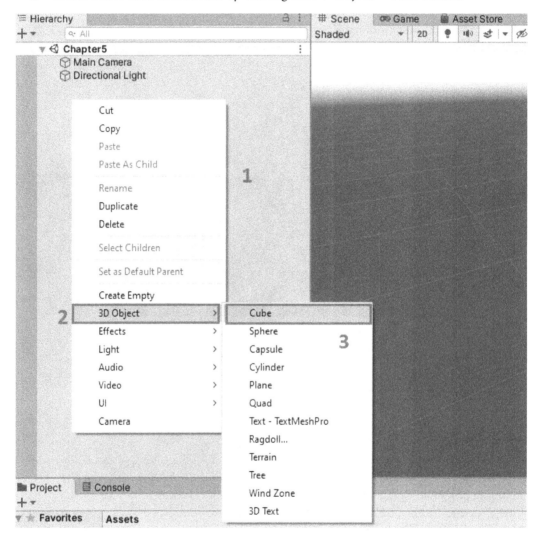

Figure 5.10 – Creating a new cube

2. Select this newly created cube and click the **Add Component** button in the Inspector window. And as you can see in *Figure 5.11*, a Box Collider has been attached to the cube:

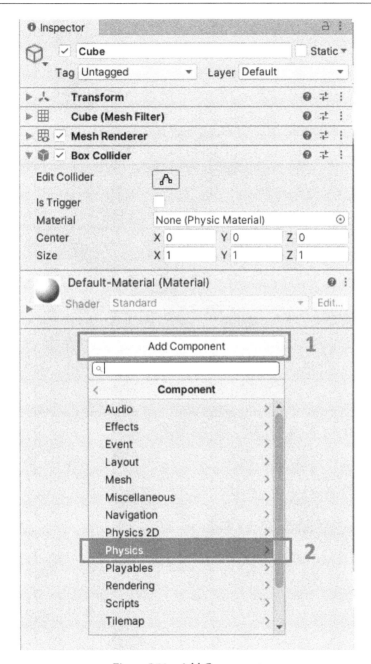

Figure 5.11 – Add Component

3.  Here, we can select the **Physics > Rigidbody** button to add a **Rigidbody** component to this cube.

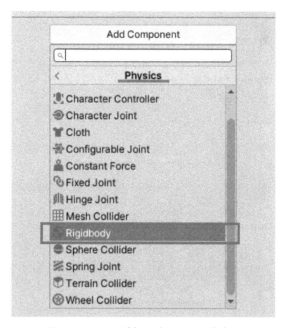

Figure 5.12 – Adding the Box Collider

Now we have added a new **Rigidbody** component, and the properties of this Rigidbody are shown in *Figure 5.13*:

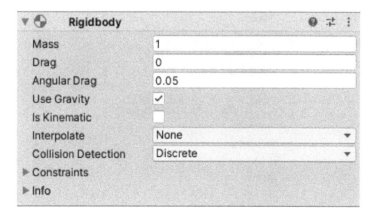

Figure 5.13 – Properties of a Rigidbody

As we can see in *Figure 5.13*, the **Use Gravity** property of the Rigidbody is checked by default, which means that this Rigidbody will apply gravity to the cube. If we run the game at this time, we will find that the cube will fall down under the influence of gravity.

In addition to the **Use Gravity** property, the Rigidbody has other properties, and we will introduce these properties below.

The first property of a Rigidbody component is **Mass**, which determines how Rigidbodies react when they collide with each other. Next is the **Drag** property, which determines how much air resistance the object is affected by when it is moving under force. By default, the value is **zero**, which means there is no air resistance when the cube is moving by force. The **Angular Drag** property is similar to the **Drag** property, the difference being that it determines how much air resistance affects the object when rotating from torque.

The **Is Kinematic** property is important because it determines whether this GameObject will be controlled by the Physics system in Unity. By default, it's disabled. If we enable it, this GameObject will no longer be driven by physics. The **Interpolate** property is useful when you find that the Rigidbody's movement is jerky. The default value of **Interpolate** is **None**, but Unity allows us to select different options for this property, such as **Interpolate** or **Extrapolate**, which, respectively, indicate that the transform is based on the transform of the previous frame for smoothing, or that the transform is smoothed based on the estimated transform of the next frame, as shown in the following screenshot:

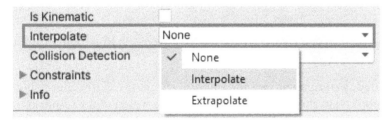

Figure 5.14 – Options of the Interpolate property

Next is the **Collision Detection** property. Sometimes, if a Rigidbody is moving too fast, causing the physics engine to not detect the collision in time, then maybe adjusting this property is a good idea. Unity also provides us with different options for **Collision Detection**; these are **Discrete**, **Continuous**, **Continuous Dynamic**, and **Continuous Speculative**.

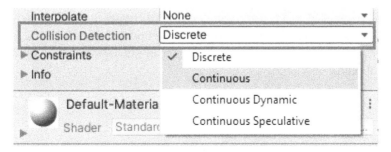

Figure 5.15 – Options of the Collision Detection property

The **Discrete** option is the default value and is used for detecting normal collisions. If you encountered issues with fast object collisions, then **Continuous** is a good choice, but you should remember that **Continuous** will impact performance compared to **Discrete**.

If you want to restrict a Rigidbody's motion, such as restricting the Rigidbody from moving in a certain direction or not being able to rotate on a certain axis, then you can do so by modifying the **Constraints** property.

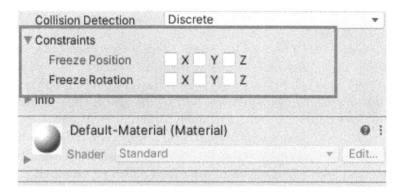

Figure 5.16 – The Constraints property

As shown in *Figure 5.16*, you can select an axis to prevent the Rigidbody from moving along it.

Through a Rigidbody component, we add physical effects to a GameObject, but sometimes we don't want the GameObject to move according to the results of the physics simulation, but just want to be able to detect the collision between two objects and trigger some events. This is a common function in games; for example, the player triggers the corresponding logic after entering a certain area. Next, we will introduce another feature provided by Unity to implement such requirements.

# Trigger

In addition to providing collision effects, colliders can also be used as triggers. However, unlike being used as a normal collider, when a trigger is enabled, there is no collision effect when Rigidbodies collide. However, the physical effect will still take effect; for example, a trigger can still fall under the influence of gravity, but it will not collide with other Rigidbodies.

When developing a Unity project, triggers are used to detect external interactions from other GameObjects and execute the code in the `OnTriggerEnter`, `OnTriggerStay`, or `OnTriggerExit` functions in the script. These three functions represent three different stages of interactions, namely, entering, staying, and exiting. We will introduce more details about those functions in the next section. For the moment, let's create a trigger by performing the following steps:

1.  Select the Cube object we created earlier to open Inspector window.

2.  Enable the **Is Trigger** property of the Box Collider component attached to this Cube object, as shown in the following screenshot:

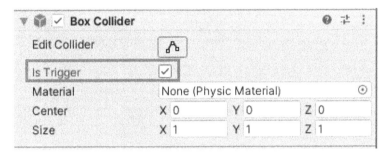

Figure 5.17 – Enabling the Is Trigger property

Now, this cube is set as a trigger, and it will no longer block other Rigidbodies. Since it is now a trigger, we can use it to create game levels. For example, when the player touches this cube, it will trigger a trap.

As a reminder, Unity also provides physical components used for 2D. If you want to develop a 2D game and need to apply physical effects to your game, then you can easily add 2D versions of these physical components in the same way.

By reading this section, we have learned some concepts of Unity's Physics system, such as colliders, Rigidbodies, and triggers. Next, we will continue to explore how to use C# scripts to interact with the Physics system.

# Scripting with the Physics system

In this section, we will explore how to interact with the Physics system via C# scripts. Similar to the previous section, we will also introduce the C# methods for colliders, triggers, and Rigidbodies, respectively. We will start with the C# methods for colliders.

# Collision methods

When a collider is not used as a trigger, collisions between Rigidbodies still occur. These three methods are called when a collision occurs, and the parameter type is the **Collision** class, which provides some information to describe the collision, such as the contact point and the impact velocity of the collision.

## OnCollisionEnter

The first method is `OnCollisionEnter`, which is called when this collider begins to touch another collider. It is useful when you want to make this object be affected by a physical collision, but also want to perform some game logic when the collision occurs. For example, when a bullet hits the target in a game, a corresponding explosion effect can be generated for it, as the following C# code snippet demonstrates:

```csharp
using UnityEngine;

public class CollisionTest : MonoBehaviour
{
[SerializeField]
private Transform _explosionPrefab;

    private void OnCollisionEnter(Collision collision)
    {
        var contact = collision.contacts[0];
        var rotation =
           Quaternion.FromToRotation(Vector3.up,
           contact.normal);
        var position = contact.point;
        Instantiate(_explosionPrefab, position, rotation);
        Destroy(gameObject);
    }
}
```

In the code snippet, we accessed the contact point data provided by the collision object and instantiated the explosion asset at that point.

## OnCollisionStay

OnCollisionStay is the second method we will explore here. As long as two objects collide, OnCollisionStay will be called once per frame. Since this method will be called during the collision of objects, it is suitable to be used to implement some logic that will last for a period of time. An interesting example of this is as follows: Suppose you are developing a helicopter game, and you want the helicopter's engine to run at 60% of its maximum strength when the skid touches the ground. In this case, we can use the following code snippet to implement this function:

```
using UnityEngine;

public class CollisionTest : MonoBehaviour
{
    private void OnCollisionStay(Collision collision)
    {
        if (collision.gameObject.name == "Ground")
        {
            //Reduce engine strength to 60%
        }
    }
}
```

## OnCollisionExit

The last method I want to introduce here is OnCollisionExit. As the name of this method implies, it will be called when this collider stops touching another collider. If some content is generated at the beginning of the object collision via OnCollisionEnter, and you want to destroy them when the object collision ends, then you should consider using OnCollisionExit:

```
using UnityEngine;

public class CollisionTest : MonoBehaviour
{
    private bool _isGrounded;

    private void OnCollisionEnter(Collision collision)
```

```
    {
        _isGrounded = true;
    }


    private void OnCollisionExit(Collision collision)
    {
        _isGrounded = false;
    }
}
```

The preceding code snippet demonstrates how to use `OnCollisionExit` to reset the `_isGrounded` field.

We have covered typical methods used in colliders. Now, we'll look at how to use triggers in a Unity project.

## Trigger methods

In fact, we still use colliders to implement triggers, and just need to check the **Is Trigger** option of the **Collider** component. At this time, the collider will no longer produce the physical collision effect, but activate trigger events.

There are three events commonly used to implement a trigger, namely, `OnTriggerEnter`, `OnTriggerStay`, and `OnTriggerExit`. These three methods are called when two GameObjects collide, and the parameter type is the `Collider` class, which provides information about other colliders involved in this collision.

### OnTriggerEnter

The first method is `OnTriggerEnter`, which is called when this collider begins touching another collider. The **Is Trigger** option should be enabled in this case. This method is useful when you want to trigger some operations on surrounding elements but don't want to produce physical collision effects. For example, you could use this to implement a trap in your game.

It is also very simple to use. We only need to include the game logic that will be triggered in the definition of this method, as shown in the following code snippet:

```
using UnityEngine;


public class TriggerTest : MonoBehaviour
{
```

```
    private void OnTriggerEnter(Collider other)
    {
        Debug.Log($"{this} enters {other}");
    }
}
```

When this GameObject collides with another GameObject, the string by means of which this GameObject enters the other GameObject will be printed in the Console window.

## OnTriggerStay

OnTriggerStay is the second method we will explore here. Similar to the OnCollisionStay method we discussed before, OnTriggerStay will be called in all frames when other colliders touch this trigger. This method is also suitable for implementing trap-like gameplay in a game; for example, the player enters a poisonous fog and will continue to be hurt:

```
using UnityEngine;

public class TriggerTest : MonoBehaviour
{
    private void OnTriggerStay(Collider other)
    {
        Debug.Log($"{this} stays {other}");
    }
}
```

Here, we also only need to put the game logic that will be triggered in the definition of the OnTriggerStay method, as shown in the preceding code snippet.

## OnTriggerExit

The last method I want to introduce here is OnTriggerExit. This method will be called when other colliders leave the trigger. This method is suitable for some tasks, such as destroying the GameObjects created when other colliders enter this trigger, resetting the state, and so on. The following code snippet shows how to destroy a GameObject in OnTriggerExit:

```
using UnityEngine;

public class TriggerTest : MonoBehaviour
```

```
{
    private void OnTriggerExit(Collider other)
    {
        Destroy(other.gameObject);
    }
}
```

# Methods of Rigidbody

The **Rigidbody** component provides us with the ability to directly interact with the Physics system in Unity. We can use the methods provided by the **Rigidbody** component in the C# script to apply a force to this Rigidbody, and we can also apply a force to a Rigidbody that simulates explosion effects.

It should be noted that, as we mentioned in *Chapter 2, Scripting Concepts in Unity*, in a script, it is recommended to use the FixedUpdate function for a physical update, so we should call Rigidbody methods in the FixedUpdate function to apply the physical effect. Now, let's explore some commonly used methods.

## AddForce

The AddForce method is one of the most commonly used methods related to physics. As its name implies, we can call this method to apply a force to the Rigidbody. The function signature of AddForce is as follows:

```
public void AddForce(Vector3 force,
    [DefaultValue("ForceMode.Force")] ForceMode mode);
```

As you can see, this method requires two parameters, namely, the force vector in world coordinates and the type of force to apply. AddForce allows us to define a force vector and choose how to apply this force to the GameObject to affect how our GameObject moves.

The first parameter, force, is a vector type that specifies the direction in which the force is applied to this object.

On the other hand, the ForceMode type parameter, mode, determines the type of force applied. ForceMode is an enum type, which defines four different types of force. By default, the AddForce method will add a continuous force to the Rigidbody, using its mass. In the following section, I will introduce the different types of force modes in detail.

## ForceMode

ForceMode is defined in the UnityEngine namespace, and we can see its definition in the following code snippet:

```
namespace UnityEngine
{
    //
    // Summary:
    //      Use ForceMode to specify how to apply a force
    //      using Rigidbody.AddForce.
    public enum ForceMode
    {
        //
        // Summary:
        //      Add a continuous force to the rigidbody,
        //      using its mass.
        Force = 0,
        //
        // Summary:
        //      Add an instant force impulse to the
        //      rigidbody, using its mass.
        Impulse = 1,
        //
        // Summary:
        //      Add an instant velocity change to the
        //      rigidbody, ignoring its mass.
        VelocityChange = 2,
        //
        // Summary:
        //      Add a continuous acceleration to the
        //      rigidbody, ignoring its mass.
        Acceleration = 5
    }
}
```

As the preceding code snippet shows, there are four types of force mode, namely, Force, Impulse, VelocityChange, and Acceleration.

Force is the default mode and in this mode, more force must be applied to push or distort objects with larger masses because it depends on the mass of the Rigidbody. It will add a continuous force to the Rigidbody.

If we choose Impulse mode as the argument, then the AddForce method will apply an instant force impulse to the Rigidbody. This mode is suitable for simulating forces from explosions or collisions. As with the Force mode, the Impulse mode also depends on the mass of the Rigidbody.

VelocityChange is the third mode here. If we select this mode, then Unity will apply the velocity change instantly with a single function call. It should be noted that the VelocityChange mode is different from the Impulse mode and the Force mode. The VelocityChange mode does not depend on the mass of the Rigidbody, which means that VelocityChange will change the velocity of each Rigidbody in the same way.

The last mode is Acceleration mode. If this mode is selected, then Unity will add a continuous acceleration to the Rigidbody. Like the VelocityChange mode, Acceleration mode also ignores the mass of the Rigidbody, which means AddForce will move every Rigidbody the same way.

So far, we have learned the different force modes available for the AddForce method. Next, let's create a new C# script and apply a force to the cube by calling AddForce:

```
using UnityEngine;

public class RigidbodyMethods : MonoBehaviour
{
[SerializeField]
private Rigidbody _rigidbody;
[SerializeField]
private float _thrust = 50f;

    private void Start()
    {
        _rigidbody = GetComponent<Rigidbody>();
    }

    private void FixedUpdate()
    {
        if (Input.GetKey(KeyCode.F))
        {
```

```
            _rigidbody.AddForce(transform.forward *
                _thrust);
        }

        if (Input.GetKey(KeyCode.A))
        {
            _rigidbody.AddForce(transform.forward *
                _thrust, ForceMode.Acceleration);
        }
    }
}
```

As shown in the code, we can apply a continuous force to the Rigidbody by pressing the *F* key on the keyboard and applying a continuous acceleration to the Rigidbody by pressing the *A* key on the keyboard.

## MovePosition

Sometimes, we just want to move our GameObjects and don't want to deal with forces. The `MovePosition` method of Rigidbody can help us to achieve this goal.

The function signature of `MovePosition` is as follows:

```
public void MovePosition(Vector3 position);
```

Here, we need a parameter position to provide the new position for the Rigidbody object to move to. To make the Rigidbody move smoothly, we often use interpolation to achieve a smooth transition between frames. Since `MovePosition` is still a method of Rigidbody, we still call it in the `FixedUpdate` function, as shown in the following code snippet:

```
using UnityEngine;

public class RigidbodyMethods : MonoBehaviour
{
[SerializeField]
private Rigidbody _rigidbody;
[SerializeField]
private float _speed = 50f;
```

```
    private void Start()
    {
        _rigidbody = GetComponent<Rigidbody>();
    }

    private void FixedUpdate()
    {
        var direction = new
          Vector3(Input.GetAxis("Horizontal"), 0,
          Input.GetAxis("Vertical"));

        _rigidbody.MovePosition(transform.position +
          direction * Time.deltaTime * _speed);
    }
}
```

Here, we get user input as the direction of movement and apply the movement to the current position. You can also see that the movement vector is multiplied by `deltaTime` and `speed`, which is for smooth movement.

After reading this section, we learned how to interact with the Physics system through C# scripts. But it's best if we implement a simple game ourselves using the physics system, and that's what we'll do in the next section! Let's move on.

# Creating a simple game based on the Physics system

We have learned the concepts of Unity's Physics system and discussed how to use C# code to interact with the Physics system. Next, we will use the knowledge we have learned to create a simple physics-based ping-pong game in Unity.

First, let's perform the following steps to create a **Plane** object as a ping-pong table:

1.  Right-click on the **Hierarchy** window to open the menu.
2.  Select **3D Object > Plane** to create a new **Plane** object in the editor.

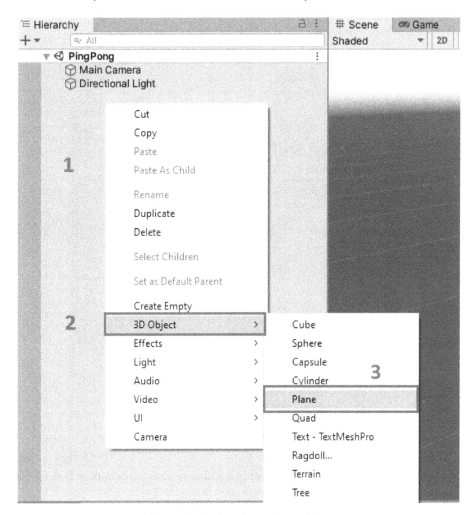

Figure 5.18 – Creating a Plane object

3.  Rename the **Plane** object to `Table`.

4.  Select **Table** to open its Inspector window, modify the **Z** value of **Scale** to 2, and we
    can see that a Mesh Collider has been added to this object by default.

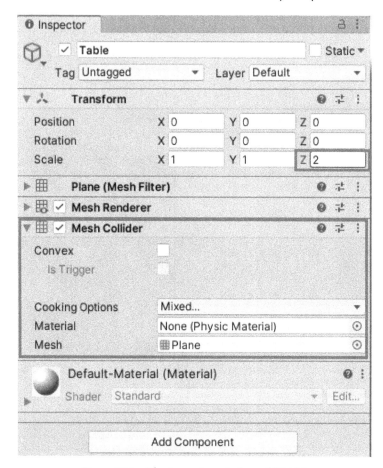

Figure 5.19 – The Inspector window of "Table"

5.  Let's create four Cube objects as walls on the table by selecting **3D Object > Cube**,
    which is similar to the process of creating a **Plane** object. By default, a Box Collider
    has been added to every Cube object.

6.  We can easily adjust the position, size, and rotation of these four Cube objects by
    using the tools in the editor to create the walls on the table.

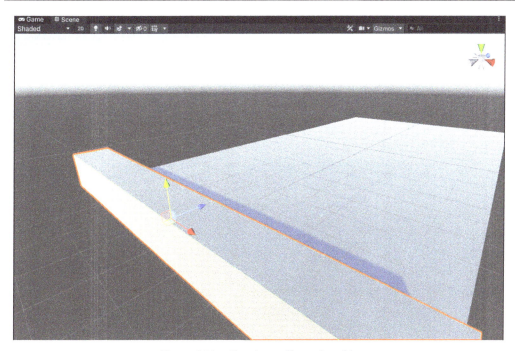

Figure 5.20 – Creating walls on the table

In order to make the table look less boring, we can apply different materials to the walls and the table. Now we have set up the ping-pong table, as shown in the following image:

Figure 5.21 – The ping-pong table

Next, we need to create two players, namely, **Player1** and **Player2**. To keep it simple, we still use two Cube objects as players:

1.  Select **3D Object > Cube** to create a new Cube object in the scene.

2.  Rename the Cube object to `Player1`.

3.  Adjust the position and size of **Player1**. For example, we can modify the **X** value of **Scale** to 3.

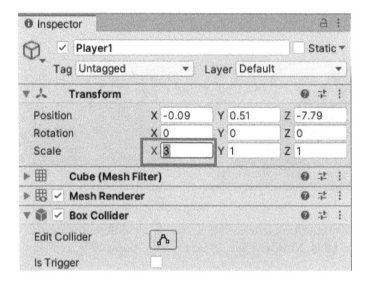

Figure 5.22 – The Inspector window of Player1

4.  Let's repeat the preceding steps to create another player.

5.  We can use different colors to identify **Player1** and **Player2** to distinguish them, as the following figure shows:

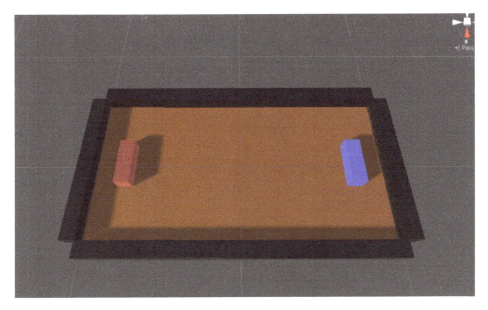

Figure 5.23 – Player1 and Player2

Now we have the **Player** objects in our simple game. Next, we will add a ping-pong ball to our game:

1.  Select **3D Object > Sphere** to create a new **Sphere** object in the scene.

2.  Rename the **Sphere** object to `Ball`.

3.  Select **Ball** to open its Inspector window. We can see that a Sphere Collider has been added to the ball by default.

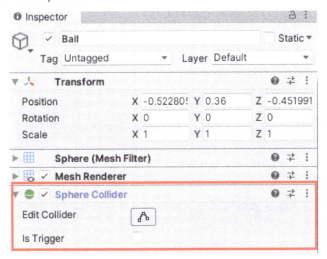

Figure 5.24 – The Sphere Collider component

4.  Then, we need to add a **Rigidbody** component to this ball by clicking the **Add Component** button and selecting **Physics > Rigidbody**.

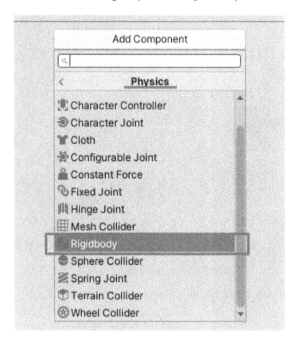

Figure 5.25 – Adding a Rigidbody component

5.  Then, we change the **Interpolate** option of this **Rigidbody** component from **None** to **Interpolate** to make the transformation smooth based on the transformation of the previous frame.

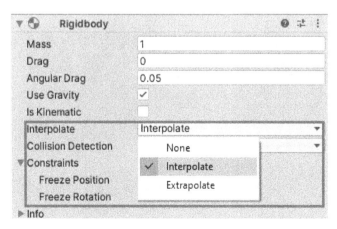

Figure 5.26 – Changing the Interpolate option

6.  Then, we also change the **Collision Detection** option of this **Rigidbody** component from **Discrete** to **Continuous Dynamic** so that we can handle the fast-moving ping-pong ball correctly.

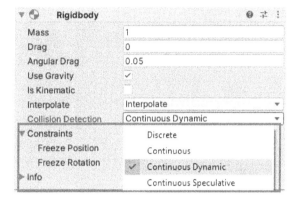

Figure 5.27 – Changing the Collision Detection option

7.  Since the real-world ping-pong ball will bounce back when it hits an obstacle, in order to simulate this bounce effect, we need to create a physic material by clicking **Create** > **Physic Material** in the Project window.

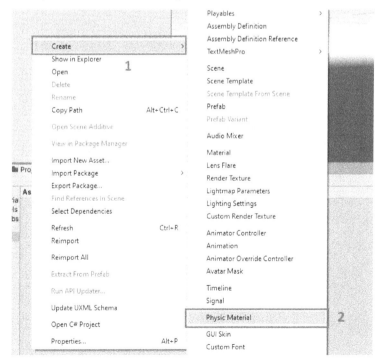

Figure 5.28 – Creating a physic material

8.  Let's select the newly created physic material to open the Inspector window and change both **Dynamic Friction** and **Static Friction** from 0.4 to 0, and **Bounciness** from 0 to 1. Also, set the **Friction Combine** option to Multiply and the **Bounce Combine** option to Maximum, as shown in the following screenshot:

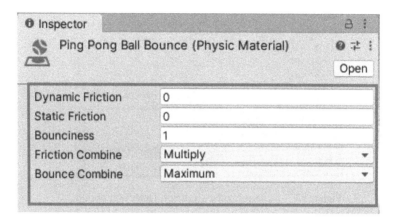

Figure 5.29 – Physic Material settings

9.  Then, assign this physic material to the **Material** option of the Sphere Collider.

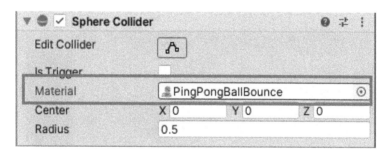

Figure 5.30 – Assigning the physic material to the Sphere Collider

Now we have set up the ping-pong ball that will be used in our game. Next, let's create a new C# script to apply force to the ball to move it:

```
using UnityEngine;


public class PingPongBall : MonoBehaviour
{
    [SerializeField] private Rigidbody _rigidbody;
    [SerializeField] private Vector3 _initialImpulse;
```

```
    private void Start()
    {
        _rigidbody.AddForce(_initialImpulse,
        ForceMode.Impulse);
    }
}
```

In this script, we are using the AddForce method and Impulse force mode that we learned about previously to apply an impulse force to the ball. The direction and magnitude of the force are provided by the _initialImpulse variable. This can be set in the editor.

Let's now attach this script to the ball and provide a value for the _initialImpulse variable.

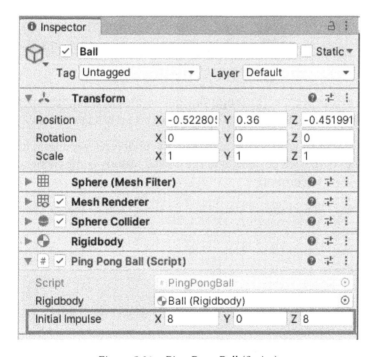

Figure 5.31 – Ping Pong Ball (Script)

As the preceding screenshot demonstrates, the value of the _initialImpulse variable is (8, 0, 8), which means we add an instant force impulse pointing to the lower-right corner of the table to the Rigidbody.

Let's play the game and see what happens.

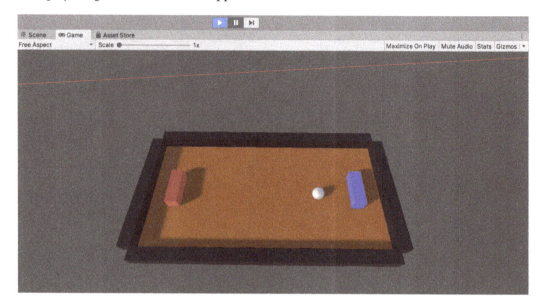

Figure 5.32 – The ball is bounced

From the picture, we can see that the ping-pong ball in the game hit the wall and bounced. Next, we will add more logic to the player objects so that we can control them in the game.

However, before we start to write C# code for our player objects, we should first add a **Rigidbody** component to each of them, and set the **Rigidbody** component settings as shown in the following screenshot:

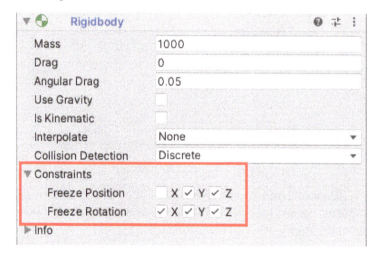

Figure 5.33 – Settings of the player's Rigidbody component

As you can see from the screenshot, we first set the mass of the **Rigidbody** component to 1000 and disabled the effect of gravity by unchecking the **Use Gravity** option.

Then, it is worth your attention that we have restricted the movement of the Rigidbody. Since the player object will only move along the x axis and will not rotate, we only keep the Rigidbody moving along the x axis without constraint.

Next, we also need to configure the controls for these two different players, as shown in the following steps:

1.  Open the **Project Settings** window by selecting **Edit > Project Settings** in the editor.

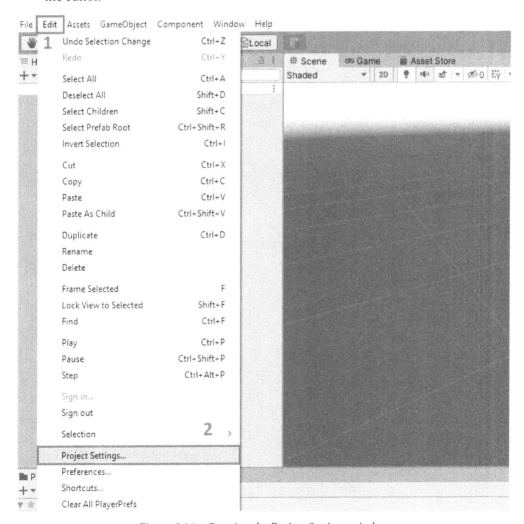

Figure 5.34 – Opening the Project Settings window

2.  Select **Input Manager** from the navigation on the left to open the **Input Manager** window.

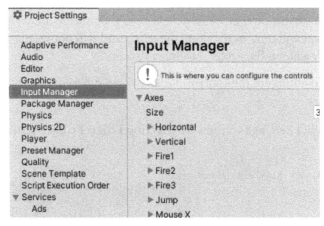

Figure 5.35 – Opening the Input Manager window

3.  We will define the input axis and related actions of player 1 and player 2 in this window to allow us to use the up and down arrow keys and the w and s keys to control the movement of these two player objects, respectively, as shown in the following screenshot:

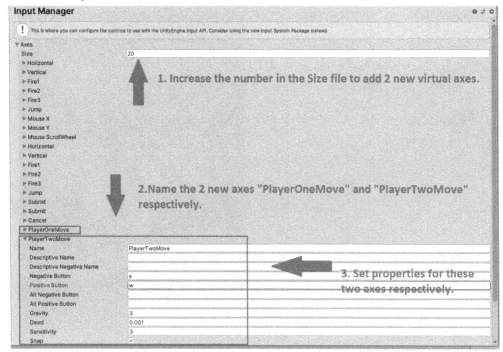

Figure 5.36 – Setting up the input controls for players

So far, we have set up the Rigidbody components and the input control needed by the player objects, and then we can write a C# script to control the player objects in our game.

Remember the `MovePosition` method we introduced before? Here, we will use this method to move the player objects:

```
using UnityEngine;

public class Player : MonoBehaviour
{
[SerializeField]
private Rigidbody _rigidbody;
[SerializeField]
private float _speed = 10f;
[SerializeField]
private bool _isPlayerOne;

    private void Start()
    {
        _rigidbody = GetComponent<Rigidbody>();
    }

    private void FixedUpdate()
    {
        var inputAxis = _isPlayerOne ? "PlayerOneMove" :
          "PlayerTwoMove";
        var direction = new
          Vector3(Input.GetAxis(inputAxis), 0, 0);
        _rigidbody.MovePosition(transform.position +
          direction * Time.deltaTime * _speed);
    }
}
```

As shown in the preceding code, this script will first determine which player the object is, get the corresponding input settings, and then determine the direction of the object's movement based on the player's input.

Now, let's attach this script to these two player objects and start the game!

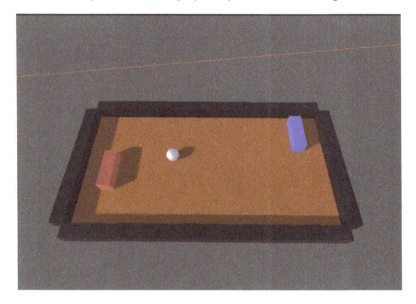

Figure 5.37 – The ping-pong game

As shown in the preceding image, we can now use the *w* and *s* keys and the *up* and *down* keys to control the movement of player 1 and player 2 and, as expected, the ping-pong ball will bounce when it hits the players.

In this section, we made a simple physics-based game, and now we will introduce how to optimize the performance of the Physics system when developing a game in Unity.

# Increasing the performance of the Physics system

Physical simulation requires a lot of calculations, especially in the case of high physical accuracy requirements. Therefore, it is very important to understand how to use Unity's Physics system correctly and reduce unnecessary computing overhead.

## The Unity Profiler

First, we should learn how to use tools to view and locate performance bottlenecks caused by the Physics system in Unity.

The **Profiler** tool in the Unity Editor is our recommended tool, which allows us to easily view various performance data and locate performance issues related to the Physics system.

Taking the ping-pong game we just made as an example, we can perform the following steps to view the performance data of this game:

1.  Start the game in the editor by clicking the **Play** button.

Figure 5.38 – Playing the game in the editor

2.  Click **Window > Analysis > Profiler** or use the keyboard shortcut *Ctrl + 7* (*Command + 7* on macOS) to open the **Profiler** window.

3.  Click the **CPU Usage** module area in the **Profiler** window to view the performance data of CPU overheads, such as the CPU time consumed by **FixedUpdate. PhysicsFixedUpdate**, as shown here:

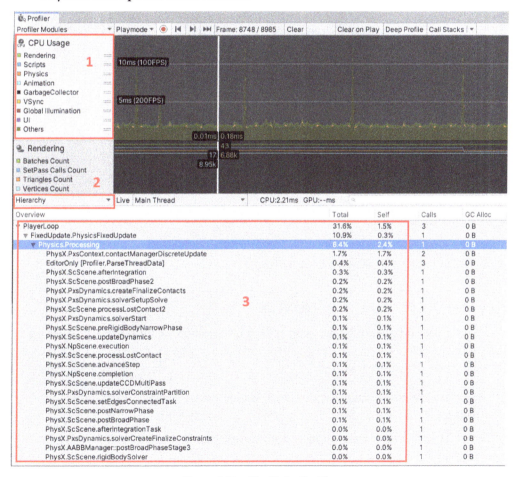

Figure 5.39 – The Unity Profiler

In addition to the **CPU Usage** module, we can also view the detailed information of the Physics system, such as the number of Rigidbodies and the number of contacts at a specific moment, as shown in the following screenshot:

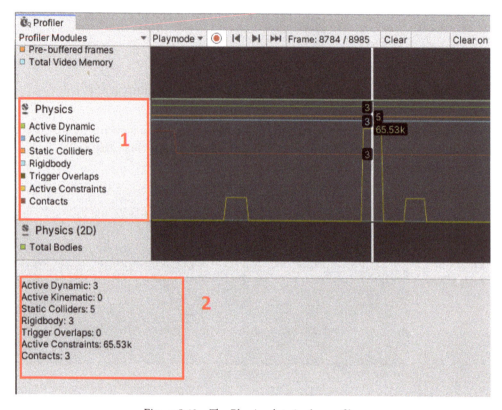

Figure 5.40 – The Physics data in the profiler

Next, we will introduce some tips for improving the performance of the Physics system.

## Increasing the fixed timestep

One idea to reduce the cost of physics computing is to reduce the number of updates per second of the Physics system. We can perform the following steps to increase this **Fixed Timestep** setting to achieve this goal:

1.  Open the **Project Settings** window by selecting **Edit** > **Project Settings** in the editor.

2.  Select **Time** from the navigation on the left to open the **Time** window.

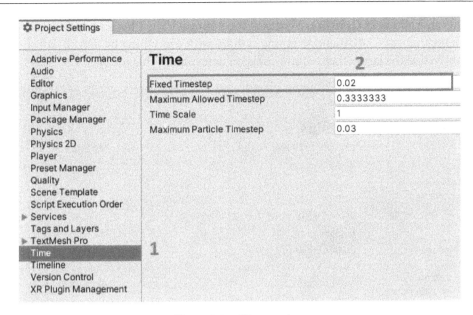

Figure 5.41 – Time settings

3.  The default value of **Fixed Timestep** is **0.02**, which means the Physics system will be updated 50 times per second. To reduce the number of updates per second, we can increase this value.

# Reducing unnecessary layer-based collision detections

Unity uses a rather inefficient physical collision detection mode by default; that is, collision detection is performed on all GameObjects. We can reduce the number of collision detections by modifying the **Layer Collision Matrix** field in the **Physics** settings of Unity and setting different layers for different GameObjects. The following steps demonstrate how to modify it:

1.  Open the **Project Settings** window by selecting **Edit** > **Project Settings** in the editor.
2.  Select **Physics** from the navigation on the left to open the **Physics** window.

3.  You can find **Layer Collision Matrix** at the bottom of the **Physics** window, and you can see in *Figure 5.42* that everything collides with everything by default. We should only enable the layers that require collision detection in this matrix.

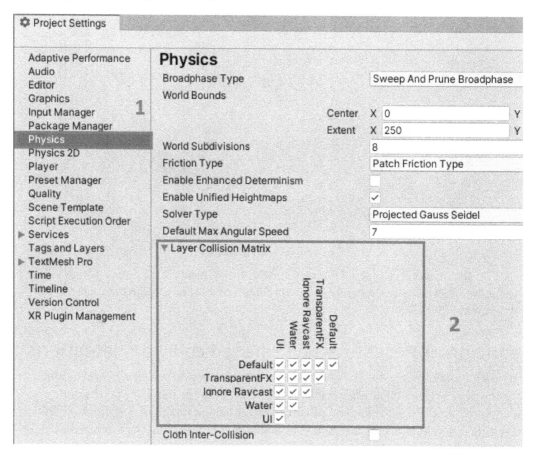

Figure 5.42 – Layer Collision Matrix

In this section, we introduced how to use Unity's Profiler tool to view the performance data of the Physics system and explored how to optimize the performance of the latter.

# Summary

In this chapter, we started by introducing the physics solutions provided by Unity, including two built-in physics solutions, **Nvidia PhysX engine** and **Box2D engine**, and Unity also provides Physics engine packages, namely, the **Unity Physics package** and the **Havok Physics for Unity package**. Then, we explored some of the most important concepts in Unity's Physics system, such as the **Collider** component, the **Rigidbody** component, and **Triggers**. We also discussed how to create a new script in Unity to interact with Unity's Physics system.

Then, we demonstrated how to implement a physics-based ping-pong game in Unity.

Finally, we explored some best practices for applying a physics simulation in Unity to optimize the performance problems caused by the Physics system.

In the next chapter, we will be discussing how to implement video and audio features in Unity.

# 6

# Integrating Audio and Video in a Unity Project

In the previous chapters, we have discussed how to use C# scripts to develop game logic in Unity, how to efficiently implement UI, how to implement animation, and how to integrate physics simulation into your game. However, one feature that is often overlooked in game development is sound. The proper use of sound effects can enhance the immersion of a game, and the background music that matches the background of the game can trigger the emotional resonance of the players. Sometimes, playing video in a game is also a way to increase the fun of a game. There is no doubt that adding video and audio to your game can make your game more lively and interesting.

In this chapter, we will introduce the following key topics:

- Concepts in Unity's audio system and video system
- Scripting with audio and video
- Things to note when using Unity to develop web applications
- Increasing the performance of the audio system

By the end of this chapter, you will be able to implement audio and video correctly and efficiently in Unity to add more realism and fun to your game.

Now, let's get started!

# Technical requirements

You can find complete code examples on GitHub in the following repository: `https://github.com/PacktPublishing/Game-Development-with-Unity-for-.NET-Developers`.

# Concepts in Unity's audio system and video system

Unity provides video and audio features, allowing your game to play videos on different platforms, and supports real-time mixing and full 3D spatial sound effects. In this section, we will introduce important concepts of the Unity audio system and video system.

## Audio clips

In order to be able to play audio in Unity, we need to import an audio file into the Unity editor first. The audio data will be saved in an **audio clip** object in Unity. We can download and import the **Ultra Sci-Fi Game Audio Weapons Pack Vol. 1** from **Unity Asset Store** at the following link: `https://assetstore.unity.com/packages/audio/sound-fx/weapons/ultra-sci-fi-game-audio-weapons-pack-vol-1-113047`. You can see this in the following screenshot:

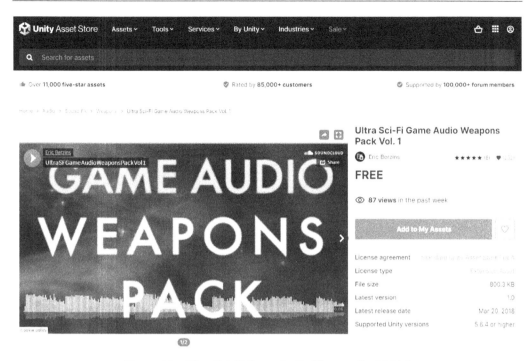

Figure 6.1 – Ultra Sci-Fi Game Audio Weapons Pack Vol. 1

The format of the audio files contained in this pack is .wav. In addition to .wav files that can be imported into Unity, Unity also supports importing files in the following formats:

- .aif
- .mp3
- .ogg
- .xm
- .mod
- .it
- .s3m

After importing these audio files, we can choose one of them to open the **Import** settings as shown in *Figure 6.2*:

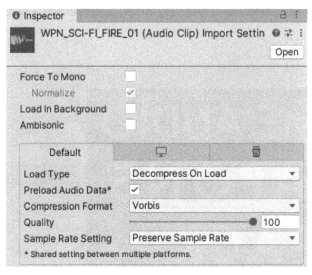

Figure 6.2 – Import settings of audio

As you can see in the **Import** settings, Unity supports mono and multichannel audio assets, up to eight channels. Unity also provides a lot of import options. Let's introduce some important options.

## Load Type

Unity provides game developers with three different ways to load audio assets at runtime. We can determine how Unity loads this audio file by modifying the **Load Type** property in the **Import** settings window.

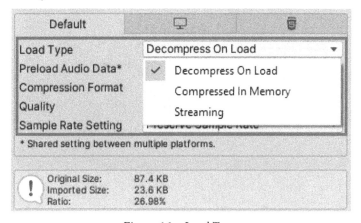

Figure 6.3 – Load Type

The three methods are as follows:

- **Decompress On Load**: This is the default value for **Load Type**. If the audio file is small, such as UI sounds or footstep sounds, we should choose this option. This is because, in this way, the audio file will be decompressed and decoded into the memory at its original size. The advantage is that it will be ready for on-demand playing with minimal CPU usage.

- **Compressed In Memory**: As a contrast with **Decompress On Load**, by choosing this method, Unity will store the compressed audio data in memory and require the CPU to decompress and decode it when playing the audio.

- **Streaming**: This is completely different from the previous two. If we choose this method, Unity will not load the audio data into the memory, but instead will stream it from disk. This method uses the least memory, but at the cost of the highest CPU usage and disk usage.

## Compression Format

In addition to the **Load Type** property just introduced, the **Compression Format** property is also very important for audio assets. Unity supports a variety of audio compression formats, and there are different formats available according to the different target platforms. For example, if the target platform is **Windows**, the following formats are available:

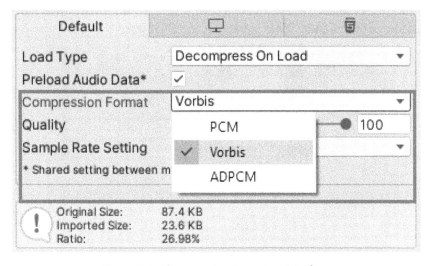

Figure 6.4 – Compression Format on Windows

On the other hand, if the target platform is **Android**, in addition to the previous formats, it also supports the MP3 format.

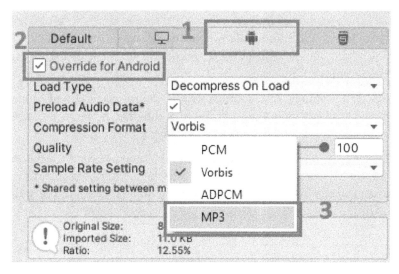

Figure 6.5 – Compression Format on Android

We will explore the different compression formats here:

- **PCM**: **Pulse-code modulation** (**PCM**) is a lossless, uncompressed format and is the standard form of digital audio in computers. It offers high quality and has a very large file size. As you can see in *Figure 6.6*, when the **PCM** format is selected, the imported size of this audio file is equal to its original size.

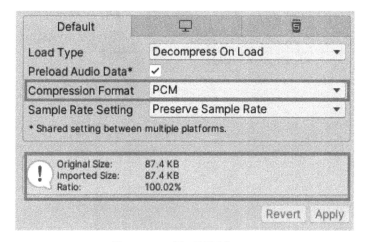

Figure 6.6 – The PCM format

- **Vorbis**: This is the default value for **Compression Format**. **Vorbis** is a very effective audio compression format. Compared with **PCM** audio, this compression produces smaller files, but the quality is lower. If we choose the **Vorbis** option, the imported size of this audio file will be much smaller than its original size. There is a **Quality** slider that allows us to adjust the compression quality.

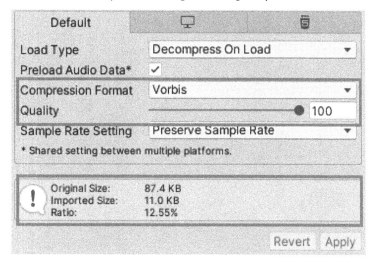

Figure 6.7 – The Vorbis format

- **ADPCM**: ADPCM is short for **adaptive differential pulse-code modulation**. Although the name is similar to PCM, it is a lossy compression format. But unlike Vorbis, its compression ratio cannot be adjusted in Unity. The compressed file size will always be 3.5 times smaller than PCM.

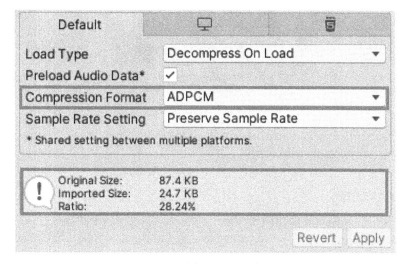

Figure 6.8 – The ADPCM format

- **MP3**: This is available on mobile platforms, such as Android. The MP3 format is similar to Vorbis, which is a very effective audio compression format. There is also a **Quality** slider that allows us to adjust the compression quality.

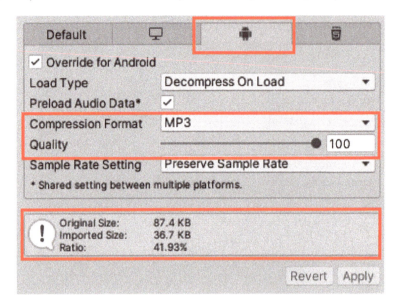

Figure 6.9 – The MP3 format

After we set the import settings for these audio files, they can be imported into the Unity editor as audio clips.

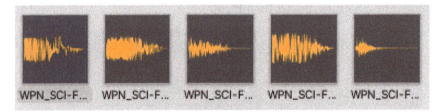

Figure 6.10 – Audio clips

As shown in *Figure 6.10*, we can find these audio clips in the **Project** window, and the icon of the audio clip will show its waveform.

## Audio Sources

In order to play the audio clip we just created in the game scene, we also need to set an **Audio Source**. Then this audio clip can be dragged to the Audio Source or used from a C# script.

Let's follow these steps to create an Audio Source first:

1.  Right-click in the **Hierarchy** window to open the menu.

2.  Choose **Create Empty** to create a new GameObject in the scene. As a reminder, the GameObject that is the Audio Source is not necessarily a static object. In many cases, the Audio Source needs to be moved, such as simulating the effect of firing a cannonball in a game. But for the sake of simplicity, we will not add movement logic to this GameObject here.

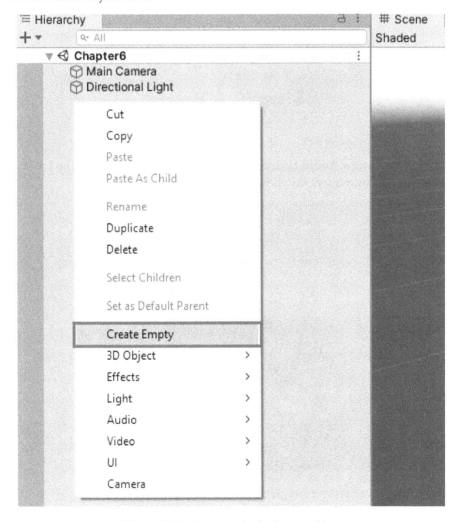

Figure 6.11 – Create an Audio Source object

3.  Select this newly created GameObject and click the **Add Component** button to open the components list.

4.  Choose **Audio | Audio Source** to add an **Audio Source** component to this GameObject.

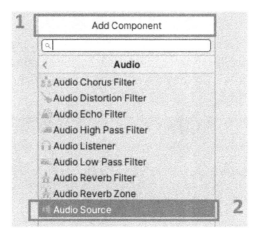

Figure 6.12 – Add an Audio Source component

Now we have created a new **Audio Source** component in our game scene. The properties of this **Audio Source** component are shown in *Figure 6.13*.

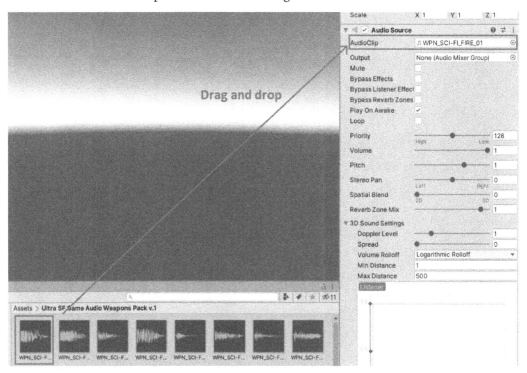

Figure 6.13 – Properties of Audio Source

We will explore some of them here:

- **AudioClip**: Here, we find the first property of **Audio Source** is a reference to **Audio Clip**. We can drag an **Audio Clip** asset to this field directly in the editor.

- **Output**: We don't have to set this property, because the output of this Audio Source will then be picked up by an Audio Listener in the scene by default. Set this property only when you want to output the sound to an Audio Mixer Group.

- **Play On Awake**: This option is enabled by default, which means the sound will start playing when the scene is loaded. If you don't want this Audio Source to emit sounds when the scene is loaded and want to control when the audio is played through code, then you can disable this option and call the `Play` method in a C# script.

In addition to the Audio Source, to emit the sound in the scene, an Audio Listener is also needed to receive the sound from the source. Next, we will discuss **Audio Listener**.

## Audio Listener

Generally speaking, you don't need to worry about the absence of **Audio Listener** in the scene, because an Audio Listener will be attached to the main camera in the scene by default when a scene is created, as shown here.

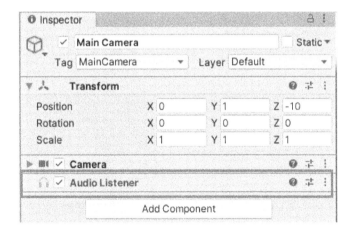

Figure 6.14 – An Audio Listener

In real life, sounds are heard by listeners, and **Audio Listener** is the representation of a listener in Unity. If you set **Audio Source** correctly in the game scene and the audio clip is available but you can't hear the sound when you run the game, then you can first check whether there is an Audio Listener in the scene. Usually, the listener is attached to the camera.

To hear the sound, we need to make sure that an Audio Listener is available, but at the same time, it should be noted that there cannot be more than one Audio Listener in the scene, otherwise you will see the following warning message in the **Console** window. So, please ensure there is always exactly one Audio Listener in the scene.

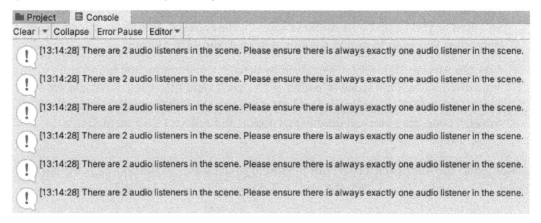

Figure 6.15 – Please ensure there is always exactly one audio listener in the scene

After introducing a few important concepts about audio in Unity, let's discuss the concepts related to video in Unity next.

## Video clips

Similar to audio clips, we also need to import external video files into the Unity editor to generate video clips. Unity supports typical file extensions for video files, such as the following:

- .mp4
- .mov
- .webm
- .wmv

After importing a video file, we can choose to open **Import settings**, as shown in *Figure 6.16*:

Figure 6.16 – Import settings of a video clip

By default, the **Transcode** option is disabled, which means that Unity will use the default settings to import this video file. If we enable this option, Unity will allow us to modify these settings, as shown in *Figure 6.17*, and we will introduce a few of them. At the bottom of the **Import settings** window, we can also directly preview the video by clicking the play button.

Now, let's check and enable the **Transcode** option and explore some of these import settings.

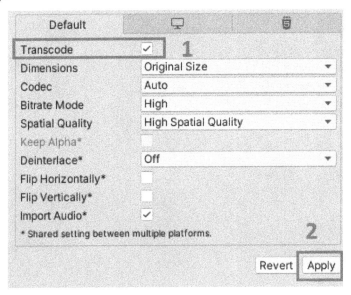

Figure 6.17 – Video import settings

- **Dimensions**: By default, Unity will not resize the original video, but if you want to resize the video file in Unity, you can change the **Dimensions** option. You will find a list of presets, such as **Half Res**, and you can also customize new sizes.

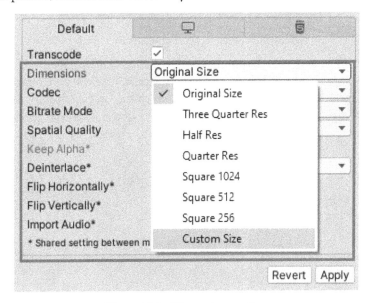

Figure 6.18 – Dimensions option

- **Codec**: Unity provides the option to transcode video clip assets into one of the following video codecs: **H264**, **H265**, and **VP8**, as shown in the following figure. **Auto** is the default value for **Codec**. Of course, you can also choose the video codec by yourself. **H264** is the best natively supported hardware-accelerated video codec.

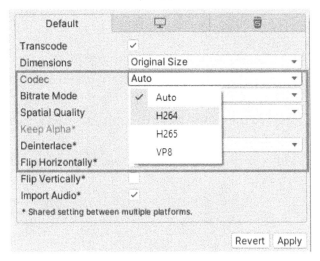

Figure 6.19 – Codec option

- **Keep Alpha**: As you can see in *Figure 6.19*, **Keep Alpha** is not an option in this case. This is because this option can only be checked when the video file contains an alpha channel. If your video file contains an alpha channel and you want to keep the alpha channel when the video is played in the game, then check this option.

- **Flip Horizontally**: As the name suggests, if this option is enabled, Unity will flip the video horizontally, switching the left side to the right side.

- **Flip Vertically**: Similar to **Flip Horizontally**, if this option is enabled, Unity will flip the video vertically to make it upside down.

- **Import Audio**: If your original video file contains audio tracks, then you can decide whether to import the audio tracks of the video by checking this option.

After setting the import settings, we can click **Apply** to transcode the video. It may take some time to complete the transcoding process.

Figure 6.20 – Transcoding the video

Now we have imported the video file into the Unity editor, next we need to set up a video player to play the video clip.

## Video Player

Let's create a Video Player by following these steps:

1. Right-click in the **Hierarchy** window to open the menu.

2. Choose **Create Empty** to create a new GameObject and rename this GameObject to `VideoPlayer`.

3. Select this newly created GameObject and click the **Add Component** button to open the components list.

4. Choose **Video | Video Player** to add a **VideoPlayer** component to this GameObject.

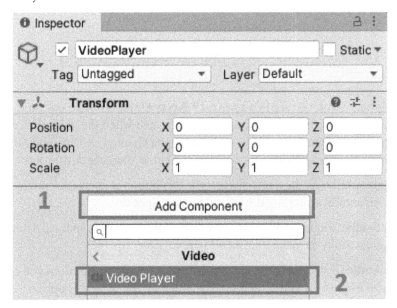

Figure 6.21 – Add a Video Player component

Now we have created a new **Video Player** in our game scene. The properties of this **Video Player** are shown in *Figure 6.22*:

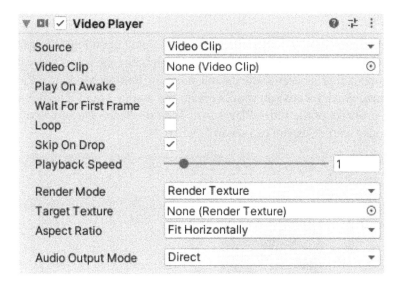

Figure 6.22 – Properties of Video Player component

Next, we will explore some of these properties:

- **Source**: In Unity, a **Video Player** can play videos from video clip assets or from a URL. By default, the **Video Player** needs a video clip asset as the video source, but we can also choose a URL as the source for video here.

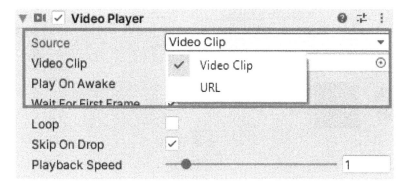

Figure 6.23 – Choose the type of video source

- **Play On Awake**: This option is enabled by default, which means the video will start playing when the scene is loaded. We can disable this option and call the `Play` method in a C# script to trigger the video playback at another point during the runtime.

- **Playback Speed**: We can increase or decrease the playback speed by adjusting this slider. The default value is 1.

# Render Mode

This is a very important setting, so we will explain it in detail. If you just set up the Video Player, drag a video clip asset to the **Source** property, and play the game, you will find that nothing will happen. This is because the default value for **Render Mode** in a Video Player is **Render Texture**, which means you should create and assign a render texture to the **Target Texture** property of the Video Player first. Then the Video Player will output the video to this render texture, as you can see in *Figure 6.24*:

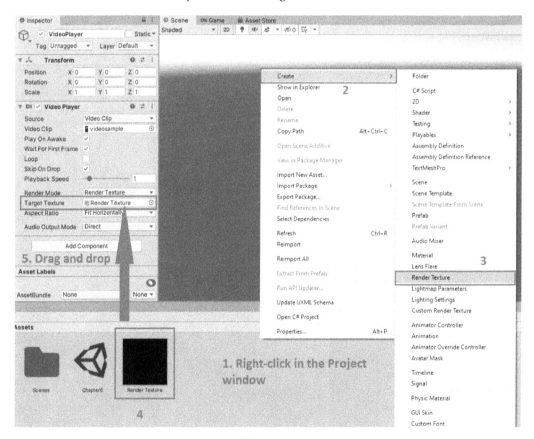

Figure 6.24 – Set the Target Texture property

However, at this stage, we only render the video to the render texture, and the video is not played in the game scene. In order to play this video in the game scene, we can create a new **Raw Image** UI element in the scene and assign this render texture to the **Raw Image** UI element.

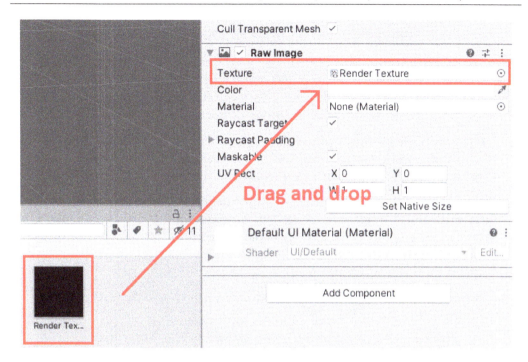

Figure 6.25 – The Raw Image UI element

Now, let's play the game again and the video plays as expected.

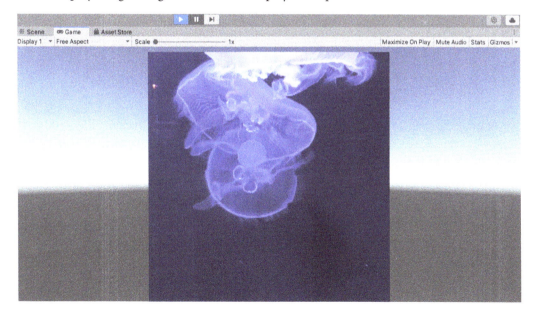

Figure 6.26 – Play the video

We can also change **Render Mode**. As you can see in *Figure 6.27*, the other options include the following:

- **Camera Far Plane**, which renders video content behind the camera's scene, allows developers to change the value of the alpha channel to make video content transparent, and can be used as a background video player.

- **Camera Near Plane**, which renders video content in front of the camera's scene, allows developers to change the value of the alpha channel to make video content transparent, and can be used as a foreground video player.

- **Material Override**: In Unity, a material is used to describe the appearance of the surface of a model. If this mode is selected, the video content will be passed into a user-specified property of the target material instead of being drawn on the screen or in a render texture. This mode is often used when making 360-degree panoramic videos in Unity.

- **API Only**, which does not render the video content, but allows developers to access the video content via an API.

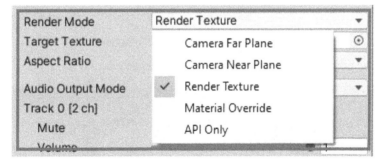

Figure 6.27 – Render Mode List

As an example, we will choose **Camera Far Plane** for **Render Mode**. Instead of a render texture, we need to provide a camera here and, as you can see in the following figure, it allows us to modify the **Alpha** value as well.

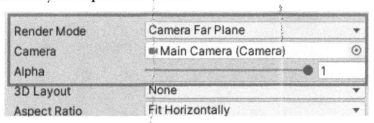

Figure 6.28 – Camera Far Plane

If we play the game, the video plays again this time.

Figure 6.29 – Play the video

In this section, we learned about some concepts of Unity's audio and video systems. Now, let's explore how to write C# code in Unity to control audio and video.

# Scripting with audio and video

In this section, we will explore how to interact with the audio and video systems via C# scripts. Similar to the previous section, we will also introduce the C# methods for **Audio Source** and **Video Player** respectively. We first start with the C# methods for **Audio Source**.

## AudioSource.Play

The first function we will introduce is the `Play` function of `AudioSource`. The function signature of `Play` is as follows:

```
public void Play();
```

It is very simple and straightforward to call this function to play an audio clip. However, if you need to deal with more complex scenarios, such as delaying the playback of an audio clip, you can call the `PlayDelayed` function, which will play the clip with a delay specified in seconds.

> **Note**
>
> There was an overloaded version of the `Play` function, which requires a `delay` parameter. However, it's deprecated now. Developers are advised to use the `PlayDelayed` function instead of the old `Play (delay)` function.

The following is the function signature of `PlayDelayed`:

```
public void PlayDelayed(float delay);
```

It requires a parameter, `delay`, which is specified in samples relative to the 44.1 kHz reference rate.

Now let's create a new C# script to first obtain a reference to the Audio Source in the scene and play the audio clip assigned to it by calling the `Play` function:

```csharp
using UnityEngine;

public class AudioPlayer : MonoBehaviour
{
[SerializeField]
private AudioSource _audioSource;

    private void Start()
    {
        if(_audioSource == null)
        {
            _audioSource = GetComponent<AudioSource>();
        }
    }

    public void OnClickPlayAudioButton()
    {
        _audioSource.Play();
    }
}
```

Then we drag this newly created script onto the Audio Source GameObject in the scene to attach this script to the GameObject as a new component.

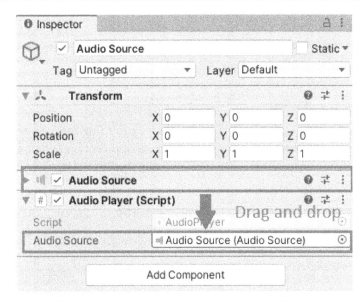

Figure 6.30 – Attaching the script to the GameObject

Here, we can manually drag the **AudioSource** component to the **Audio Source** field of the audio player component to obtain a reference to the **AudioSource** component, as shown in *Figure 6.30*. If you forget to assign a value to it, then you can use the GetComponent<AudioSource>() function to get the **AudioSource** component in the code as well.

Next, we will create a UI button in the scene and bind the button with the OnClickPlayAudioButton function so that when the button is clicked, the **Audio Source** will play the audio clip.

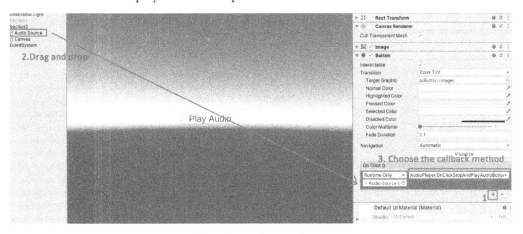

Figure 6.31 – Create a button

Now we can run the game and click the button to play the sound effect in the scene. This function is very useful when implementing audio effects; for example, when the player fires a gun, the sound of the bullet can be played, and so on.

## AudioSource.Pause

An Audio Source can be used to play background music. In some cases, we would like the background music to be paused, such as when the player enters a different scene or triggers a new plot. At this point, we can consider using the `Pause` function to pause playing the background music clip.

The function signature of `Pause` is very simple, as follows:

```
public void Pause();
```

We can create another function for the `AudioPlayer` class we created earlier:

```
public void OnClickPauseAudioButton()
{
    _audioSource.Pause();
}
```

Since the assets pack we downloaded earlier only contains sound effects with a short duration, in order to demonstrate the function of pausing background music, we can download and import **Free Music Tracks For Games** from the Unity Asset Store at the following link: `https://assetstore.unity.com/packages/audio/music/free-music-tracks-for-games-156413`.

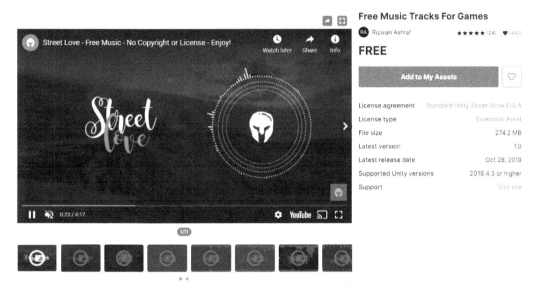

Figure 6.32 – Free Music Tracks For Games

Then replace the sound effect clip referenced by AudioSource with a new background music clip. Next, we will create another UI button and bind the button with the newly created OnClickPauseAudioButton function.

Now, we can run the game. If you click the first button, the background music will play; if you click the second button, we can pause the music.

AudioSource also provides an UnPause function to unpause the paused playback and an isPlaying property to check whether the current audio clip is playing.

The following is the function signature of UnPause:

```
public void UnPause();
```

We can use them to implement a more flexible function of pausing and continuing music playback as in the following code snippet:

```
public void OnClickPauseAudioButton()
{
    if(_audioSource.isPlaying)
    {
        _audioSource.Pause();
    }
    else
```

```
        {
            _audioSource.UnPause();
        }
    }
```

In this way, we can click the second button to pause the music playback, and click again to continue playing the music.

## AudioSource.Stop

In some cases, you may want the background music of the game to stop and then start from the beginning, instead of pausing and continuing to play. The Stop function of AudioSource is a suitable solution here.

The function signature of Stop is also very simple, as shown in the following code snippet:

```
public void Pause();
```

Let's create another function in the C# script to stop the background music and start playing from the beginning:

```
public void OnClickStopAndPlayAudioButton()
{
    if(_audioSource.isPlaying)
    {
        _audioSource.Stop();
    }
    else
    {
        _audioSource.Play();
    }
}
```

And we will also create a third UI button and bind the button with the OnClickStopAndPlayAudioButton function.

Run the game and click this button and the background music starts to play. Click again to stop the background music, and if you click for a third time, the background music will start to play from the beginning.

# VideoPlayer.clip

By default, a `VideoPlayer` component will play the video clip it refers to. However, it's a common requirement that we should be able to change the video when the game is running instead of creating many different Video Player instances. So, we can just modify the clip property of `VideoPlayer` via C# code:

```csharp
using UnityEngine;
using UnityEngine.Video;

public class VideoManager : MonoBehaviour
{
[SerializeField]
private VideoPlayer _videoPlayer;
[SerializeField]
private VideoClip _videoClip;

    void Start()
    {
        if (_videoPlayer == null)
        {
            _videoPlayer = GetComponent<VideoPlayer>();
        }
    }

    public void OnClickChangeVideoClip()
    {
        _videoPlayer.clip = _videoClip;
    }
}
```

In this case, we create a new C# script called `VideoManager`, which will get a reference to the target `VideoPlayer` component and a reference to the video clip asset. There is also a function called `OnClickChangeVideoClip`, which will later be bound to a UI button to change the video clip being played.

Compared to setting an Audio Source, setting a Video Player is slightly more complicated, because we also need to select a **Render Mode** option for **Video Player**. For simplicity, here we select the **Camera Near Plane** option and use **Main Camera** in the scene to render each frame of the video clip, as shown in *Figure 6.33*.

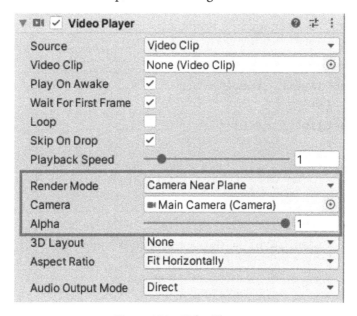

Figure 6.33 – Video Player

Then, we also need to assign the newly created script **VideoManager** to the same GameObject.

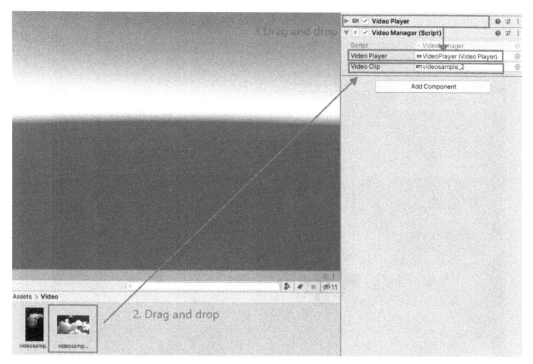

Figure 6.34 – Video Manager

As you can see in *Figure 6.34*, we not only assigned a reference to **Video Player** to the **VideoManager** script but also assigned a reference to a video clip asset to it.

The third thing is to create a new UI button and bind the button with the `OnClickChangeVideoClip` function we mentioned earlier.

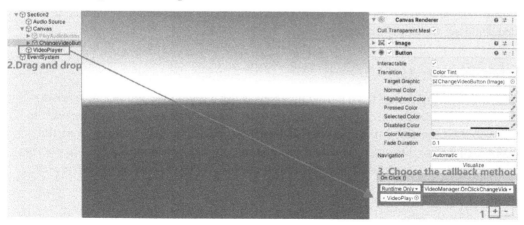

Figure 6.35 – UI button

Let's play the game in the editor and click the button to change the video clip.

Figure 6.36 – Change the video clip

As shown in *Figure 6.36*, the video clip of the **Video Player** component is changed to the video clip asset we want it to play.

# VideoPlayer.url

Sometimes, playing videos from video clip assets is not a good idea. For example, we do not want to increase the size of the game due to the inclusion of video files, or we want to develop WebGL-based games, and WebGL does not support video clip assets. Then, the use of a URL to provide video resources becomes an obvious solution. So, let's add another function called OnClickSetVideoURL to let the Video Player in the game scene play the video pointed to by the URL:

```
[SerializeField] private string _videoURL;
...
    public void OnClickSetVideoURL()
```

```
    {
        _videoPlayer.url = _videoURL;
    }
```

And we also need to create a new UI button and bind the button with the
`OnClickSetVideoURL` function.

Figure 6.37 – Set Video URL

Run the game and click the **Set Video URL** button to play the video from the URL,
as shown in the preceding figure.

> **Note**
>
> Unity does not support playing videos from YouTube, so you can host your
> video resources on other platforms, such as the Azure cloud.

## VideoPlayer.Play

In the previous two examples, whether we set the video clip asset or the video URL, the Video Player will automatically play the video. This is because we have enabled the **Play On Awake** option by default, as shown in *Figure 6.38*.

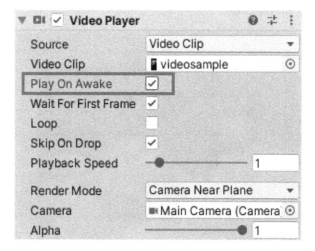

Figure 6.38 – Play On Awake

Usually, we prefer to be able to control when to play the video ourselves. Therefore, it is a good idea to disable this option and use C# code in a script to control playback, as shown in the following code block:

```
public void OnClickPlay()
{
    _videoPlayer.Play();
}
```

Here, we will create the third UI button and bind the button with the `OnClickPlay` function.

Figure 6.39 – Play Video

This time, if we run the game and click the **Change Video Clip** button or the **Set Video URL** button, there will be no video playing automatically. We also need to click the **Play Video** button to call the `Play` function of **Video Player** to play the video, as shown in *Figure 6.39.*

# VideoPlayer.frame and VideoPlayer.frameCount

Speaking of controlling video playback, the video progress bar is a useful feature. We can also implement a video progress bar in Unity. Next, let's discuss how to use the `frame` and `frameCount` properties of **Video Player** to implement a video progress bar.

The `frameCount` property is read-only and provides the number of frames in the current video content. On the other hand, the `frame` property can be modified and provides the frame index of the current frame. Therefore, we should first create a UI slider and then modify the value of the slider according to the value of `frame` and `frameCount` for a **VideoPlayer** component, as shown in *Figure 6.40*.

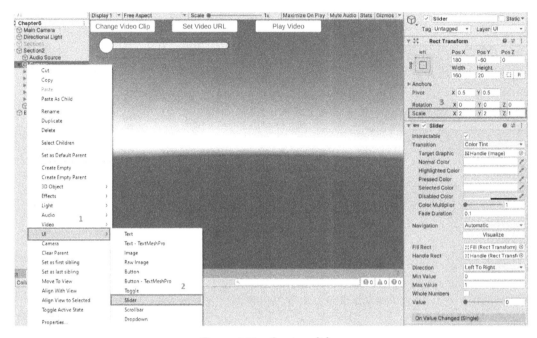

Figure 6.40 – Create a slider

We also need to modify the C# script to obtain a reference to the slider and update the value of the slider based on the value of `frame` and the value of `frameCount`:

```
using UnityEngine;
using UnityEngine.Video;
using UnityEngine.UI;


public class VideoManager : MonoBehaviour
{
    [SerializeField] private VideoPlayer _videoPlayer;
    [SerializeField] private VideoClip _videoClip;
    [SerializeField] private string _videoURL;
    [SerializeField] private Slider _progressBar;
```

```
void Start()
{
    if (_videoPlayer == null)
    {
        _videoPlayer = GetComponent<VideoPlayer>();
    }
}

private void Update()
{
    _progressBar.value = (float)_videoPlayer.frame /
        (float)_videoPlayer.frameCount;
}

public void OnClickChangeVideoClip()
{
    _videoPlayer.clip = _videoClip;
}

public void OnClickSetVideoURL()
{
    _videoPlayer.url = _videoURL;
}

public void OnClickPlay()
{
    _videoPlayer.Play();
}
}
```

In this case, we are using the UnityEngine.UI namespace because we need to access the UI slider from our code. And we implemented the Update function to update the value of the slider.

Let's run the game and play the video as before.

Figure 6.41 – The progress bar

We can see that as the video plays, the progress bar is also updated.

In this section, we explored and demonstrated the use of C# code to control audio and video, such as how to play audio and video, pause audio and video, and implement a progress bar via C# code.

However, if you are using Unity to develop a web application, then you may encounter other problems. Let's continue to explore.

# Things to note when using Unity to develop web applications

Unity is a cross-platform game engine, which means that we can deploy games that use the same code base and resources on different platforms, including WebGL. However, if you are using Unity to develop games for the web platform, here are some notes about implementing a video player.

# URL

First of all, the `VideoPlayer.clip` property is not supported on WebGL, which means that you can implement your video player solution by playing the video content in the video clip assets in the editor. However, once you build and deploy your web application to the server and run it, the video will not be played, even if the required video assets are packaged and deployed together.

Figure 6.42 – WebGL

As shown in *Figure 6.42*, when we run the web app and click the **Play Video** button, nothing will happen.

In this case, we have to provide a video source via the `VideoPlayer.url` property instead. If the video file has been hosted on another cloud platform, then we can directly use the method introduced in the previous section to play the video pointed to by the URL. In addition, `VideoPlayer.url` also supports local absolute or relative paths. Therefore, we can also build and deploy video files and other content of the game together. It should be noted that in this case, we no longer use Unity's video clip assets, but directly use the original video files, and put these video files in a folder called `StreamingAssets`.

> **Note**
>
> **StreamingAssets** is a special folder name of a Unity project. Files in this folder are available in their original format.

Here, we can create a new folder in the root of the project, rename it **StreamingAssets**, then put the original video file in this folder.

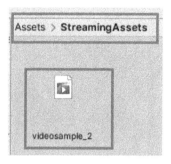

Figure 6.43 – StreamingAssets folder

As you can see in *Figure 6.43*, the video file is in its original format and has not been converted into a Unity video clip asset.

Next, let's create another C# script to demonstrate how to make **Video Player** load this video file and play it in the browser:

```csharp
using System.IO;
using UnityEngine;
using UnityEngine.Video;

public class VideoManagerForWeb : MonoBehaviour
{
    [SerializeField] private VideoPlayer _videoPlayer;
    [SerializeField] private string _videoFileName;

    void Start()
    {
        if (_videoPlayer == null)
        {
            _videoPlayer = GetComponent<VideoPlayer>();
        }
    }
```

```
    public void OnClickSetVideoURL()
    {
        _videoPlayer.url =
            Path.Combine(Application.streamingAssetsPath,
            _videoFileName);
    }
}
```

In this script, we access the Application.streamingAssetsPath property to get the path to the folder at runtime and assign the path to the url property of VideoPlayer.

Now, instead of running the game in the editor, we build and deploy it as a web application, and then run it in the browser.

Figure 6.44 – Play the video in the broswer

This time the video played as expected in the browser.

# Frame rate

When developing WebGL applications with Unity, another thing you should pay attention to is the **frame rate** of the video. In Unity, the frame rate is expressed as frames per second.

Let's print the video length, video frame count, and video frame rate information of the sample video we are using in the editor.

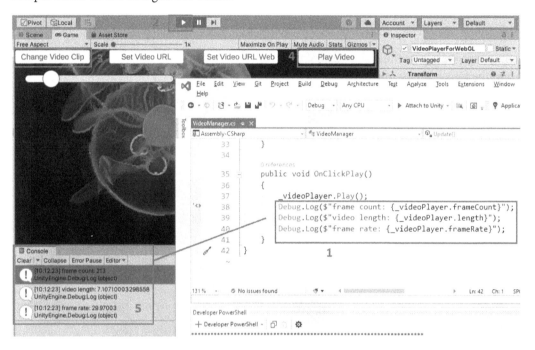

Figure 6.45 – Frame rate

As you can see here, the frame count of this video is 213, the video length is 7.1 seconds, and the frame rate is 30 FPS.

However, since the underlying implementation on the WebGL platform, that is, the JavaScript API for HTML5 `<video>`, does not disclose frame rate information, the frame rate is always assumed to be 24 FPS, even if the real frame rate of the video is 30 FPS. Therefore, frames/second of a video is always 24, which should be paid attention to when implementing the video progress bar for WebGL.

In this section, we discussed the things that need to be paid attention to when using Unity to develop video functions due to some limitations of the web platform. Next, we will explore how to use the profiler tool provided by Unity to locate the performance problems caused by audio and how to solve them.

# Increasing the performance of the audio system

In game development, the importance of audio is often overlooked. Sometimes this is also reflected in performance optimization. Game developers usually invest more effort in other performance areas, such as performance optimization for graphics rendering. But as games become more and more complex, audio can also cause performance problems, such as greater memory usage and so on. In this section, we will explore how to optimize audio performance in Unity.

## The Unity Profiler

First, we should learn how to use the Unity Profiler tool to view and locate performance bottlenecks caused by the audio system in Unity:

1.  Click **Window | Analysis | Profiler** or use the keyboard shortcut *Ctrl + 7* (*command + 7* on macOS) to open the **Profiler** window.

2.  Click the **Audio** module area in the **Profiler** window to view the performance data of the audio system. You can find out how many Audio Sources are playing, the number of audio clips being used, and the amount of memory being used for audio, and so on, as shown in *Figure 6.46*:

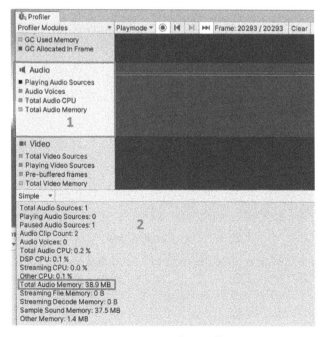

Figure 6.46 – Audio Profiler

As shown in *Figure 6.46*, the value of **Total Audio Memory** is 38.9 MB, which is very bad because, currently, only one Audio Source is playing sound. Therefore, we can click on the drop-down menu labeled **Simple** and switch to the **Detailed** view.

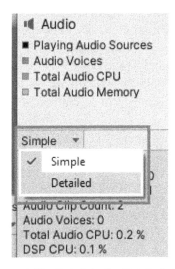

Figure 6.47 – Switch to the Detailed view

3.  We can get more information about the audio system and identify the specific audio asset that occupies 38.9 MB of memory.

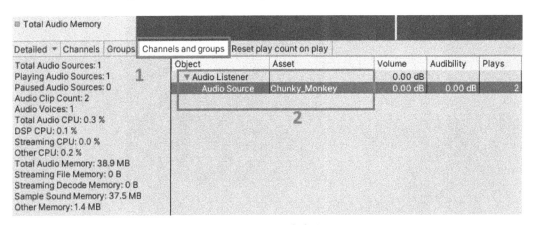

Figure 6.48 – Detailed view

Next, we will introduce how to reduce the memory occupied by this audio resource.

# Using Force To Mono to save memory

If we inspect this audio asset, we will find the audio asset is stereo, as shown in *Figure 6.49*.

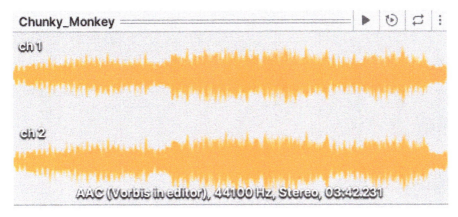

Figure 6.49 – The audio clip

However, since there is only one Audio Source in the game scene, which means that the sound is emitted from one point, the effect of using stereo is lost here, but the memory consumption is twice that of mono. Therefore, if the game does not require stereo and needs to reduce memory overhead, we can just enable the **Force To Mono** option in the audio clip's import settings to convert the stereo audio clip to a mono audio clip.

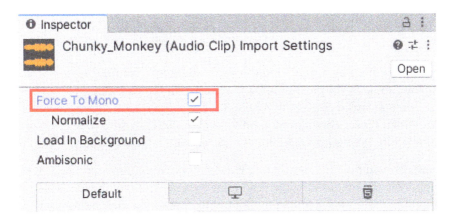

Figure 6.50 – Enable Force To Mono

Then let's play the audio again. This time we find that the memory consumption of this audio clip has dropped from 38.9 MB to 20.2 MB, which is almost halved.

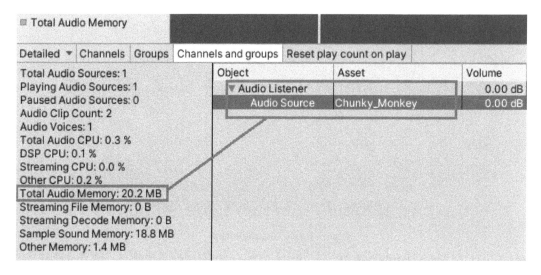

Figure 6.51 – Memory consumption has dropped

In this section, we introduced how to use Unity's Profiler tool to view the performance data of the audio system and explored how to optimize the performance of the audio system.

## Summary

In this chapter, we started by introducing the audio and video features provided by Unity, then we explored some of the most important concepts in the Unity audio system and video system, such as **audio clip** assets, **Audio Source** components, **Audio Listener** components, **Video Player** components, and so on. We also discussed how to create a new script in Unity to interact with Unity's audio system and video system.

Then we demonstrated how to implement a video for the web platform because WebGL does not support Unity's **video clip** assets and due to underlying implementation reasons, the video frame rate is always assumed to be 24 FPS. These need to be paid attention to.

Finally, we explored how to view and locate performance bottlenecks caused by the audio system in Unity.

In the next chapter, we will introduce the mathematics of computer graphics in Unity.

# Part 3:
# Advanced Scripting
# in Unity

In this section, we will introduce advanced topics in Unity, such as the Scriptable Render Pipeline, the **Data-Oriented Technology Stack (DOTS)**, and serialization in Unity. Additionally, we'll also cover how to use the Microsoft Azure cloud for assets management and hosting player data in the cloud.

This part includes the following chapters:

# 7

# Understanding the Mathematics of Computer Graphics in Unity

Mathematics is a topic that is often discussed in game development. Although Unity has provided game developers with many helper functions to reduce the complexity of using mathematics in Unity, it is still necessary to have some basic mathematical knowledge about computer graphics, such as coordinate systems, vectors, matrices, and quaternions.

In this chapter, we will explore the following key topics:

- Getting started with coordinate systems
- Working with vectors
- Working with the transformation matrix
- Working with quaternions

By the end of this chapter, you will have mathematical knowledge of computer graphics and know how to use vectors, matrices, quaternions, and Euler angles in scripts correctly.

Now, let's get started!

# Getting started with coordinate systems

Like many files, most model files are binary files. When a game engine, such as Unity, needs to render a model, the data of the model, such as the vertex array of the model and the index of the vertex array, will be extracted and processed through the render pipeline of the game engine.

> **Note**
> You can find more information about the render pipeline in computer graphics at `https://www.khronos.org/opengl/wiki/Rendering_Pipeline_Overview`.

A graphics render pipeline mainly includes two functions: one to convert the 3D coordinates of an object into 2D coordinates in the screen space and the other to color each pixel of the screen. Finally, the 3D model is rendered on the 2D screen.

In the process of the render pipeline, a lot of coordinate system conversion work will be involved, as you can see in *Figure 7.1*. So, it's an important topic and we will introduce information about coordinate systems in this section:

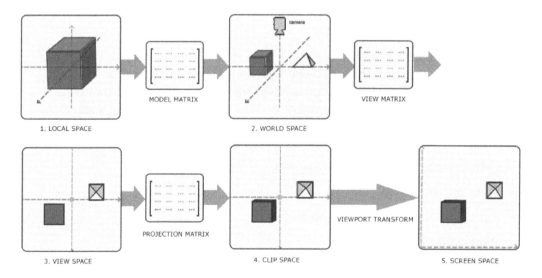

Figure 7.1 – Coordinate transformation process (CC BY 4.0)

# Understanding left-handed and right-handed coordinate systems

A coordinate system is a geometric system that typically uses numbers to determine the position of a point in space.

In mathematics, there are many different types of coordinate systems, such as the **number line coordinate system**, **Cartesian coordinate system**, and **polar coordinate system**. In computer graphics, the **Cartesian coordinate system** is the most commonly used.

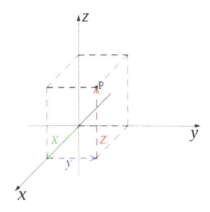

Figure 7.2 – Cartesian coordinate system

The Cartesian coordinate system is also very common in our daily lives, that is, the $x$ axis, $y$ axis, and $z$ axis are used to describe the coordinate information of the object. When used to describe 3D space, the Cartesian coordinate system can be either a **left-handed coordinate system** or a **right-handed coordinate system**. As their names imply, we can actually distinguish between the two by using the left hand and the right hand.

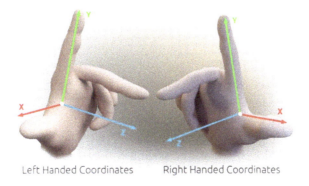

Left Handed Coordinates          Right Handed Coordinates

Figure 7.3 – Coordinate systems (CC BY-SA 3.0)

As shown in *Figure 7.3*, we can distinguish between the left-handed coordinate system and the right-handed coordinate system by visualizing the thumb pointing to the *x* axis, the index finger to the *y* axis, and the middle finger to the *z* axis.

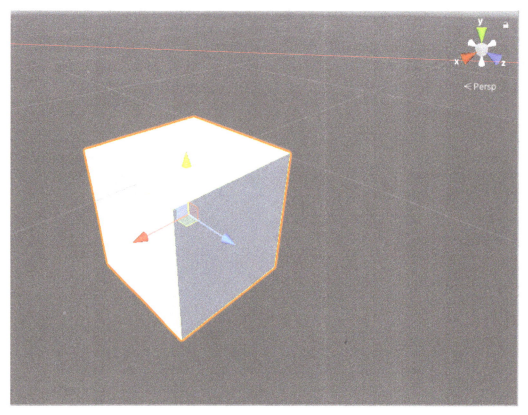

Figure 7.4 – Left-handed coordinate system in Unity

If we look in the Unity Editor, we can see that Unity uses the left-handed coordinate system, as shown in *Figure 7.4*.

## Local space

Coordinate space is the space where 3D positions and transformations exist within the coordinate system, such as **local space** and **world space**. In Unity, we often work with local space or world space. Local space is related to the concept of the **parent-child relationship**, which means it uses the origin and axes of the GameObject's parent node in the hierarchy of GameObjects. The position, rotation, and scaling of the parent GameObject will affect the local space defined by it. Therefore, this is useful not when we are dealing with the transformation of a single GameObject but of a group of GameObjects.

For example, in *Figure 7.5*, the five cube objects are all children of the GameObject named **LocalSpace**:

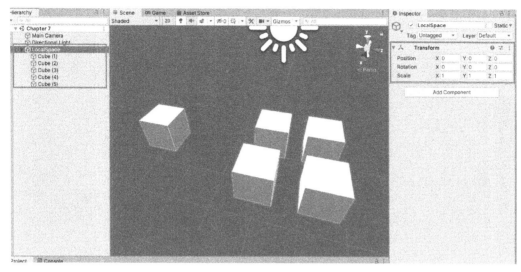

Figure 7.5 – LocalSpace parent object

We can see that the **Position** and **Rotation** values of the parent GameObject are 0. Now, let's move this parent object down 2 units along the *y* axis and also rotate it 45 degrees around the *y* axis.

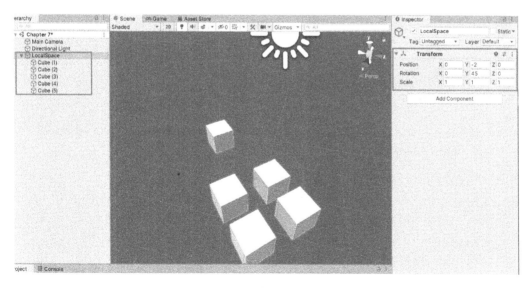

Figure 7.6 – LocalSpace parent object

As shown in *Figure 7.6*, all these cubes have moved down 2 units along the *y* axis and rotated 45 degrees around the *y* axis. However, if we look at the position and rotation of individual cubes in the **Inspector** window, we can see that these values have not changed. This is because, currently, they are in local space defined by their parent object, and the position and rotation of a single cube relative to its parent object have not changed.

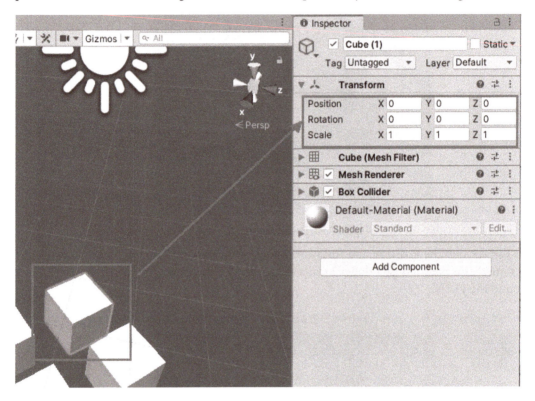

Figure 7.7 – Local space child object

We can change the local position, local rotation, and local scale of a child object through C# code at runtime, as shown:

```
public class LocalSpaceTest : MonoBehaviour
{
    private Vector3 _localPosition = new Vector3(-2, 0, 0);
    private Vector3 _localScale = new Vector3(1, 2, 1);
    private Transform _transform;
```

```
    private void Start()
    {
        _transform = gameObject.transform;
        _transform.localPosition = _localPosition;
        _transform.localScale = _localScale;
    }
}
```

Attach this script to the child object named **Cube (1)** and run the game. We can see in the following screenshot that the child object has moved 2 units along the *x* axis relative to the parent object and is magnified 2 times along the *y* axis relative to the parent object:

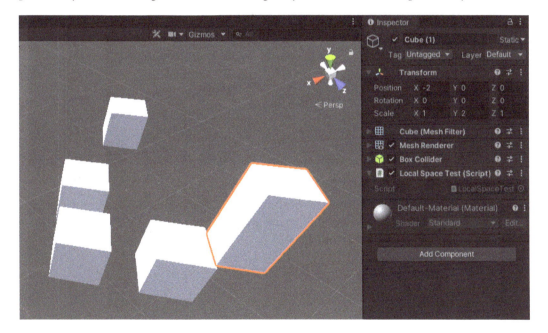

Figure 7.8 – Changing the local postion and local scale

In this section, we discussed local space. Next, we will explore world space.

# World space

Unlike local space, which is defined by a parent GameObject, world space is the coordinate system for the entire Scene. The center of the Scene is the origin of world space.

Let's create a new Cube object in the Scene and this time, this new cube is not a child of other GameObjects.

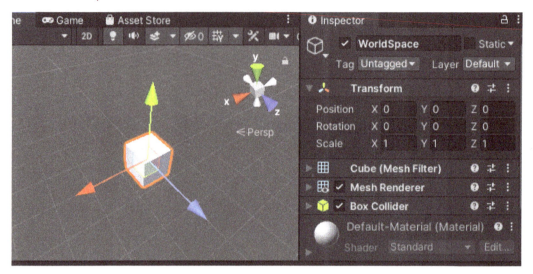

Figure 7.9 – World space

As shown in *Figure 7.9*, when the position of the cube is 0, the cube is located in the center of the Scene. If we change the $x$ value of the cube position from 0 to 1, then the cube will advance 1 unit along the $x$ axis of world space.

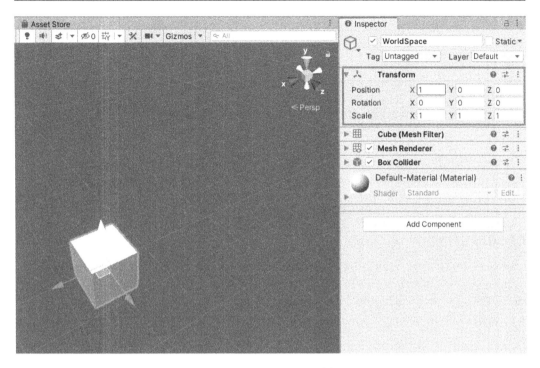

Figure 7.10 – Moving in world space

We can also modify the position, rotation, and scale of a GameObject in world space in a C# script. The following code snippet shows how to do this:

```
using UnityEngine;

public class WorldSpaceTest : MonoBehaviour
{
    void Start()
    {
        transform.position = new Vector3(0, 1, 0);
    }
}
```

The `position` property is the world space position of the transform. In addition to directly modifying the `position` or `rotation` properties, we can also call the following method to modify the `position` and `rotation` properties of the object at the same time:

```
public void SetPositionAndRotation(Vector3 position, Quaternion
    rotation);
```

We can see that this method requires a `Vector3`-type parameter and a `Quaternion`-type parameter. We will introduce vectors and quaternions later in the *Working with vectors* and *Working with quaternions* sections, respectively.

## Screen space

As we mentioned at the beginning of this section, the coordinate system can be used to determine a point in space. This refers to not only 3D space but also 2D space. Screen space is the space defined by the viewer's screen. It means that the screen space projects the content onto the screen.

In screen space, the coordinates are in 2D; **(0,0)** is the lower-left corner and (screen.width, screen.height) is the upper-right corner, as shown in the following screenshot:

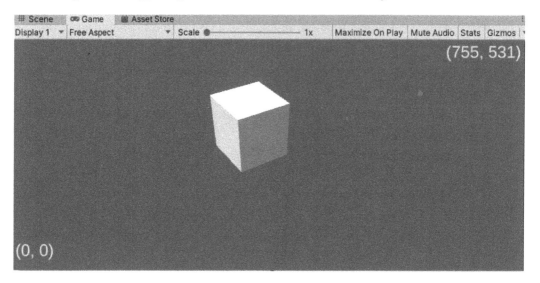

Figure 7.11 – Screen space

2D elements are often described in screen space, and the most common is the UI. Another common use of screen space is to get the position of the mouse input. The reason is obvious: the mouse moves on the screen. The following code snippet demonstrates how to get the position of the mouse in a C# script:

```
using UnityEngine;

public class ScreenSpaceTest : MonoBehaviour
{
    void Update()
    {
        Vector2 mousePosition = Input.mousePosition;
        Debug.Log($"Mouse Position: {mousePosition}");
    }
}
```

The mousePosition property of the Input class will return the current mouse position in screen space, and the preceding code will print the mouse position to the **Console** window, as shown in *Figure 7.12*:

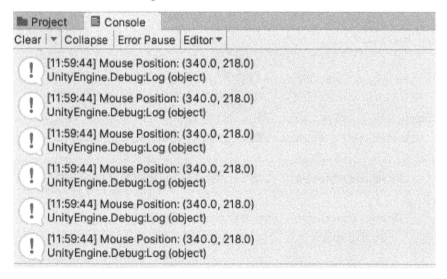

Figure 7.12 – Mouse position

After obtaining the screen space position of the mouse, we can use the methods provided by Unity's Camera class to convert the screen space position to the world space position. In addition, Unity allows us to create a ray that goes from the camera through a screen point to the game world. This can help us deal with a common situation in games where we need to know what the player is clicking on in the 3D game world, even though the player can only actually click on a 2D screen.

The method signatures of some methods are as follows:

```
public Ray ScreenPointToRay(Vector3 pos);
public Vector3 ScreenToWorldPoint(Vector3 position);
```

As we just mentioned, the ScreenPointToRay method is very useful because it returns a Ray instance from the camera pointing to the mouse position in the world space. I hope you still remember the Collider component in the physics system we introduced in the previous chapter because we can use this method to cast a ray to the collider and get the details of the collider, and it can also be used to draw a line in the Scene view of the Unity Editor to help with debugging.

Next, we will modify the previous code to implement a function that can detect whether there is a collider at the mouse click position and draw a red line in the Scene view if there is a collider:

```
Ray _ray;

private void FixedUpdate()
{
  _ray =
    Camera.main.ScreenPointToRay(Input.mousePosition);

  if (Physics.Raycast(_ray, out RaycastHit hit, 50))
  {
      Debug.DrawLine(_ray.origin, hit.point,
        Color.red);
  }
}
```

As the code snippet shows, we are calling the `ScreenPointToRay` method to create a ray pointing in the direction of the mouse from the location of the main camera in the scene, and then using this ray to detect colliders in the Scene by calling `Physics.Raycast`, and finally calling `Debug.DrawLine` to draw a red line in the Scene view, as shown in the following screenshot:

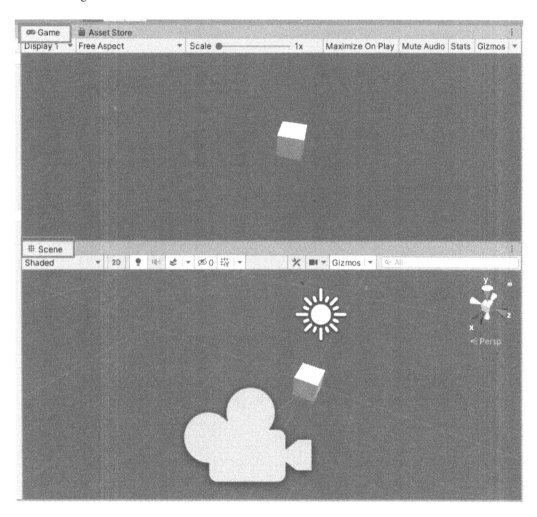

Figure 7.13 – Drawing a red line

In *Figure 7.13*, the top is the game view, which is the window where the game is running, and the bottom is the Scene view, which is the window for debugging.

We have introduced you to coordinate systems in this section. Next, we will discuss another very important topic: vectors.

# Working with vectors

In game development, we use vectors to define directions and positions. As shown in the following figure, we draw a line between two points to represent a vector. In this case, the vector starts from the origin, which is point **B (0, 0)** on the graph, to point **A (6, 2)**:

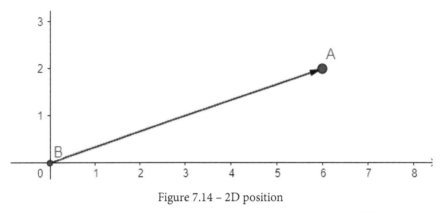

Figure 7.14 – 2D position

We can see this vector is made up of two components, namely *x* and *y*. They represent the distance from the origin along the *x* axis and the *y* axis. Therefore, this vector can be used to define the position of point **A** relative to the origin in the space. In addition to the position of point **A**, we can also calculate the length of the distance between these two points, and we call it the **magnitude**. The magnitude of a 2D vector is the square root of `(x*x+y*y)`.

In Unity, we will use the Vector2 structure to represent 2D vectors and points. The magnitude property of Vector2 returns the value of the magnitude of this 2D vector.

3D vectors are similar to 2D vectors, but we also need to consider the value of the *z* axis. The magnitude of a 3D vector is the square root of `(x*x+y*y+z*z)`.

Unity also provides the `Vector3` structure to represent 3D vectors and points. If you look at the Inspector window of a GameObject in the Scene, you will find that the **Position**, **Rotation**, and **Scale** properties of the object are all Vector3 types, as shown in the following figure:

| Transform | | | | | | | |
|---|---|---|---|---|---|---|---|
| Position | X | 0 | Y | 0 | Z | 0 |
| Rotation | X | 0 | Y | 0 | Z | 0 |
| Scale | X | 1 | Y | 1 | Z | 1 |

Figure 7.15 – Transform of a GameObject

# Vector addition

Since vectors can be used to describe positions, they can also be used to describe positions that change over time. A moving object has a velocity, which is the speed of the object in a given direction.

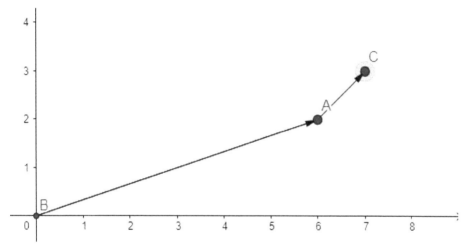

Figure 7.16 – Vector addition

As shown in *Figure 7.16*, suppose an object is currently located at point **A** and its velocity is (1, 1) per minute, which means the object will move in a direction that is 1 unit further on the *x* axis and 1 unit further on the *y* axis. So, we will add its current position vector to its velocity vector to calculate where it will end up after 1 minute:

```
(6, 2) + (1, 1) = (7, 3)
```

The new position of this object after 1 minute is (7, 3).

# How to subtract vectors

Vector subtraction and vector addition are very similar. We can reverse the direction of the second vector and use vector addition. Let's still use the previous example. Suppose a moving object is currently located at point A and its velocity is (-1, -1) per minute, which means the object will move in a direction that is -1 unit further on the $x$ axis and -1 unit further on the $y$ axis.

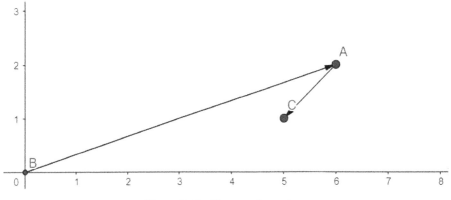

Figure 7.17 – Vector subtraction

Let's add its current position vector to its velocity vector to calculate where it will end up after 1 minute again:

```
(6, 2) - (1, 1) = (6, 2) + (-1, -1) = (5, 1)
```

The new position of this object after 1 minute is (5, 1).

In Unity, we can add vectors and subtract vectors in C# code, as follows:

```
private void Start()
{
    var vector1 = new Vector3(1, 1, 1);
    var vector2 = new Vector3(1, 2, 3);
    var addVector = vector1 + vector2;
    var subVector = vector1 - vector2;

    Debug.Log($"Addition: {addVector}");
    Debug.Log($"Subtraction: {subVector}");
}
```

In the previous code snippet, we created two 3D vectors, (1, 1, 1) and (1, 2, 3). Then, we added and subtracted them respectively and printed the results to the **Console** window.

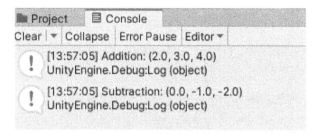

Figure 7.18 – Adding vectors and subtracting vectors

In order to move objects in Unity, knowledge of vectors is needed. But sometimes we don't have to directly calculate the result of vector addition or subtraction in the code. This is because Unity provides us with the Transform.Translate function to move objects. Of course, we still need to pass a vector parameter to provide velocity:

```
private void Update()
{
    transform.Translate(_speed * Time.deltaTime *
        Vector3.forward);
}
```

The preceding code snippet demonstrates how to move an object by calling the Transform.Translate function.

## Dot product

In addition to vector addition and vector subtraction, 3D vector operations commonly used in game development also include **dot product** operations and **cross product** operations. We will introduce them separately in this section and the next section.

First, we will explore the dot product in Unity. The dot product or scalar product takes two vectors and returns a single scalar value.

Suppose there are two 3D vectors, named *vector1* and *vector2*; the calculation process of the dot product is very simple, as shown:

```
scalar value = (x1 * x2) + (y1 * y1) + (z1 * z2)
```

In game development, vector dot product operations are often used to find out whether these two vectors are perpendicular to each other. If the result of their dot product operation is 0, the two vectors are perpendicular to each other. If the result is positive, the angle between the two vectors is less than 90 degrees. If the result is negative, the angle between the two vectors is greater than 90 degrees.

Next, we can create two 3D vectors in the Unity Editor to demonstrate how to use the vector dot product operation.

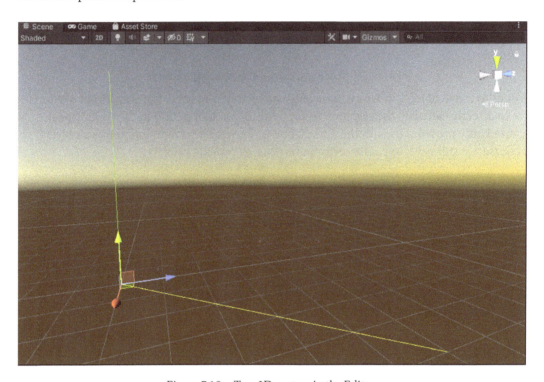

Figure 7.19 – Two 3D vectors in the Editor

As shown in *Figure 7.19*, the green line represents the first vector, which is $(0, 5, 0)$, and the yellow line represents the other vector, which is $(5, 0, 5)$. The result of the dot product operation of these two vectors is as follows:

```
0 = (0 * 5) + (5 * 0) + (0 * 5)
```

At the same time, we can see in *Figure 7.19* that these two vectors are perpendicular.

If the first vector is (0, 5, 5), the result of the dot product operation of these two vectors will be as follows:

```
25 = (0 * 5) + (5 * 0) + (5 * 5)
```

As shown in *Figure 7.20*, the two vectors are not perpendicular this time, and the included angle is less than 90 degrees:

Figure 7.20 – Two 3D vectors in the Editor

If the first vector is (0, 1, -1), the result of the dot product operation of these two vectors will be as follows:

```
-5 = (0 * 5) + (1 * 0) + (-1 * 5)
```

As shown in *Figure 7.21*, the two vectors are not perpendicular this time, and the included angle is greater than 90 degrees:

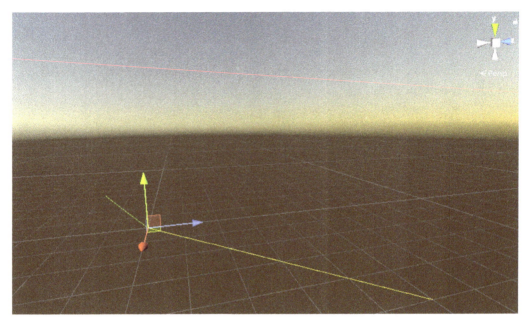

Figure 7.21 – Two 3D vectors in the Editor

Unity provides us with a function to calculate the result of the dot product of two 3D vectors, as follows:

```
public static float Dot(Vector3 lhs, Vector3 rhs);
```

It is a static function and we can call it directly in our C# code:

```
public class VectorTest : MonoBehaviour
{
    private Vector3 _vectorA = new Vector3(0, 1, -1);
    private Vector3 _vectorB = new Vector3(5, 0, 5);

    private void Start()
    {
        var result = Vector3.Dot(_vectorA, _vectorB);
        Debug.Log(result);
    }
}
```

# Cross product

On the other hand, the cross product takes two vectors as well but returns another vector instead of a single scalar value. This vector is perpendicular to both of the original two vectors.

$$\begin{bmatrix} b_x \\ b_y \\ b_z \end{bmatrix} \times \begin{bmatrix} c_x \\ c_y \\ c_z \end{bmatrix} = \begin{bmatrix} b_y c_z - c_y b_z \\ c_x b_z - b_x c_z \\ b_x c_y - c_x b_y \end{bmatrix}$$

Figure 7.22 – Cross product (CC BY-SA 4.0)

Compared with the dot product, the calculation process of the cross product is more complicated. The preceding figure demonstrates this process.

Unity also offers another helpful function to calculate the result of the cross product of two 3D vectors, as follows:

```
public static Vector3 Cross(Vector3 lhs, Vector3 rhs);
```

It is a static function and we can call it directly in our C# code:

```
    void FixedUpdate()
    {
        var vector1 = new Vector3(0, 1, 0);
        var vector2 = new Vector3(1, 0, 1);
        Debug.DrawLine(Vector3.zero, vector1, Color.green);
        Debug.DrawLine(Vector3.zero, vector2,
            Color.yellow);
        var resultVector = Vector3.Cross(vector1, vector2);
        Debug.DrawLine(Vector3.zero, resultVector,
            Color.cyan);
    }
```

In this code snippet, we calculate the result of the cross product of vector1 and vector2 and at the same time, we also draw these three vectors in the Unity Editor, as shown in the following figure:

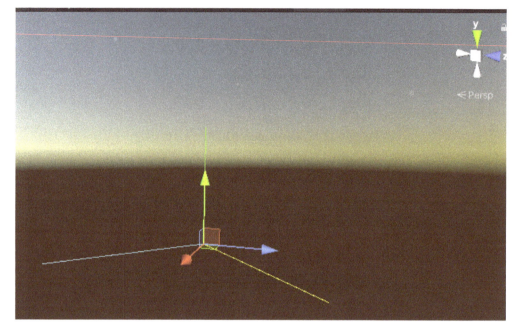

Figure 7.23 – Cross product

In this section, we introduced vectors and explored how to use vectors correctly in Unity scripts. Next, let's continue to explore another important concept in computer graphics, namely matrices.

# Working with the transformation matrix

In game development, the **transformation matrix** is also a common term. Specifically, we use the transformation matrix to encode transformations, including translation, rotation, and scaling transforms.

Unity provides us with the Matrix4x4 struct in C# to represent a standard 4x4 transformation matrix.

$$\begin{bmatrix} 1 & 0 & 0 & 0 \\ 0 & 1 & 0 & 0 \\ 0 & 0 & 1 & 0 \\ 0 & 0 & 1 & 0 \end{bmatrix}$$

Figure 7.24 – A 4x4 matrix

As shown in *Figure 7.24*, a transformation matrix is a grid of numbers. Although it is a common term, we rarely use this matrix directly in scripts. This is because the calculation of the matrix is relatively cumbersome, and Unity, as an easy-to-use game engine, has encapsulated the complex calculations in the `Transform` class for us, and we only need to call some functions. Therefore, in this section, we only give a brief introduction to the transformation matrix.

Before we start, you should know that transformations include translation, rotation, scaling, and these operations can be represented as matrices. We will discuss them one by one in the following subsections.

## Translation matrix

We can move an object by using a translation matrix. The following diagram shows a translation matrix and how to move the original vector by multiplying the translation matrix:

$$\begin{bmatrix} 1 & 0 & 0 & T_x \\ 0 & 1 & 0 & T_y \\ 0 & 0 & 1 & T_z \\ 0 & 0 & 0 & 1 \end{bmatrix} \cdot \begin{pmatrix} x \\ y \\ z \\ 1 \end{pmatrix} = \begin{pmatrix} x + T_x \\ y + T_y \\ z + T_z \\ 1 \end{pmatrix}$$

Figure 7.25 – Translation matrix (CC BY 4.0)

Let's create a C# script and demonstrate how to move an object by using a matrix directly in Unity:

```
using UnityEngine;

public class MatrixTest : MonoBehaviour
{
    void Start()
    {
        var translationMatrix = new Matrix4x4(
            new Vector4(1, 0, 0, 0),
            new Vector4(0, 1, 0, 0),
            new Vector4(0, 0, 1, 0),
            new Vector4(3, 2, 1, 1)
        );
```

```
var newPosition =
    translationMatrix.MultiplyPoint
    (transform.position);

    transform.position = newPosition;
    }
}
```

As you can see in this code snippet, we used four `Vector4` instances to create an instance of the `Matrix4x4` struct. It should be noted here that each `Vector4` we used to create the matrix represents a column of the matrix, not a row. Therefore, the code creates a matrix, as shown in the following figure:

$$
\begin{pmatrix}
1 & 0 & 0 & 3 \\
0 & 1 & 0 & 2 \\
0 & 0 & 1 & 1 \\
0 & 0 & 0 & 1
\end{pmatrix}
$$

Figure 7.26 – Creating a matrix

Then, we calculated the new position of the object by calling the `MultiplyPoint` function of `Matrix4x4`, where the parameter is the original position of the object. Finally, we set the position of the object to this new vector.

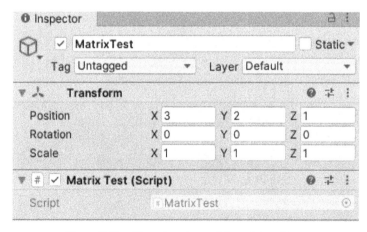

Figure 7.27 – Changing the position of the object

If we create an object at the origin and run this script, the result will be that the object is moved to the point (3, 2, 1), as shown in *Figure 7.27*.

# Rotation matrix

Similarly, a matrix can also be used to rotate an object, that is, a rotation matrix. This time, we also need to create an instance of `Matrix4x4` in the C# script, but instead of calling its constructor, we call this function:

```
public static Matrix4x4 Rotate(Quaternion q);
```

The `Rotate` function is a static function of `Matrix4x4`, and it creates and returns a rotation matrix. This function requires a quaternion-type parameter. We will introduce quaternions in the next section.

Now, let's write some code to rotate the object by using `Matrix4x4`:

```
var rotation = Quaternion.Euler(0, 90, 0);
var rotationMatrix = Matrix4x4.Rotate(rotation);
var newPosition =
    rotationMatrix.MultiplyPoint(transform.position);
transform.position = newPosition;
```

This code will move the point from its original position to a place rotated 90 degrees around the *y* axis.

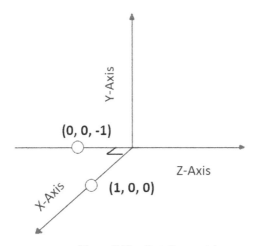

Figure 7.28 – Rotation matrix

Let's set the original position of this object to $(1, 0, 0)$ and then run the code. The new position of this object should be $(0, 0, -1)$, as shown in the preceding figure.

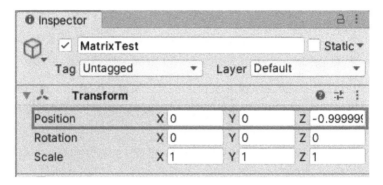

Figure 7.29 – The real result

After running the code, we can see that the real result is consistent with what we expected.

## Scaling matrix

When we scale a vector, we will keep its direction unchanged and change the length by the amount we want to scale. We can also use a scaling matrix to scale a point away from the origin. You can imagine that a model is composed of many vertices. When we scale a model, we actually extend or shrink the positions of the vertices that make it up.

Unity also provides us with the following function to directly create a scaling matrix in a C# script:

```
public static Matrix4x4 Scale(Vector3 vector);
```

The **Scale** function is a static function of `Matrix4x4`, and it creates and returns a scaling matrix:

```
private void ScalingMatrixTest()
{
    var scale = new Vector3(3, 2, 1);
    var scalingMatrix = Matrix4x4.Scale(scale);
    var newPosition =
      scalingMatrix.MultiplyPoint(transform.position);
    transform.position = newPosition;
}
```

In order to demonstrate how to apply a scaling matrix to a point, we created the preceding code snippet. As you can see in the code, we created a new `Vector3` to present the scaling factors. Then, we created a scaling matrix by calling the `Matrix4x4.Scale` function and finally, we applied this matrix to a point.

Let's create a new GameObject in the Scene and locate this GameObject at the position $(1, 1, 0)$.

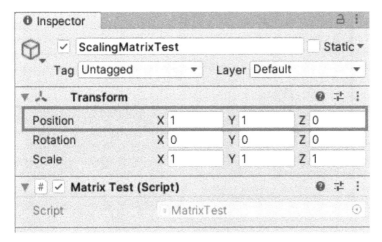

Figure 7.30 – GameObject at $(1, 1, 0)$

Then, attach this script to it and run the script.

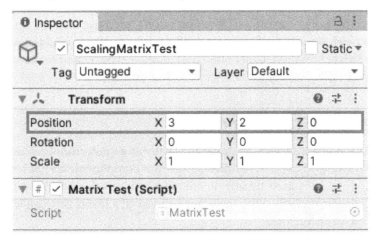

Figure 7.31 – Apply the scaling matrix to this object

As shown in *Figure 7.31*, the new position of this object is (3, 2, 0). This is because this scaling matrix increases the point three times along the *x* axis from its original position and two times along the *y* axis.

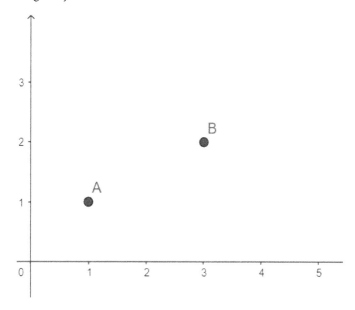

Figure 7.32 – Scaling a point

As we mentioned at the beginning of this section, in Unity development, matrix operations are relatively low-level operations. Unity has provided us with many functions to cover up the complexity of matrices. Developers don't often use matrices directly, but as an important concept, we still need to understand some concepts around them. However, when it comes to object rotation, Unity often uses another type to save rotation data. If you are interested in this, let's continue to explore **quaternions**.

# Working with quaternions

In Unity, the rotation of a transform is stored internally as a **quaternion**, which has four componenets, namely *x*, *y*, *z*, and *w*. However, these four components do not represent angles or axes, and we developers usually do not need to access them directly. You may be confused because if you look at the Inspector window of a transform, you will find the rotation is displayed as a Vector3.

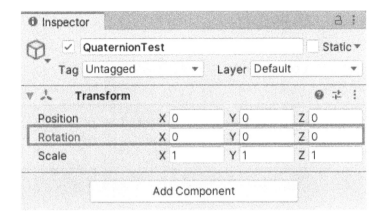

Figure 7.33 – Rotation property in the Inspector window

This is because although Unity uses quaternions to store rotations internally, in addition to quaternions, rotations can also be represented by three angle values of *x*, *y*, and *z*, namely **Euler angles**.

Therefore, for the convenience of developers to edit, Unity displays the value of the equivalent Euler angle in the Inspector.

So, why doesn't Unity use Euler angles to store rotations directly? It is composed of three axes angles and is in a format that is easy for humans to read. This is because the Euler angle is affected by the **gimbal lock**, which means that the "degree of freedom" is lost.

On the other hand, using quaternion rotation will not cause the gimbal lock issue. Therefore, Unity uses quaternions to store rotations internally. But what you have to remember is that the four components of a quaternion do not represent angles, so we will not modify the value of a component individually, and it is very complicated to modify a quaternion directly. Fortunately, Unity provides us with many built-in C# functions in the Quaternion struct to manage quaternion rotations easily. It is our best choice to use the Quaternion structure and its functions to manage the rotation values in Unity.

We can divide these functions into three groups according to their purpose, namely creating rotations, manipulating rotations, and working with Euler angles. Let's explore them next.

## Creating rotations

The first function we will introduce is LookRotation and the function signature of this function is as follows:

```
public static Quaternion LookRotation(Vector3 forward,
    [DefaultValue("Vector3.up")] Vector3 upwards);
```

This is a static function; you can pass in parameters to specify the forward and upward direction for it, and it will return the correct rotation value according to the passed-in parameters.

In the following example, we set up a Scene in which there are two objects, named `target` and `player`, and created a new C# script called `LookAtScript.cs`. We then attached this script to the player object, as shown in *Figure 7.34*. The blue cube represents the player and the red sphere represents the target object:

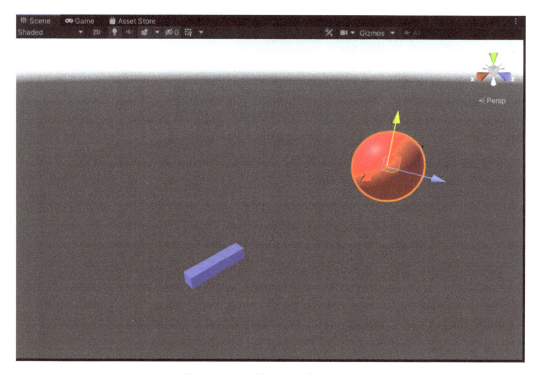

Figure 7.34 – Objects in the Scene

In the following script, we demonstrate how to implement the function that the player always faces the target object no matter where the target moves:

```csharp
using UnityEngine;


public class LookAtTest : MonoBehaviour
{
    [SerializeField] private Transform _targetTransform;
```

```
    private void Update()
    {
        if (_targetTransform == null) return;
        var dir = _ targetTransform.position -
          transform.position;
        transform.rotation = Quaternion.LookRotation(dir);
    }
}
```

First, we calculated the direction from the player to the target. Next, we called the Quaternion.LookRotation function to calculate the rotation value.

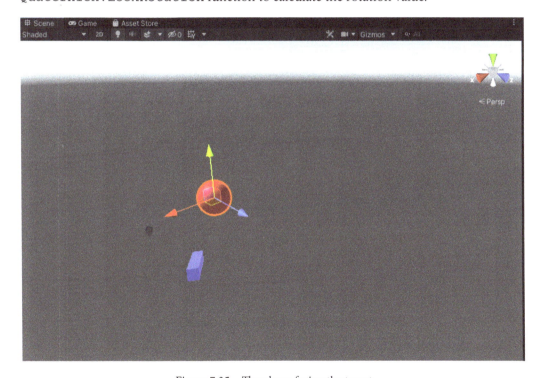

Figure 7.35 – The player facing the target

Finally, we moved the target object and the player also moved to face the target, as shown in *Figure 7.35*.

# Manipulating rotations

There are some functions that are used to manipulate rotations and `Quaternion`. `Slerp` is one of them. The following is the function signature of it:

```
public static Quaternion Slerp(Quaternion a, Quaternion b,
    float t);
```

This is a static function. The result of calling `Quaternion.Slerp` is that the object will start to rotate, slower, then faster in the middle.

In the following example, we still use the Scene we set up earlier, this time creating a new C# script called `OrbitScript.cs`. Then, we will attach this script to the player object to implement a gravity orbit effect.

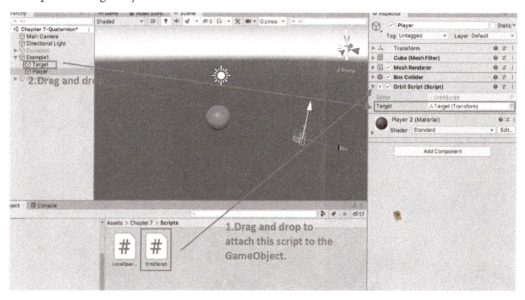

Figure 7.36 – Attaching the script to the GameObject

The code of `OrbitScript.cs` is as follows:

```
using UnityEngine;

public class OrbitScript : MonoBehaviour
{
    [SerializeField] private Transform _target;

    void Update()
```

```
    {
        if (_target == null) return;

        var dir = _target.position - transform.position;

        var targetRotation = Quaternion.LookRotation(dir);
        var currentRotation = transform.localRotation;

        transform.localRotation =
          Quaternion.Slerp(currentRotation, targetRotation,
          Time.deltaTime);
        transform.Translate(0, 0, 5 * Time.deltaTime);
    }
}
```

In this script, we reused some code from LookAtScript.cs. We also first calculated the angle of the player toward the target. But unlike the previous script, we did not directly modify the player's rotation, but saved the target rotation and the player's current rotation with two temporary variables, namely targetRotation and currentRotation. Then, the Quaternion.Slerp function was called to make the player gradually turn to the target, which is also the key to achieving the effect of gravity orbit. Finally, we called the transform.Translate function to keep the player moving forward.

Figure 7.37 – Running the game

If we run the game, we will find that the player will move around the target and turn to face the target, as shown in *Figure 7.37*.

# Working with Euler angles

If in some cases you prefer to use Euler angles instead of quaternions, Unity allows you to convert Euler angles to a quaternion, but you should not retrieve Euler angles from a quaternion and apply it to the quaternion after modifying it, because this may cause problems.

Quaternion.Euler is one of these functions that we can use to convert Euler angles into quaternions. The following is its function signature:

```
public static Quaternion Euler(Vector3 euler);
```

This function requires a Vector3-type parameter, which provides the angle around the *x* axis, the angle around the *y* axis, and the angle around the *z* axis. Based on this data, this function returns the corresponding quaternion rotation.

The following code snippet demonstrates how to use Euler angles in the script correctly:

```
using UnityEngine;

public class EulerAnglesTest : MonoBehaviour
{
    private float _xValue;

    private void Update()
    {
        _xValue += Time.deltaTime * 5;
        var eulerAngles = new Vector3(_xValue, 0, 0);
        transform.rotation = Quaternion.Euler(eulerAngles);
    }
}
```

In the code, we created Euler angles that rotate around the $x$ axis, and then called the `Quaternion.Euler` function to convert Euler angles into quaternions.

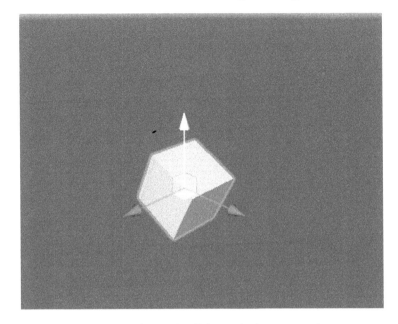

Figure 7.38 – Converting Euler angles into quaternions

Attach this script to a cube and run the game. You will find the cube rotates around the $x$ axis.

In this section, we introduced you to quaternions and explored how to use quaternions correctly in Unity scripts. It should be noted that in Unity, rotation can not only be represented by quaternions, but also by Euler angles. When Euler angles are used to represent rotation, its format is easy for humans to read, but due to the influence of the gimbal lock, Unity still uses quaternions to save rotations internally.

# Summary

This chapter first introduced the concept of the coordinate system in computer graphics, and then discussed the coordinate system used by Unity. Then, we discussed the concept of vectors and how to perform vector operations such as vector addition, vector subtraction, dot product, and cross product in Unity.

We also introduced the concept of a matrix and demonstrated how to use a matrix to translate, rotate, and scale in Unity.

Finally, we explored how to create rotations and manipulate rotations in quaternions, and demonstrated how to use Euler angles in the script correctly.

By reading this chapter, you should now have a bit more mathematical knowledge about computer graphics. In the next chapter, we will introduce the Scriptable Render Pipeline in Unity.

# 8
# The Scriptable Render Pipeline in Unity

In the *Chapter 7, Understanding the Mathematics of Computer Graphics in Unity*, we learned about the mathematics used in computer graphics. This knowledge is general computer graphics knowledge, and all 3D software and game engines use those mathematical concepts. For a game engine, rendering is one of the most important functions. In this chapter, we will specifically explore the rendering functions provided by Unity.

The following key topics will be explored in this chapter:

- An introduction to the Scriptable Render Pipeline
- Working with Unity's Universal Render Pipeline
- The Universal Render Pipeline shaders and materials
- Increasing performance of the Universal Render Pipeline

By the end of this chapter, you will understand what the **Scriptable Render Pipeline** is and how to enable the Scriptable Render Pipeline-based **Universal Render Pipeline** and **High Definition Render Pipeline** in your project. You will also know how to use the **Universal Render Pipeline Asset** to configure your render pipeline and how to use the **Volume framework** to apply post-processing effects to your game. You will also know how to create a custom shader and material that can be used in the Universal Render Pipeline, how to use Unity's **Frame Debugger** tool to view the information of the rendering process, and how to use the **SRP Batcher** to reduce the number of draw calls in your project.

It sounds exciting! Now, let's get started!

# An introduction to the Scriptable Render Pipeline

Since its first release in 2004, the Unity game engine has grown into the world's most widely used real-time content creation platform. A large number of games are developed using the Unity game engine. At the same time, Unity is being rapidly applied to the content design and production process of traditional industries, including VR, AR, and MR simulation applications, architectural design display, automobile design and manufacturing, and even film and television animation production. The development of real-time rendering technology based on computer graphics is an important reason for the rapid growth and widespread use of the Unity engine.

Before the Unity 2018 version, developers could only use the **built-in render pipeline** provided by Unity. Since the Unity engine itself is a closed source engine, the built-in render pipeline in Unity is like a black box for developers, and developers cannot know the specific logic implementation of rendering inside the Unity engine. Furthermore, games developed using Unity's built-in render pipeline will use the same set of rendering logic on different platforms. It is very difficult for developers to customize the render pipeline for different platforms.

With the release of the Scriptable Render Pipeline, developers can view its code directly on GitHub and use C# code to control the rendering process, customizing a unique rendering pipeline for their games or applications.

> **Note**
> You can find the code of the Scriptable Render Pipeline on GitHub:
> `https://github.com/Unity-Technologies/Graphics`.

The Scriptable Render Pipeline is a toolbox provided by Unity for developers, through which developers can freely implement specific rendering functions in Unity. For the convenience of developers, there are two pre-build render pipelines based on the Scriptable Render Pipeline available, namely the Universal Render Pipeline and the High Definition Render Pipeline.

By using Scriptable Render Pipeline-based pre-built render pipelines, we can directly modify a specific function in the render pipelines without having to implement a new pipeline from scratch, simultaneously obtaining excellent rendering results and continuous updates. Therefore, when using Unity to develop games, there are three ready-made render pipelines to choose from:

- The legacy built-in render pipeline
- The Universal Render Pipeline
- The High Definition Render Pipeline

Of course, you can also choose to develop your own render pipeline based on the Scriptable Render Pipeline.

The following are some open source projects on GitHub made using the Universal Render Pipeline or the High Definition Render Pipeline, which you can download and use.

# The Fontainebleau Demo

The first one is the **Fontainebleau Demo** project, made with the High Definition Render Pipeline:

Figure 8.1 – The Fontainebleau Demo

As shown in *Figure 8.1*, this project allows you to walk in a forest in first-person mode. You can find the project here: `https://github.com/Unity-Technologies/FontainebleauDemo`.

## The Spaceship Demo

The second project I want to introduce is the **Spaceship Demo** project. This is a playable AAA first-person mode demo, as shown in *Figure 8.2*:

Figure 8.2 – The Spaceship Demo

In this open source project, you can see how to implement GPU-accelerated particle effects, such as realistic flames, smoke, and electrical spark visual effects. You can find the project here: `https://github.com/Unity-Technologies/SpaceshipDemo`.

## The BoatAttack Demo

If you choose to use the Universal Render Pipeline as the render pipeline of your game, then this open source project is worth checking out:

Figure 8.3 – The BoatAttack Demo

As shown in *Figure 8.3*, this is a boat racing game, made using the Universal Render Pipeline. You can find the project here: `https://github.com/Unity-Technologies/BoatAttack`.

In addition to open source projects on GitHub, Unity also provides developers with free resources on Unity's **Asset Store** for developers to learn and use.

Here are some samples.

# The Heretic: Digital Human

The first one I want to share is the **The Heretic: Digital Human** project on the Asset Store:

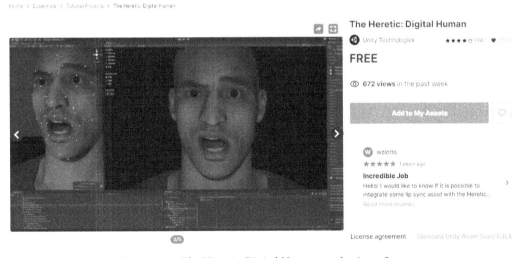

Figure 8.4 – The Heretic: Digital Human on the Asset Store

As shown in *Figure 8.4*, this is a free project that shows how to make a digital human with real skin, eyes, eyebrows, and so on. You can find the project here: `https://assetstore.unity.com/packages/essentials/tutorial-projects/the-heretic-digital-human-168620`.

## The Heretic: VFX Character

The second project is from **The Heretic** on the Asset Store as well, and it's free to download:

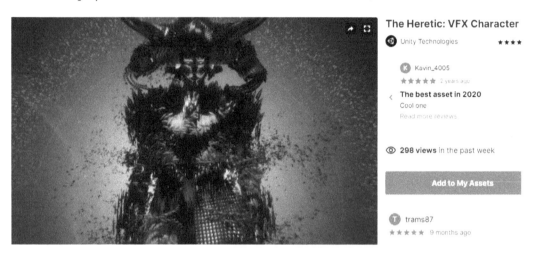

Figure 8.5 – The Heretic: VFX Character

As shown in *Figure 8.5*, this project demonstrates how to create a VFX-based character with the High Definition Render Pipeline in Unity. You can find the project here: `https://assetstore.unity.com/packages/essentials/tutorial-projects/the-heretic-vfx-character-168622`.

Well, after introducing a lot of projects based on Scriptable Render Pipeline, which are open source and free, are you more interested in this render pipeline now? If so, then we will briefly introduce the two pre-built render pipelines based on the Scriptable Render Pipeline, namely the Universal Render Pipeline and the High Definition Render Pipeline.

# Universal Render Pipeline

The Universal Render Pipeline is a pre-built render pipeline based on the Scriptable Render Pipeline in Unity. As its name implies, this render pipeline can be used on all platforms supported by Unity. Different pipelines cannot be mixed, so once you choose to use the Universal Render Pipeline, the built-in render pipeline and the High Definition Render Pipeline will not be enabled.

Unity uses the legacy built-in render pipeline by default, but you can enable the Universal Render Pipeline in your project in different ways.

If you want to develop a new project, then you can use the **3D Sample Scene (URP)** project template provided by **Unity Hub** to create a new Universal Render Pipeline project:

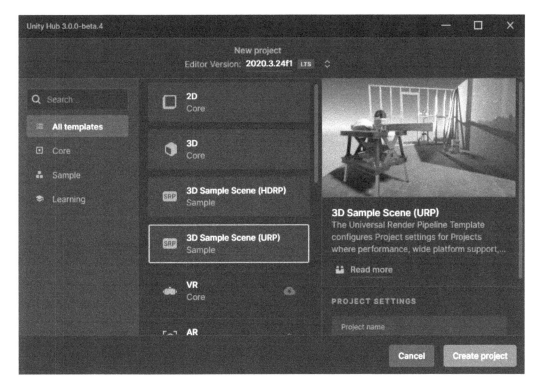

Figure 8.6 – The 3D Sample Scene (URP) project template

As shown in *Figure 8.6*, the **3D Sample Scene (URP)** project template configures project settings to use the Universal Render Pipeline in the project:

Figure 8.7 – A new URP project

After waiting for the new project to be created, you can view the sample scene rendered using the Universal Render Pipeline, as shown in *Figure 8.7*.

However, if you want to switch an existing project from the built-in render pipeline to the Universal Render Pipeline, recreating a new project using the Universal Render Pipeline is not suitable for your project. At this point, choosing to use Unity's Package Manager to install the Universal Render Pipeline is a more suitable option:

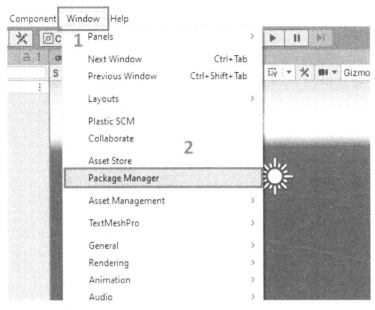

Figure 8.8 – Opening the Package Manager window

The Package Manager window can be opened by clicking on **Window | Package Manager** in the Unity Editor toolbar, as shown in the previous figure:

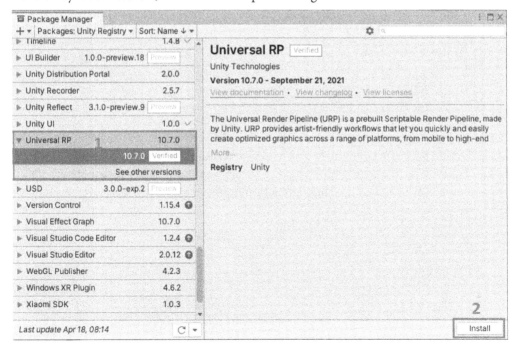

Figure 8.9 – Package Manager

As shown in *Figure 8.9*, you can find the **Universal RP** package in the packages list and install it in your project.

In this section, we briefly introduced the Universal Render Pipeline and how to install it in your project. A more detailed introduction on how to use it will be covered in the *Working with Unity's Universal Render Pipeline* section. Next, let's continue our journey to briefly explore the High Definition Render Pipeline and how to install it in your project.

## The High Definition Render Pipeline

The High Definition Render Pipeline is another pre-built render pipeline based on the Scriptable Render Pipeline in Unity. Unlike the Universal Render Pipeline, it does not support all platforms supported by Unity, only supporting high-end platforms. The following table shows the platforms supported by the High Definition Render Pipeline:

| Microsoft | Sony | macOS | Linux | Google |
|---|---|---|---|---|
| Xbox One | PlayStation 4 | macOS using Metal graphics | Linux with Vulkan | Stadia |
| Xbox Series X and Xbox Series S | PlayStation 5 | | | |
| Windows and Windows Store, with DirectX 11 or DirectX 12 and Shader Model 5.0 | | | | |
| Windows with Vulkan | | | | |

Figure 8.10 – The platforms supported by the High Definition Render Pipeline

As shown in the previous table, the High Definition Render Pipeline is currently mainly used for platforms such as consoles or desktop computers. If you are developing a mobile-oriented project, then the High Definition Render Pipeline is not a suitable choice.

Since we want to cover as many usage scenarios as possible, this chapter will mainly focus on the use of the Universal Render Pipeline – hence the brief introduction to the High Definition Render Pipeline. However, if you want to try it or really need to use the High Definition Render Pipeline, installing it is very similar to installing the Universal Render Pipeline, as described earlier.

Firstly, if you are starting a new project, you can use the **3D Sample Scene (HDRP)** project template provided by Unity Hub to create a new High Definition Render Pipeline project.

As shown in *Figure 8.11*, the **3D Sample Scene (HDRP)** project template configures project settings to use the High Definition Render Pipeline in the project:

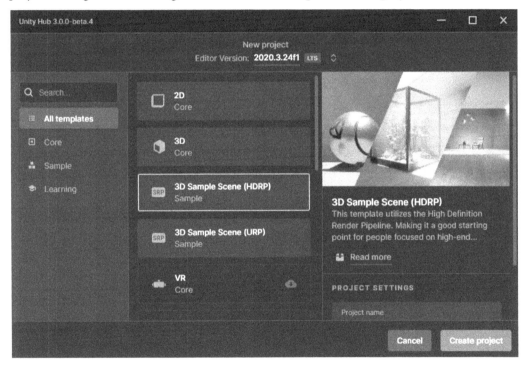

Figure 8.11 – The 3D Sample Scene (HDRP) template

After waiting for the new project to be created, you can view the sample scene rendered using the High Definition Render Pipeline, as shown in *Figure 8.12*:

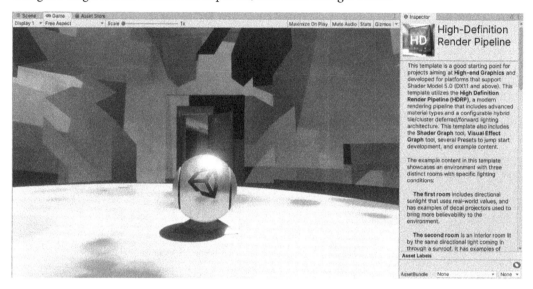

Figure 8.12 – A new HDRP project

Of course, you can also install the **High Definition RP** package from the Package Manager in Unity:

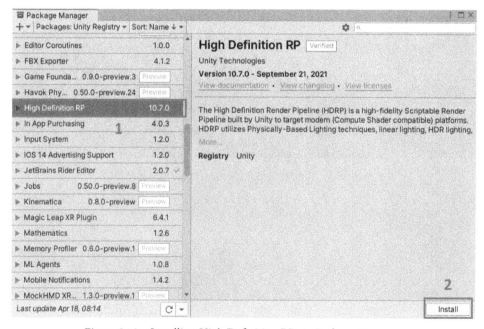

Figure 8.13 – Installing High Definition RP via Package Manager

As shown in *Figure 8.13*, you can find the **High Definition RP** package in the packages list and install it to your project.

In this section, we briefly introduced the High Definition Render Pipeline and how to install it in your project.

By reading this section, *An Introduction to the Scriptable Render Pipeline*, you should now have an understanding of what the Scriptable Render Pipeline is and how to install a Scriptable Render Pipeline-based Universal Render Pipeline and High Definition Render Pipeline in your project. Next, we will discuss in detail how to use the Universal Render Pipeline correctly in your project.

Let's get started!

# Working with Unity's Universal Render Pipeline

The Universal Render Pipeline is widely used by Unity developers. It is not only used to develop games for PC or video game consoles; you can also use it to develop mobile games.

We can create a new Universal Render Pipeline project through the Unity Hub project template. Through the project template, Unity will automatically set up all the render pipeline resources for us. The project also contains a sample scene, as shown in *Figure 8.14*. You can find a camera, a directional light, a spot light, a post-process volume, reflection probes, and some models in this scene:

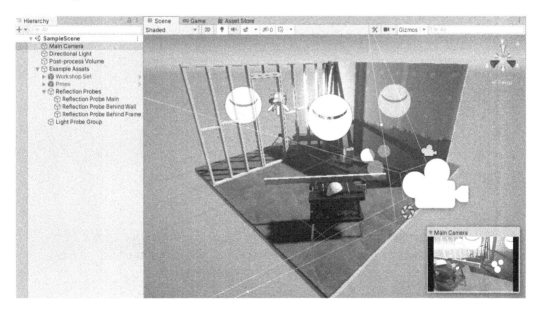

Figure 8.14 – The sample scene of the Universal Render Pipeline

For starters, this sample scene is a good starting point. We will use it to explain how to use the Universal Render Pipeline.

# Exploring the sample scene

Let's first explore this sample scene. As you can see in *Figure 8.14*, this scene is not complicated, but it contains most of the functions of the Universal Render Pipeline. We will introduce these components in the scene separately.

## The main camera

Let's start with the main camera in the scene. We can select the **Main Camera** in the **Hierarchy** window to open the **Inspector** window of it, as shown in the following figure:

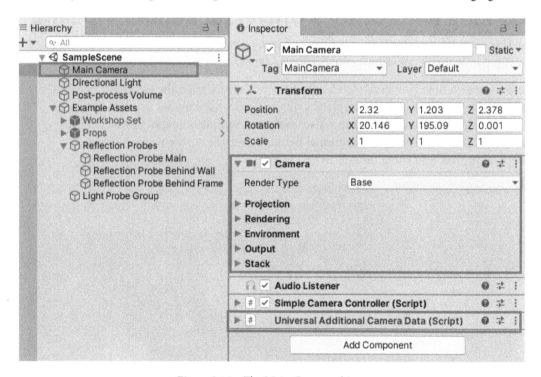

Figure 8.15 – The Main Camera object

There is a **Camera** component attached to the **Main Camera** object, which provides all the functions related to the camera object. You can set the background, culling mask, anti-aliasing setting, perspective settings of the camera, and so on.

Another component that you need to be aware of is the **Universal Additional Camera Data** component, which you can find at the bottom of *Figure 8.15*. If you are using the Universal Render Pipeline, Unity does not allow you to remove it from the camera because this component is used to store data internally.

## The directional light

There is only one directional light in this sample project, which is used to simulate the sunlight. You can modify the color, intensity, and shadow effect of the light by modifying the settings of the **Light** component attached to the light object in the scene. You can also modify the rotation property of the **Transform** component of the light object to adjust the direction of the directional light, as shown in *Figure 8.16*.

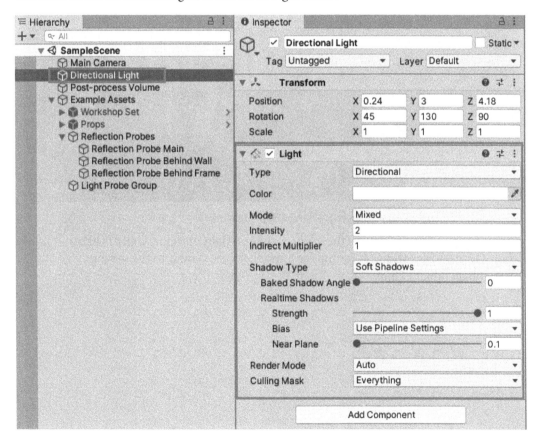

Figure 8.16 – The Directional Light object

In this example, the intensity value of this light is **2**, and soft shadows are used.

## The Spot Light

There are four types of lights in Unity, which are **Directional Light**, **Point Light**, **Spot Light**, and **Area Light**. In this sample scene, in addition to a directional light used to simulate the sunlight, there is also a spot light used to simulate a spotlight:

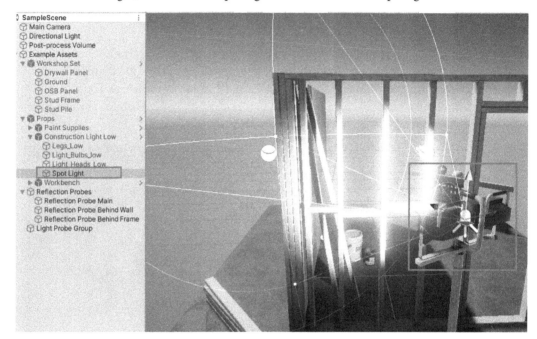

Figure 8.17 – The Spot Light object

As shown in the preceding figure, the effect of the spotlight object in Unity is like spotlights in the real world. The settings of a spot light are similar to the settings of a directional light in Unity:

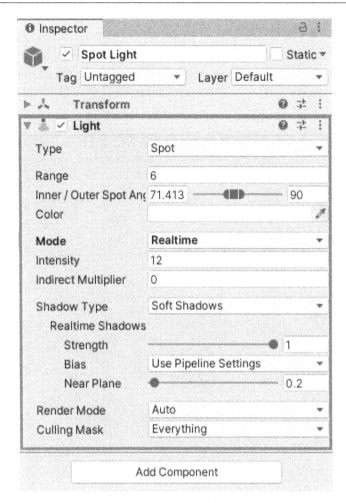

Figure 8.18 – The Spot Light settings

You still can modify the color, intensity, and shadow effect of the spot light by modifying the settings of the **Light** component attached to the **Spot Light** object, and you can also modify the range and the inner/outer spot angle of this spot light, as shown in *Figure 8.18*.

## The Post-process Volume

Now, let's take a look at the Post-process Volume object in the sample scene. In game development, post-processing is a technique that is often used to add various effects to a rendered image, common effects such as tone mapping, depth of field, bloom, anti-aliasing, and motion blur:

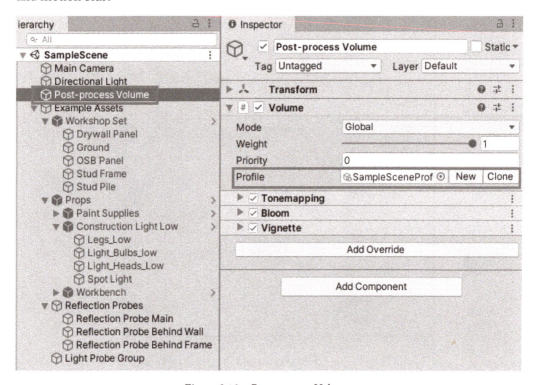

Figure 8.19 – Post-process Volume

The Universal Render Pipeline provides the **Volume** component and the **Volume Profile** object to manage different post-processing effects applied to rendered images, as shown in *Figure 8.19*. One advantage of using the **Volume** component is that component and specific settings can be decoupled. All settings on the **Volume** component come from the associated **Volume Profile** object. We will discuss the **Volume Profile** object in detail later.

In this sample scene, the **Tonemapping**, **Bloom**, and **Vignette** effects are applied. If you're curious about the original rendered image without post-processing, let's see what happens when we disable this post-process volume:

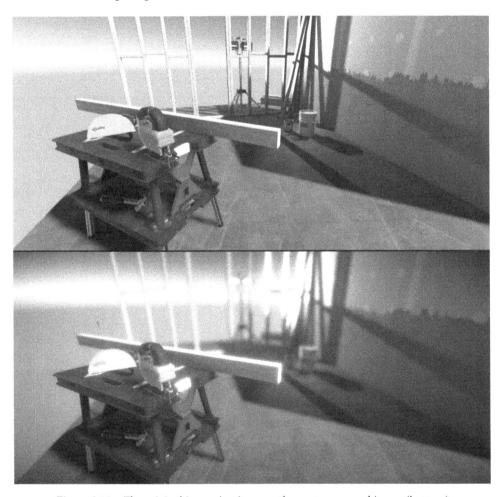

Figure 8.20 – The original image (top) versus the post-processed image (bottom)

*Figure 8.20* shows a comparison of the original image and the post-processed image of the sample scene.

# The reflection probes

A **reflection probe** can provide efficient reflection information for related models in a scene by sampling the scene around itself so that the surface of the model in the scene has a realistic reflection effect:

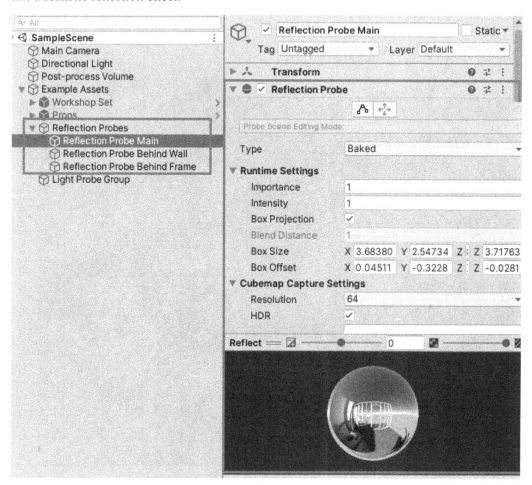

Figure 8.21 – Reflection probes in the scene

In this sample scene, we can see there are three reflection probes as child objects of the GameObject named **Reflection Probes**, as shown in the preceding figure.

If we select one of these reflection probes, then the corresponding reflection probe will be displayed in the scene view and show the reflection information, as shown in *Figure 8.22*:

Figure 8.22 – Viewing the reflection information in the scene

Since the reflection probes in different positions will obtain different reflection information, in order to use the reflection information correctly, they need to be placed in the proper place. While the definition of "proper place" varies from scene to scene, a general guideline is that you should place reflection probes near any large objects in the scene that will be significantly reflected. For example, place reflection probes in the areas around the center and corners of walls in the scene. Of course, this doesn't mean ignoring all the smaller objects in the scene. For example, a campfire in a scene may be a small object compared to a wall, but reflecting the fire from the campfire is just as important to create a realistic rendering of the scene.

# The Universal Render Pipeline asset

Since this new sample project is created using the Universal Render Pipeline template, Unity has automatically set everything up for us to make the Universal Render Pipeline work properly. However, if your project is using the built-in render pipeline for development and you want to switch to using the Universal Render Pipeline, or if your project has been developed using the Universal Render Pipeline but you want to use another render pipeline, it is necessary to know how to set up it in Unity.

> **Note**
>
> In this chapter, we are using the **forward rendering path** in the Universal Render Pipeline. The so-called **rendering path** refers to a series of operations related to lighting and shading. Unity's built-in render pipeline provides different rendering paths, such as the forward rendering path and the **deferred rendering path**. After version 12.0.0 of the Universal Render Pipeline, developers can also use the deferred rendering path in the pipeline, but that is beyond the scope of this chapter. If you are interested in this topic, you can find out more at `https://docs.unity3d.com/Packages/com.unity.render-pipelines.universal@12.0/manual/rendering/deferred-rendering-path.html` and `https://docs.unity3d.com/Packages/com.unity.render-pipelines.universal@12.0/manual/urp-universal-renderer.html#rendering-path-comparison`.

We will walk you through the following steps to learn how to set up a render pipeline for your project in Unity:

1. Let's start with the **Project Settings** window. You can open this window through **Edit | Project Settings** in the Unity Editor toolbar.

2. Next, click the **Graphics** item in the category list on the left to open the **Graphics** Settings panel, as shown in *Figure 8.23*:

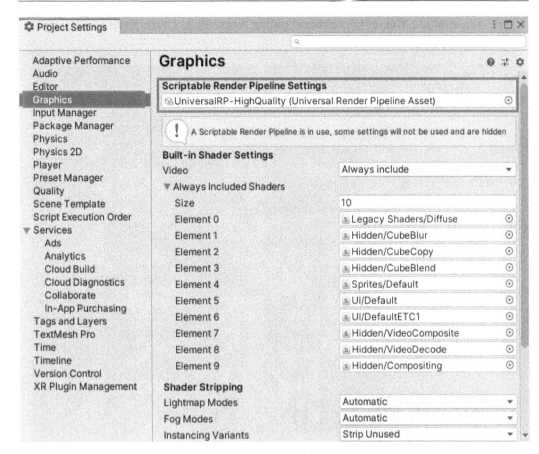

Figure 8.23 – The Graphics settings

Let's take a look at the **Scriptable Render Pipeline Settings** property in detail:

- The **Scriptable Render Pipeline Settings** property of the **Graphics** settings is associated with an object of the **Universal Render Pipeline Asset** type named **UniversalRP-HighQuality**, which is automatically created when this project is created using the template.

- If the **Scriptable Render Pipeline Settings** property of the **Graphics** settings is set to none, then Unity will use the default built-in render pipeline.

3.  You can find this **Universal Render Pipeline Asset** object in the **Assets > Settings** folder of the project:

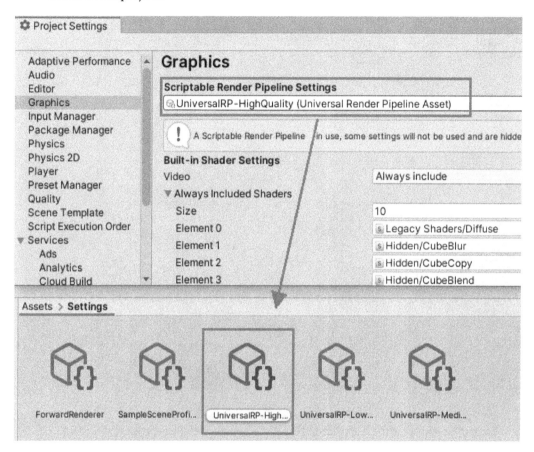

Figure 8.24 – The Universal Render Pipeline Asset objects

4.  Then, select the **UniversalRP-HighQuality** object to open the **Inspector** window so that we can check the detailed information of this **Universal Render Pipeline Asset** object. As shown in *Figure 8.25*, a **Universal Render Pipeline Asset** object provides various settings for the current Universal Render Pipeline, such as rendering functions and rendering quality:

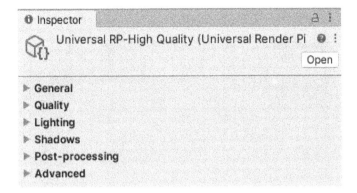

Figure 8.25 – The Inspector window of this Universal Render Pipeline Asset object

- Next, let's walk through these settings. We can configure the general settings of the render pipeline in the **General** section, as shown in *Figure 8.26*. For example, if the **Depth Texture** option is enabled, you can access the depth map generated by the render pipeline from your shader code:

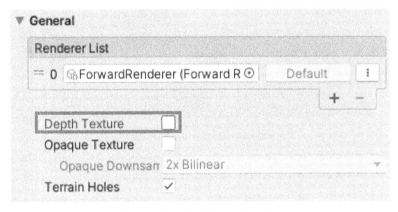

Figure 8.26 – The General settings

---

Note

In game development, **depth textures** are used to represent the depth information of objects in 3D space from the camera's perspective.

---

- We can also control the global rendering quality:

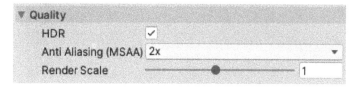

Figure 8.27 – The Quality settings

In *Figure 8.27*, we enable the global **HDR** option, and the global **Anti Aliasing** setting is **2x**. We can also modify the rendering resolution by adjusting the **Render Scale** slider.

- As a very important factor for real-time rendering, the lighting of the Universal Render Pipeline can also be configured in the **Lighting** section, as shown in *Figure 8.28*:

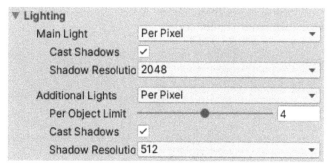

Figure 8.28 – The Lighting settings

The main light in the settings panel is the brightest directional light in the game scene. You can decide whether to enable it and whether to allow it to cast shadows.

- In *Figure 8.29*, you can find the **Shadows** settings under the **Lighting** settings. You can modify the parameters here to adjust what the shadows look like in Unity:

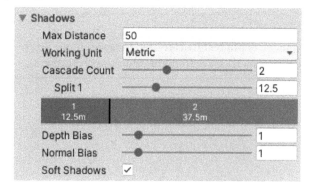

Figure 8.29 – The Shadows settings

- Finally, let's explore the **Advanced** settings:

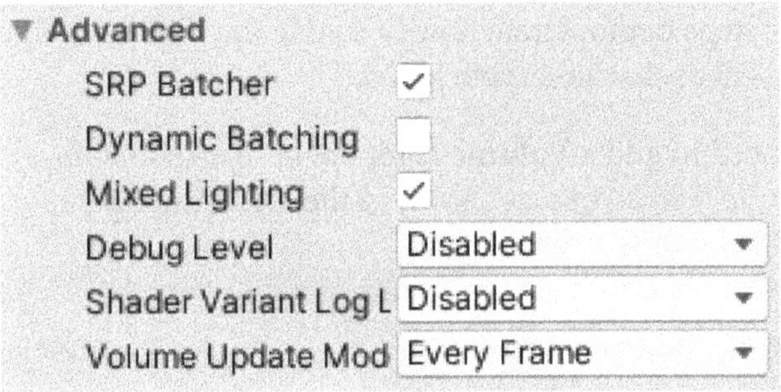

Figure 8.30 – The Advanced settings

Here, we can check the **SRP Batcher** option to enable the SRP Batcher function to improve the performance of the Universal Render Pipeline, which we will explain in detail in the *Increasing performance of the Universal Render Pipeline* section. We can also modify the level of the **Debug** log output, as shown in *Figure 8.30*.

> **Note**
> If there is no Universal Render Pipeline Asset in your project, you can create a new one by clicking **Assets** > **Create** > **Rendering** > **Universal Render Pipeline** > **Pipeline Asset** in the Unity Editor toolbar.

In this section, we introduced the Universal Render Pipeline Asset in Unity and how to switch between different render pipelines by changing the **Scriptable Render Pipeline Settings** property of the **Graphics** settings. Next, we will explore another important asset in the Universal Render Pipeline, namely the Volume Profile.

# The Volume framework and post-processing

The **Volume** framework is provided to game developers by the Scriptable Render Pipeline. By using this framework, developers can decouple a component from the specific settings of the component. Those render pipelines based on the Scriptable Render Pipeline, such as the Universal Render Pipeline and the High Definition Render Pipeline, use this framework.

As we mentioned before, the Universal Render Pipeline uses the **Volume** component and the **Volume Profile** object to manage different post-processing effects applied to rendered images. The following steps demonstrate how to enable the Volume framework and apply some post-processing effects to the sample project:

1. First of all, we need to add a **Volume** component to a GameObject in the scene to enable the Volume framework, as shown in the following figure:

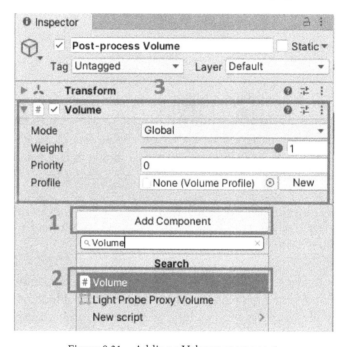

Figure 8.31 – Adding a Volume component

2. In *Figure 8.31*, the **Profile** property of this **Volume** component is **0**, so we can either create a new **Volume Profile** file by clicking the **New** button below it or assign an existing **Volume Profile** file to it. Here, we will create a new **Volume Profile** file:

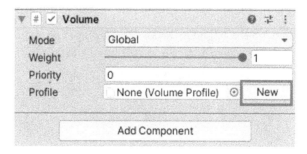

Figure 8.32 – Creating a new Volume Profile file

3. Click the **Add Override** button to open the **Volume Overrides** panel, and click the **Post-processing** item to open the **Post-processing** overrides list:

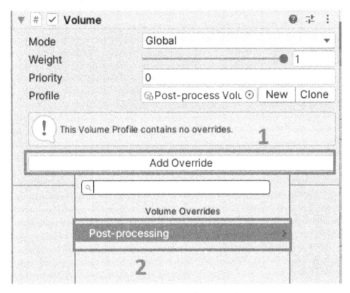

Figure 8.33 – Add Override

4. In *Figure 8.34*, you can see lots of post-processing effects in the **Post-processing** overrides list. You can choose the effects that you want to apply to the rendered image, such as **Bloom**:

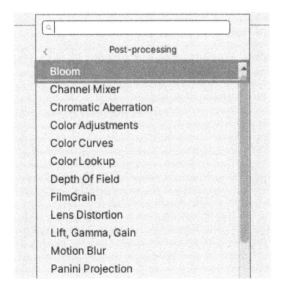

Figure 8.34 – Post-processing effects

5.  Finally, let's modify the configuration of the **Bloom** effect; check the **Threshold** and **Intensity** options and set their values to **0.9** and **4** respectively, as shown in the following figure:

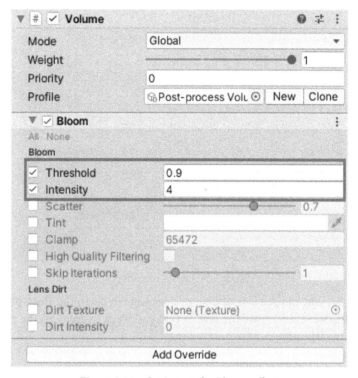

Figure 8.35 – Setting up the Bloom effect

After completing the preceding steps, switch to the game view. We can see the game scene in *Figure 8.36* after applying the **Bloom** effect:

Figure 8.36 – The applied Bloom effect image (top) versus the original image (bottom)

In this section, we started by exploring the sample scene that is included in the template to learn the functions of the Universal Render Pipeline. Then, we introduced how to switch between different render pipelines and the Universal Render Pipeline Assets. Finally, we demonstrated how to use the Volume framework to implement post-processing in the Universal Render Pipeline. The next stop of our journey is to explore the shaders and materials that are important for rendering in Unity.

# The Universal Render Pipeline shaders and materials

**Shaders** and **materials** are essential for rendering models in Unity. Shaders are used to provide algorithms to calculate the color of each pixel. A material provides various parameters for the shader associated with it to determine how to render the model, such as providing texture as the input of the shader and defining how the shader samples the texture:

Figure 8.37 – Materials and shaders

If we select a model in the scene, such as the safety hat model, the material settings will be displayed in the **Inspector** window, as shown in *Figure 8.37*.

# Commonly used shaders

Each material can be associated with a specified shader, and the parameters required by this shader are displayed in the **Inspector** window. The commonly used shader when using the Universal Render Pipeline is **Universal Render Pipeline/Lit**, and this safety hat model is rendered using this shader as well. By adjusting various parameters, the **Universal Render Pipeline/Lit** shader can be used to render different material surfaces, such as metal, glass, and wood.

> **Note**
>
> The shader with **Lit** in its name implies that this shader will perform lighting calculations. The shader with **Unlit** in the name means that the shader does not consider the lighting factor when calculating the color of a pixel.

We can change the shader associated with a material by selecting a different one through the **Shader** drop-down window, as shown in *Figure 8.38*:

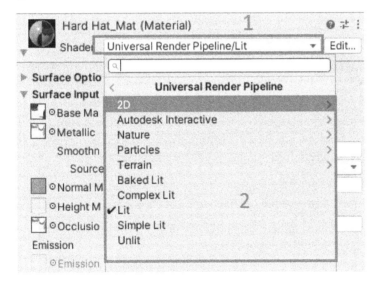

Figure 8.38 – Shaders

Once we have determined the shader associated with the material, we can then provide various parameters for this specific shader through this material:

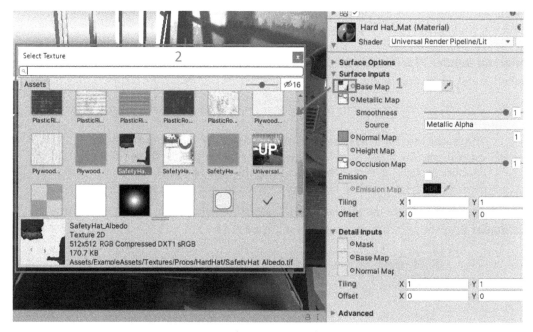

Figure 8.39 – The parameters of the shader

In *Figure 8.39*, developers can specify various maps in the **Surface Inputs** section for the **Universal Render Pipeline/Lit** shader. The textures associated with these map parameters are used to provide different information for the shader. We will explain in detail as follows:

- **Base Map** is used to provide the base color of the surface to the shader.

- **Metallic Map** is used to provide metallic workflow information to the shader to determine how "metal-like" the surface is.

- **Normal Map** is used to add more details to the surface of the model that do not exist on the original model.

- **Occlusion Map** is used to provide information to the shader to simulate shadows from ambient lighting.

The parameters required by different shaders may be different, and due to different shader algorithms, the final rendering results are also different:

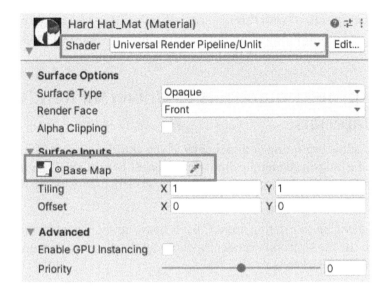

Figure 8.40 – The Unlit shader

For example, if we change the shader associated with this material to **Universal Render Pipeline/Unlit**, then only **Base Map** remains in the **Surface Inputs** section to provide the base color for the surface:

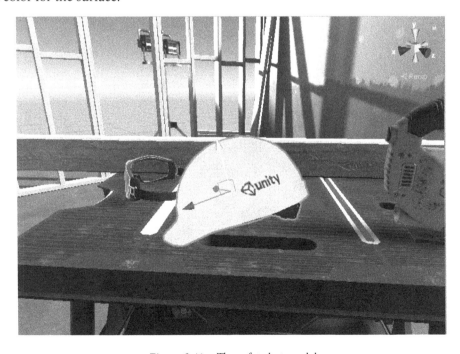

Figure 8.41 – The safety hat model

The safety hat model rendered with this material will only display the basic color and will no longer be affected by any lighting. You can see the difference between the safety hat model and the surrounding models in the preceding screenshot.

# Upgrading project materials to Universal Render Pipeline materials

As we mentioned at the beginning of this chapter, if you choose to use the Universal Render Pipeline, the built-in render pipeline will no longer be available. This includes not only the built-in render pipeline itself but also the shaders used with the built-in render pipeline.

Therefore, when changing an existing project that is being developed using the built-in render pipeline to use the Universal Render Pipeline, developers often encounter a problem known as "material errors":

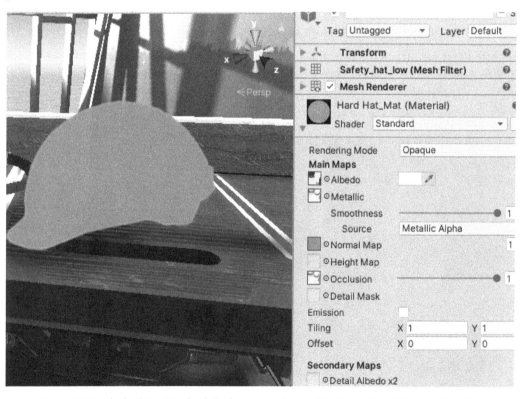

Figure 8.42 – The built-in Standard shader cannot be used in the Universal Render Pipeline

For example, we can change the shader used to render the safety hat model from **Universal Render Pipeline/Unlit** to the built-in **Standard** shader. Then, you can see that the safety hat model displays a pink color, which means that there is an error in the material, as shown in the preceding figure.

Therefore, if your project is developing using the built-in render pipeline but you need to switch to using the Universal Render Pipeline, then in order to ensure that the Universal Render Pipeline can work correctly, you need to upgrade the existing materials to Universal Render Pipeline materials.

You can manually modify the shaders associated with the existing materials, such as replacing the built-in **Standard** shader with the **Universal Render Pipeline/Lit** shader:

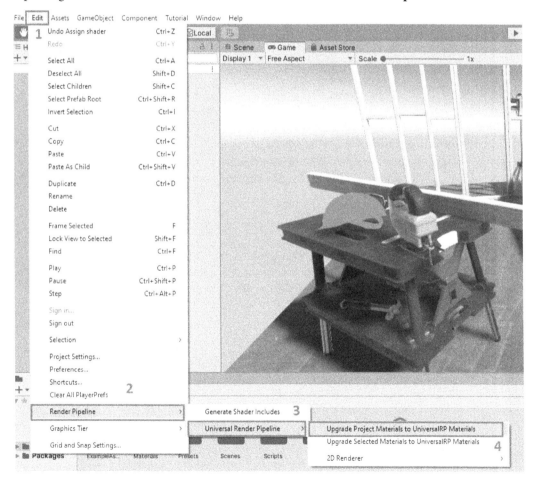

Figure 8.43 – Upgrading the project materials

On the other hand, Unity also provides a function for developers to upgrade existing materials to Universal Render Pipeline materials automatically. You can find it by clicking **Edit** > **Render Pipeline** > **Universal Render Pipeline** > **Upgrade Project Materials to UniversalRP Materials** in the Unity Editor toolbar, as shown in the preceding screenshot.

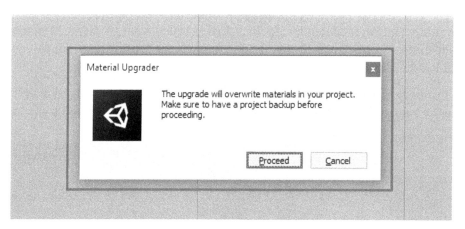

Figure 8.44 – Material Upgrader

Then, the **Material Upgrader** window will pop up. As shown in *Figure 8.44*, the changes cannot be undone, so if you want to upgrade all of the materials in your project to Universal Render Pipeline materials and have backed up the project, click the **Proceed** button.

> **Note**
>
> This **Material Upgrader** tool can only upgrade the built-in shaders to Universal Render Pipeline shaders but not the custom shaders created by developers. Therefore, the custom shaders still need to be modified manually.

## Creating a shader and a Shader Graph

Sometimes, you may want to create a new shader to implement some custom features that can be used with the Universal Render Pipeline. There are two ways to do it – you can either create a new shader file or a Shader Graph file.

# Creating a new shader file

First of all, even if our project uses the Universal Render Pipeline, we can still use the legacy way to create custom shader files using shader templates in the built-in render pipeline in Unity. As shown in *Figure 8.45*, we can click **Assets** > **Create** > **Shader** to create a new shader:

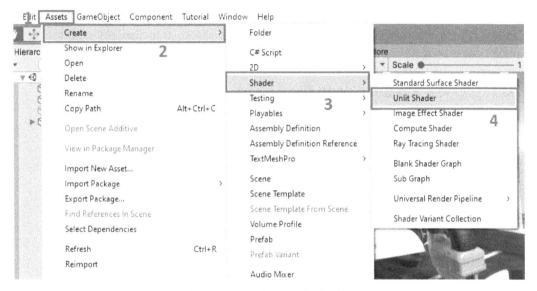

Figure 8.45 – Create a shader file

Some built-in shader templates are listed, such as **Standard Surface Shader**, **Unlit Shader**, and **Image Effect Shader**. Here, we choose the **Unlit Shader** item to create a new shader that does not consider lighting factors and name this shader CustomShader:

Figure 8.46 – CustomShader

Then, a shader file is created in our project, as shown in the preceding figure. You can open a shader source file in your IDE by double-clicking it, and then you can use Unity's **ShaderLab language** to write shader code that defines how Unity calculates the color rendered for each pixel.

> **Note**
>
> How to write shaders in Unity's ShaderLab language is beyond the scope of this chapter, but if you're interested, you can find more information at `https://docs.unity.cn/Packages/com.unity.render-pipelines.universal@7.7/manual/writing-custom-shaders-urp.html`.

In addition to creating a shader file, we can also create a new custom Shader Graph file to render these models in the scene. Let's continue.

## Creating a new Shader Graph file

Compared with the legacy way of creating a shader file, creating a new **Shader Graph** file is easier. The Shader Graph feature was introduced to Unity for the first time in Unity 2018. When developing a Shader Graph file, you don't need to write shader code but use the visualization node to develop directly.

We can still create a new unlit shader, but this time, we will use Shader Graph instead of a shader file, as shown in the following steps below:

1. As shown in *Figure 8.47*, click **Assets** > **Create** > **Shader** > **Universal Render Pipeline** > **Unlit Shader Graph** in the Unity Editor toolbar to create a new Shader Graph file and name it `CustomShaderGraph`:

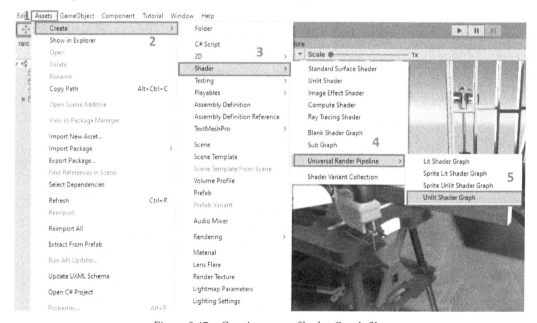

Figure 8.47 – Creating a new Shader Graph file

2.  A new Shader Graph file is created, and its suffix is `.shadergraph`, as shown in the following figure:

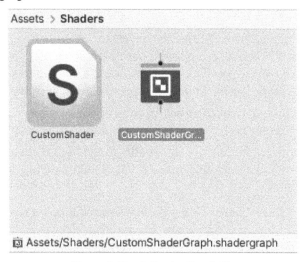

Figure 8.48 – A Shader Graph file

3.  Double-click this file, and this time, the Shader Graph file will not be opened in an IDE but a visual node editor, displayed directly in the Unity Editor, as shown in *Figure 8.49*:

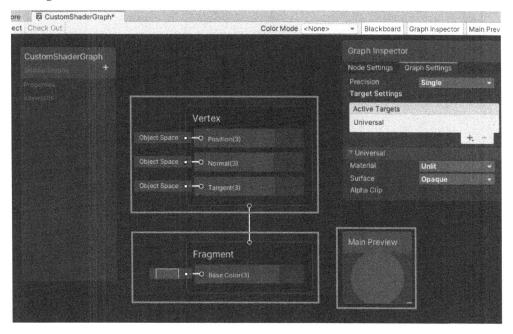

Figure 8.49 – The Shader Graph editor

This visual node editor is a lot to take in, so let's walk through it in more detail:

- In Unity, a shader usually consists of two parts, namely the **Vertex** program and the **Fragment** program.
- The **Vertex** program is usually used to convert the 3D coordinates of the vertices of the model into 2D coordinates in the screen space. We already introduced the knowledge of coordinate systems in the previous chapter. In this example, there are three nodes in the vertex program, namely **Position**, **Normal**, and **Tangent**.
- Alternatively, the **Fragment** program determines the color of the pixels, and in this example, the **Fragment** program only has one node, named **Base Color**.
- You can also preview the result of this shader in the **Main Preview** window in the lower-right corner, as shown in *Figure 8.49*. This shader we just created here will render pixels in blue.

Now that we've created a new Shader Graph file and opened it in the Shader Graph editor, which allows us to edit, add, and delete nodes, we will next take a look at how to edit a node in a Shader Graph file. Let's go!

## Editing the properties of a node in Shader Graph

We can edit the properties of an existing node in the Shader Graph file. As we mentioned earlier, there is a node named **Base Color**, so let's edit this node as follows:

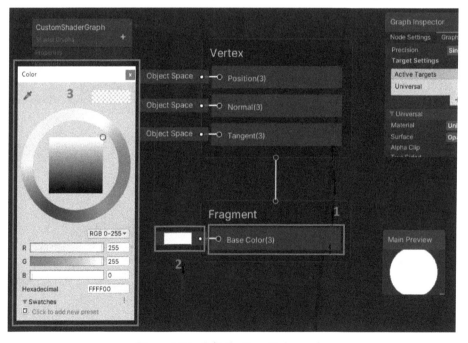

Figure 8.50 – Edit the Base Color node

1.  In the Shader Graph editor, select the **Base Color** node in the **Fragment** section.

2.  Click the color input of this node to open the color picker window.

3.  Select the color you want to use in the color picker window – in this case, we chose yellow for the **Base Color** node. The **Main Preview** window shows us what's happening to the shader, as shown in *Figure 8.50*.

As you can see, it is very easy to modify an existing node; in addition to modifying a node, we can also create a new one to provide more data to the shader. Let's continue.

## Adding a new node in Shader Graph

Developing a shader usually involves sampling a texture and returning a color value for the shader to use. Let's perform the following steps to add a new node to add the ability to sample textures to our example shader:

1.  Right-click in the Shader Graph editor and select **Create Node** from the pop-up menu:

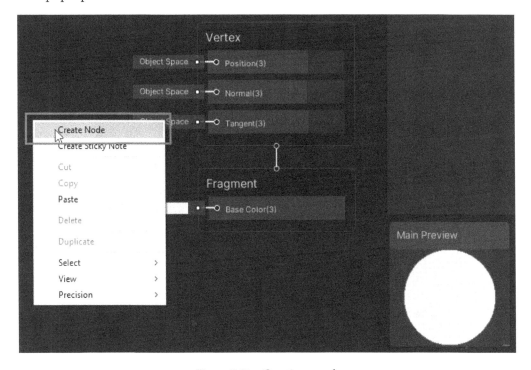

Figure 8.51 – Creating a node

2. Enter texture in the search bar in the top-left corner of the **Create Node** window, and then select **Sample Texture 2D** item in the results list:

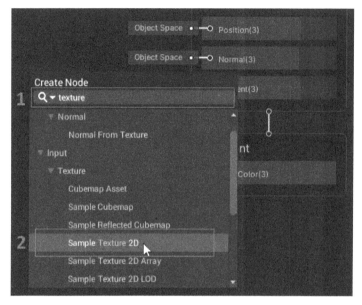

Figure 8.52 – Selecting the Sample Texture 2D node

3. As shown in *Figure 8.53*, a new **Sample Texture 2D** node is created. Click the **Texture** slot of this node to provide the texture asset:

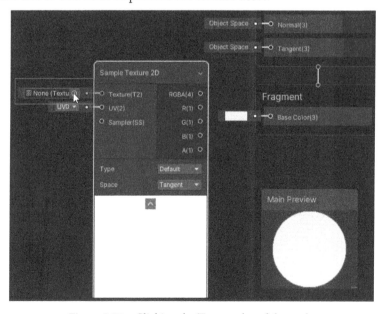

Figure 8.53 – Clicking the Texture slot of the node

4. Select a texture from the pop-up **Select Texture** window:

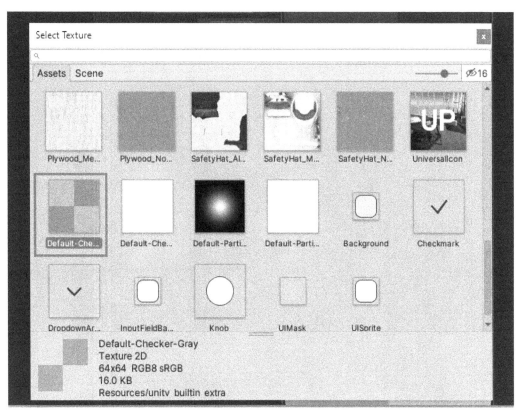

Figure 8.54 – Selecting a texture

5.  Then, as shown in *Figure 8.55*, the **Sample Texture 2D** node samples the texture from its **Texture** input and gets the texture's color:

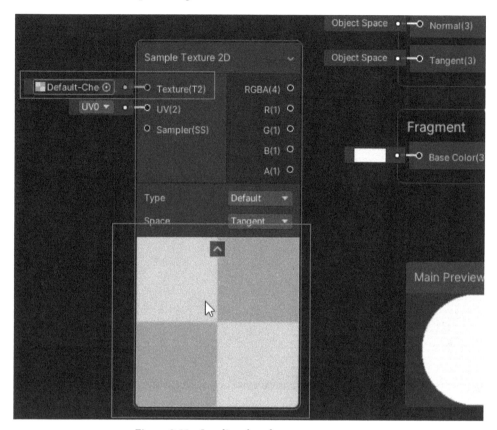

Figure 8.55 – Loading data from a texture asset

Now, we have added a new node to the Shader Graph file, but the color obtained from texture sampling is still stored in the **Sample Texture 2D** node. Next, we need to connect it with the **Base Color** node in the **Fragment** section so that the shader can render the pixels with the colors obtained from this texture.

## Connect two nodes in Shader Graph

We can pass data such as color from one node to another by connecting two nodes in a Shader Graph file, so let's connect the **Sample Texture 2D** node with the **Base Color** node using the following steps:

1.   Click the radio button next to the **RGBA(4)** output, as shown in *Figure 8.56*:

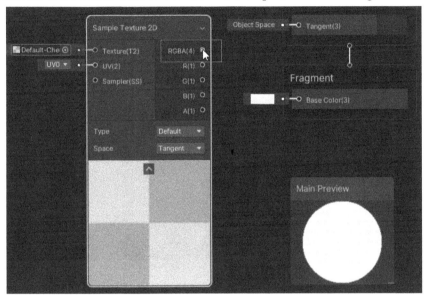

Figure 8.56 – Clicking the radio button

2.   After that, a line that can be dragged freely will appear:

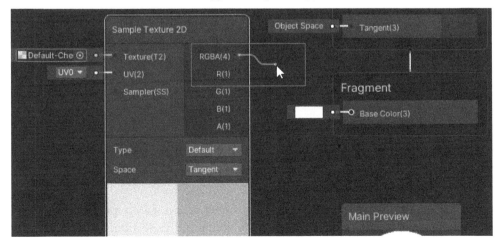

Figure 8.57 – A line will appear

3.  Drag this line to the color input of the **Base Color** node. As shown in *Figure 8.58*, we connected these two nodes, and the **Main Preview** window shows us that the shader has rendered the pixel using the color obtained from the texture:

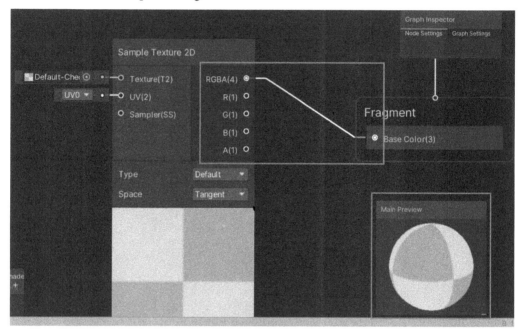

Figure 8.58 – Connecting two nodes

Now you should know how to create a Shader Graph file and how to modify, add, and connect nodes in it. As developers, we do not need to write shader code when using Shader Graph, but Unity will automatically generate shader code based on the content of the Shader Graph file.

In this section, we introduced knowledge related to Universal Render Pipeline shaders and materials, then demonstrated how to upgrade a built-in material to Universal Render Pipeline material, and finally, explored how to create a custom shader that can be used in the Universal Render Pipeline. Next, we will continue to discuss how to find performance issues and improve performance of the Universal Render Pipeline.

# Increasing performance of the Universal Render Pipeline

Rendering is a major function of a game engine. Therefore, it is very important to understand how to use Unity's render pipeline efficiently. In this section, the topic we will discuss is performance.

# The Frame Debugger

First, we should learn how to use tools to view and locate performance bottlenecks caused by rendering in Unity.

The **Frame Debugger** tool in the Unity Editor is our recommended tool, which allows us to easily view the entire process of rendering a frame in Unity.

Let's follow the following steps to see how Unity's render pipeline renders a frame of your game:

1.  Start the game in the editor by clicking the **Play** button:

Figure 8.59 – Playing the game in the editor

2.  Click **Window** > **Analysis** > **Frame Debugger** in the Unity Editor toolbar to open the **Frame Debugger** window, as shown in *Figure 8.60*:

Figure 8.60 – Opening the Frame Debugger

3.  In the **Frame Debug** window, click the **Enable** button to take a snapshot of the current frame of your game, as shown in *Figure 8.61*:

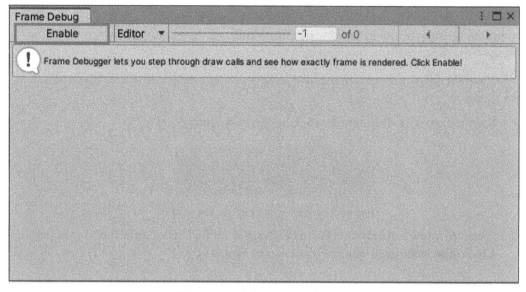

Figure 8.61 – The Frame Debugger

4.  In *Figure 8.62*, we can see there are **109** draw calls, which call to the graphics APIs, such as **OpenGL**, **Direct3D**, and **Vulkan**, to draw objects. We can also select a specific draw call to view the detailed information of it:

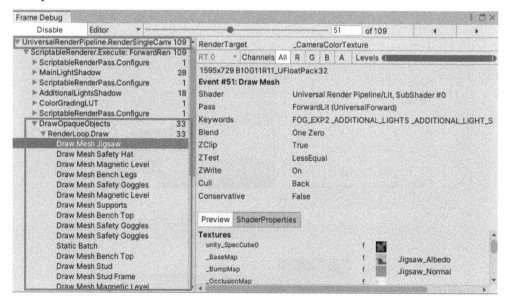

Figure 8.62 – Viewing the draw call information

Through the Frame Debugger tool, we can understand the entire rendering process and view the information of a specific draw call, which provides us with insight to determine what should be done to improve rendering performance. For example, in *Figure 8.62*, we can see that **33** draw calls are used to render opaque objects. Therefore, reducing the count of draw calls here is what we should do. Next, we will introduce how to use the SRP Batcher to do it.

## The SRP Batcher

The **SRP Batcher** is a feature provided by the Scriptable Render Pipeline, so every render pipeline based on the Scriptable Render Pipeline can use this feature to reduce the number of draw calls and improve rendering performance.

In order to ensure that the SRP Batcher can work correctly in your project, you need to ensure two things. The first is to enable the **SRP Batcher** function of the Universal Render Pipeline, and the second is to ensure that the shaders in your project are compatible with the SRP Batcher.

Let's first make sure that the SRP Batcher is enabled in the render pipeline. As we mentioned in *The Universal Render Pipeline Asset* subsection, we can enable it by checking the **SRP Batcher** option in the **Advanced** settings of the Universal Render Pipeline Asset file that our project is using, as shown in *Figure 8.63*:

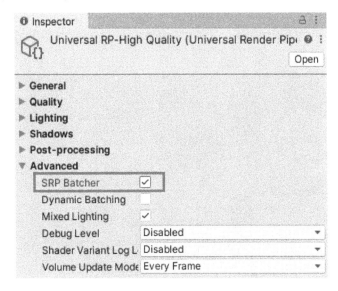

Figure 8.63 – Enabling the SRP Batcher

Next, let's check whether the shaders we are using to render these opaque objects are compatible with the SRP Batcher:

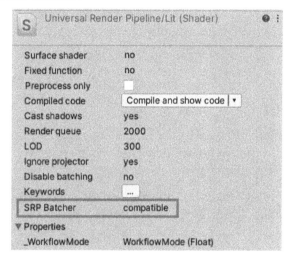

Figure 8.64 – The SRP Batcher compatibility status of the shader

We can find the SRP Batcher compatibility status of the shader in the **Inspector** window, as shown in the preceding figure. Here, the **Universal Render Pipeline/Lit** shader is used, which is compatible with the SRP Batcher.

Now, let's run the game and check the Frame Debugger again:

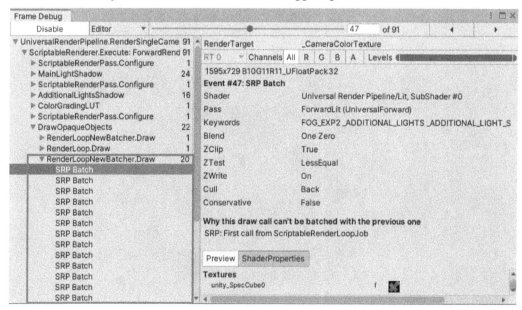

Figure 8.65 – The number of draw calls is reduced

As you can see in *Figure 8.65*, the total number of draw calls has been reduced from **109** to **91**, the number of draw calls used to render opaque objects has been reduced from **33** to **20**, and each draw call is marked as **SRP Batch**.

In this section, we started by introducing how to use Unity's Frame Debugger tool to view the entire rendering process and the information of a specific draw call. Then, we also explored how to reduce the number of draw calls and improve rendering performance by using the SRP Batcher.

# Summary

This chapter introduces three ready-made render pipelines to choose from in Unity, namely the legacy built-in render pipeline and two pre-made render pipelines based on the Scriptable Render Pipeline – the Universal Render Pipeline and the High Definition Render Pipeline. At the same time, we also introduced some open source projects that use these render pipelines for you to learn and use.

Then, we discussed how to use the Universal Render Pipeline in Unity by first exploring a sample scene, and then we explained how to use the Universal Render Pipeline Asset to configure your render pipeline and the Volume framework to apply post-processing effects to your game.

We also introduced the concept of shaders and materials, demonstrated how to upgrade a built-in material to Universal Render Pipeline material, and explored how to create a custom shader that can be used in the Universal Render Pipeline.

Finally, we explored how to use Unity's Frame Debugger tool to view the information of the rendering process and how to use the SRP Batcher to reduce the number of draw calls.

By reading this chapter, you should now understand how to work with the Universal Render Pipeline correctly in Unity. In the next chapter, we will introduce how to use the **Data-Oriented Technology Stack** (**DOTS**) in Unity.

# 9

# The Data-Oriented Technology Stack in Unity

The Unity engine is a very developer-friendly engine. When developing game logic, Unity's **GameObject-Components** architecture can help developers develop functions quickly, and adding a new behavior to a GameObject in Unity just requires attaching the corresponding component to it. However, with today's games becoming more complex, this approach, while very developer-friendly, especially to those familiar with traditional **Object-Oriented Programming** (**OOP**) models, is not ideal for game performance and project maintainability.

Therefore, Unity introduced the **Data-Oriented Technology Stack** (**DOTS**) to allow developers to write game code using an alternative programming philosophy that is data-oriented rather than object-oriented. It also introduces multithreading capabilities to optimize the performance of the game.

The following key topics will be part of our learning path in this chapter:

- DOTS overview
- Multithreading and the C# Job System in Unity

- Working with ECS in Unity
- Using C# and the Burst compiler

By reading this chapter, you will learn what DOTS is and the difference between data-oriented design and traditional object-oriented design. You will also find out how to use Unity's **C# Job System** to implement multithreading to improve game performance, how to use Unity's **Entity Component System** (**ECS**) to write game logic code in a data-oriented way, and how to use the **Burst compiler** to optimize the generated native code for Unity games.

## Technical requirements

The example project for this chapter is already available on GitHub. You can find it here: `https://github.com/PacktPublishing/Game-Development-with-Unity-for-.NET-Developers/tree/main/Chapter9-DOTS`.

## DOTS overview

DOTS is a new programming pattern in Unity and a topic that has been discussed a lot in the Unity developers community in recent years.

Figure 9.1 – The Megacity demo based on DOTS

If you have previous .NET programming experience, you will be familiar with the **Object-Oriented Programming** (**OOP**) pattern. OOP is widely adopted in the software industry, and developing games with Unity was no exception until Unity introduced DOTS. There's no doubt that OOP is an old habit for many programmers. Therefore, before discussing why Unity introduced DOTS, we will first talk about the problems that may be encountered when using OOP in Unity development.

## Object-oriented design pattern versus DOTS

First of all, let's talk about the concepts of OOP. We can find some useful information on Wikipedia. These concepts include **object/class**, **inheritance**, **interface**, **information hiding**, and **polymorphism**. The following link provides detailed explanations: `https://en.wikipedia.org/wiki/Object-oriented_design`.

If we focus on the object/class and inheritance, we will find that these two concepts are the biggest difference between OOP and DOTS.

Let's start with an **object/class**. In a traditional OOP pattern, a class is a tightly coupled set of data and behavior that acts on that data. Here, we have an example:

```
using UnityEngine;

public class Monster : MonoBehaviour
{
    # region Data

    private string _name;
    private float _hp;
    private Vector3 _position;
    private bool _isDead;

    #endregion

    #region Behavior
    public void Attack(Monster target){}
    public void Move(float speed){}
    public void Die(){}
    #endregion
}
```

As you can see in the code, we have a class called `Monster` and it has some data for its position, health, name, and whether or not it's dead. In addition, this class can also behave like a real object on its own data. Each `Monster` object can attack the target, move itself, or die.

So far, everything is perfect; the objects in the program are like objects in the real world, as if they have a life of their own, which is also in line with human experience. Next, let's discuss the **inheritance** concept of OOP. We can extend the data and behavior of a class and reuse some of its code via inheritance.

Suppose in this example we realize that not all monsters attack other monsters; some may attack humans. From a programming perspective, humans and monsters have a lot in common: position, health, and whether or not it's dead. But some monsters may not be killed, and humans cannot fly to move. In real life, we have a superset of monsters and humans, namely, creatures. Let's put the data that monsters and humans share in the `Creature` class so that they can both have this data, but we don't have to type them again, as shown:

```
public class Creature : MonoBehaviour
{
    #region Data

    private string _name;
    private float _hp;
    private Vector3 _position;
    private bool _isDead;

    #endregion
}
public class Monster : Creature
{
    # region Data

    private bool _canFly;

    #endregion

    #region Behavior
```

```
public void Attack(Creature target){}
public void Move(float speed){}
public void Die(){}


#endregion
}
```

Basically, if we keep going with this idea, we will end up with some complex class diagrams where you have a bunch of different creatures, such as monsters, humans, animals, and plants. We haven't even considered performance and we have already found that OOP can give us a lot of trouble.

Now, let's see how OOP also misuses the hardware. With the development of technology, processor hardware is getting faster and faster, but a point that is often overlooked is that if the data cannot be submitted from memory to the processor fast enough, then no matter how fast the processor is, it will not work as fast as expected. **Cache**, which is located closer to the processor core, is smaller, faster memory. When the processor issues a memory access request, it will first check whether there is data in the cache. If it exists (this is also called a cache hit), the data is returned directly without accessing the main memory; if it does not exist, the corresponding data in the main memory must be loaded into the cache first and then returned to the processor. CPUs typically use a hierarchy with multiple cache levels; for example, in a two-level cache system, the **L1 cache** is close to the processor side, and the **L2 cache** is close to the memory side.

CPU caches are designed around several assumptions. The reason why the caches are effective is mainly that the access to the memory when the program runs is characterized by locality. This locality includes both **spatial locality** and **temporal locality**. That is, the pieces of data we need to perform a series of related operations may be very close to each other in memory, or the data we just used for an operation may soon be used again for another operation. Taking advantage of this locality, caches can achieve extremely high hit rates.

However, OOP often misuses the hardware. Let's still use the `Monster` class as an example. Assuming a `Monster` object will occupy 56 bytes of memory, we iterate over a list of monsters and call their `Move()` function to change the monster's position property.

The pseudocode is as follows:

```
public void Update()
{
    for (var i = 0; i < _monsters.count; i++)
    {
        _monsters[i].Move(speed);
```

```
        }

    }
```

This code actually modifies a set of Vector3 data every frame, but how is this data allocated in memory? The following diagram shows how monster objects are allocated in memory when a 64-byte cache line is split up into 8-byte chunks:

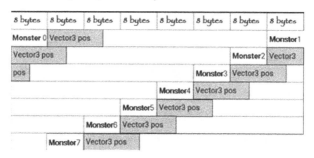

Figure 9.2 – The data layout in memory (OOP)

From the diagram, we can see that the position data that will be modified at every frame is discontinuous in memory, which means that our game cannot effectively use high-speed memory, that is, the caches.

Now, let's look at the new programming pattern in Unity, DOTS.

Unlike OOP, DOTS' philosophy is to design for data rather than objects, focusing on prioritizing and organizing data to make its memory access as efficient as possible. Let's still use the Monster class as an example to see how its data is allocated in memory when using DOTS.

| 8 bytes | 8 bytes | 8 bytes | 8 bytes | 8 bytes | 8 bytes | 8 bytes | 8 bytes |
|---|---|---|---|---|---|---|---|
| Vector3 pos | Vector3 pos | Vector3 pos | Vector3 pos | Vector3 pos | Ve |
| ctor3 pos | Vector3 pos | Vector3 pos | Vector3 pos | Vector3 pos | Vector3 |
| po s | Vector3 pos | Vector3 pos | Vector3 pos | Vector3 pos | Vector3 pos |
| Vector3 pos | Vector3 pos | Vector3 pos | Vector3 pos | Vector3 pos | Ve |
| ctor3 pos | Vector3 pos | Vector3 pos | Vector3 pos | Vector3 pos | Vector3 |
| po s | Vector3 pos | Vector3 pos | Vector3 pos | Vector3 pos | Vector3 pos |
| Vector3 pos | Vector3 pos | Vector3 pos | Vector3 pos | Vector3 pos | Ve |
| ctor3 pos | Vector3 pos | Vector3 pos | Vector3 pos | Vector3 pos | Vector3 |
| po s | Vector3 pos | Vector3 pos | Vector3 pos | Vector3 pos | Vector3 pos |

Figure 9.3 – The data layout in memory (DOTS)

Do you remember? When moving a monster, we actually only need 12 bytes of position data for the monster, so the code only needs to load and process the position data of all the monsters to move them. Using DOTS allows us to pack all of this position data into an array and allocate memory more efficiently, as shown in the previous figure. Placing data in a contiguous array in memory improves data locality, which results in extremely high hit rates for caches, which improves code performance.

So, how does Unity's DOTS make developers' code run more efficiently? Well, DOTS in Unity is not just a change of programming paradigm from object-oriented to data-oriented; it actually includes a series of new technology modules, namely the following:

- The C# Job System

- ECS

- The Burst compiler

Each of them consists of one or more Unity packages. We can install the corresponding functions through Unity's Package Manager. Next, we will briefly introduce these three modules, respectively.

# C# Job System

By using the C# Job System, we can write efficient asynchronous code in Unity that takes full advantage of the hardware.

Figure 9.4 – Tech demo using the C# Job System

The preceding figure shows a demo project developed using Unity's C# Job System, showing thousands of "soldiers" attacking the enemy in the scene. You can find this project on GitHub: `https://github.com/Unity-Technologies/UniteAustinTechnicalPresentation`.

We will discuss the C# Job System in detail in the *Multithreaded and C# Job System in Unity* section.

## ECS

The full name of **ECS** is **Entity Component System**. It is the core part of DOTS in Unity and is built around using data-oriented design, which is very different from the object-oriented design you may be used to.

Figure 9.5 – The Megacity demo

The preceding figure shows Unity's impressive tech demo using ECS called **Megacity**, which developers can download here: `https://unity.com/megacity`.

Figure 9.6 – Megacity download page

We'll cover ECS in detail in the *Working with ECS in Unity* section.

# The Burst compiler

The Burst compiler in Unity is an advanced compiler technology. Unity projects made with DOTS can use Burst technology to improve their runtime performance. Burst works on a subset of C# called **High-Performance C# (HPC#)** and applies advanced optimization methods under the LLVM compiler framework to generate efficient binaries, which achieves efficient use of device energy.

We will introduce how to use it in your project in a later section, *Using C# and the Burst compiler*.

This section introduces DOTS-related knowledge, such as what technology modules DOTS contains, how its design philosophy differs from traditional OOP, and what problems it solves. However, DOTS is not a replacement for OOP; it just provides another efficient programming pattern for game developers in Unity. For example, you can still use the C# Job System to implement multithreaded programming in the traditional Unity GameObject-Components style, rather than maintaining thread pools yourself. Well, next, let's explore how to implement efficient multithreaded programming in Unity.

# Multithreading and the C# Job System in Unity

Asynchronous programming is very common when developing .NET projects. But unlike what many people who are familiar with .NET development think, Unity's support for asynchronous programming was not friendly at first.

## Coroutines

Before Unity 2017, if a game developer wanted to handle asynchronous operations, a common scenario was waiting for a network response. The ideal solution was to use **coroutines** in Unity.

We can start a coroutine in Unity as follows:

```
void Start()
{
    var url = "https://jiadongchen.com";
    StartCoroutine(DownloadFile(url));
}


private static IEnumerator DownloadFile(string url)
{
    var request = UnityWebRequest.Get(url);
    request.timeout = 10;
    yield return request.SendWebRequest();
    if (request.error != null)
    {
        Debug.LogErrorFormat("error: {0}, url is: {1}",
            request.error, url);
        request.Dispose();
        yield break;
    }

    if (request.isDone)
    {
        Debug.Log(request.downloadHandler.text);
        request.Dispose();
        yield break;
    }
}
```

As you can see in the code, we use the `StartCoroutine` function to start a coroutine, and inside the coroutine, we can pause the execution by using `yield` statements. However, coroutines are still inherently single-threaded, just spread-out tasks across multiple frames, rather than multithreaded.

## async/await

Unity introduced the `async/await` operator in Unity 2017, allowing game developers to use `async/await` in their games to write asynchronous code, but it's still not like a normal .NET/C# program. This is because the Unity engine manages these threads by itself, and most of the logic runs on Unity's main thread, which includes not only the C# code as scripts but also the engine's C++ code. We can use the **Unity Profiler** tool to view the CPU timeline. As shown in the following screenshot, the Unity engine runs scripts in the main thread by default:

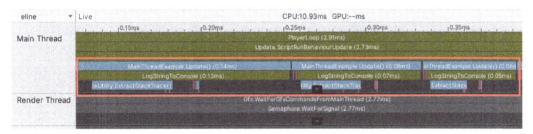

Figure 9.7 – The Timeline of CPU

There are 50 GameObjects in this scene, and each of them is attached with a **MainThreadExample** script. You can see that the **Update** functions in these 50 scripts are executed one by one.

You can multithread different types of tasks; for example, doing some Vector3 math in a separate thread is no problem. But as long as the task needs to access the transform or GameObject outside Unity's main thread, the program will throw an exception.

Let's look at an example. The purpose of the following code is to change the scale of the GameObject and use `async/await` to perform the operation in another thread:

```
using System.Threading.Tasks;
using UnityEngine;

public class AsyncExceptionTest : MonoBehaviour
{
```

```
    private async void Start()
    {
        await ScaleObjectAsync();
    }

    private async Task<Vector3> ScaleObjectAsync()
    {
        return await Task.Run(() => transform.localScale = new
            Vector3(2, 2, 2));
    }
}
```

Attach this script to a GameObject in the scene, then click the **Play** button in the Unity editor to run the script. The result of the operation is that the scale of the GameObject has not changed, and a **UnityException: get_transform can only be called from the main thread** exception is thrown, as shown in the following screenshot:

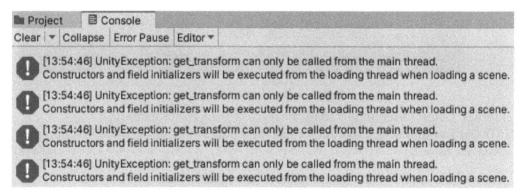

Figure 9.8 – Exception

So, you should take care of this and not access transforms or GameObjects from threads other than Unity's main thread.

As we mentioned earlier, we can do math in a separate thread. So, in order to make the previous code work correctly, we can just calculate the scale value in different threads, access the Transform component, and modify the localScale property of it in Unity's main thread:

```
    private async Task ScaleObjectAsync()
    {
        var newScale = Vector3.zero;
```

```
    await Task.Run(() => newScale = CalculateSize());
    transform.localScale = newScale;
}

private Vector3 CalculateSize()
{
    Debug.Log("Threads");
    return new Vector3(2, 2, 2);
}
```

This time, everything is going well and if we view the Unity Profiler again, we can find the timeline of these threads in the **Scripting Threads** section, as shown in the following screenshot:

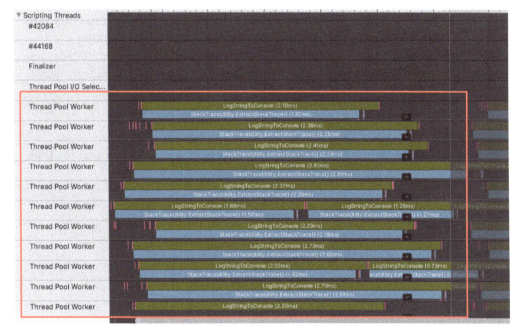

Figure 9.9 – Scripting threads

However, as a developer, there are still many challenges with writing thread-safe and efficient code even in C#, such as the following:

- Thread-safe code is hard to write.
- Race conditions, where the result of a computation depends on the order in which two or more threads are scheduled.
- Inefficient context switching; is time-consuming when switching threads.

The C# Job System in Unity is a solution that focuses on solving these challenges so that we can enjoy the benefits of multithreading to develop games. Next, let's explore how to use the C# Job System in our Unity projects.

# Working with the C# Job System

The **Job System** was originally the internal thread management system of the Unity engine, but with the growth of developers' demands for multithreaded programming in Unity, Unity introduced the C# Job System, which allows developers to write multithreaded parallel processing code painlessly in C# scripts to improve games' performance. Game developers do not need to implement complex thread pools themselves to keep each thread running properly. The C# Job System is integrated with Unity's native Job System, and C# script code and Unity engine's C++ code share threads.

This form of cooperation allows game developers to write code in the way required by the Job System; the Unity engine handles multithreading for game developers and developers no longer have to worry about problems that may be encountered when writing multithreaded code, because the C# Job System will not create any managed threads, but instead use Unity's worker threads on multiple cores, giving them tasks, which are called **jobs** in Unity.

## Installing the Jobs package

In order to install and enable the **Job System** in your project, you need to install the **Jobs** package first, as shown in the following screenshot:

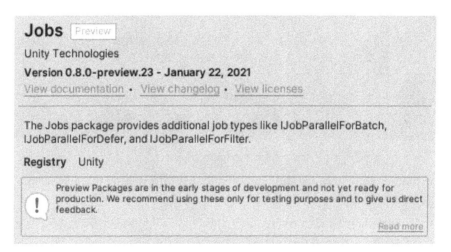

Figure 9.10 – The Jobs package

However, the **Jobs** package is currently still in the preview state, as shown in the preceding screenshot, and the Unity Package Manager does not display packages in the preview state by default. So, if you can't find the **Jobs** package, then you need to follow these steps to allow showing the package in the preview state:

1.   Open the **Project Settings** window by clicking the **Edit | Project Settings...** item in the Unity editor toolbar, as shown in the following screenshot:

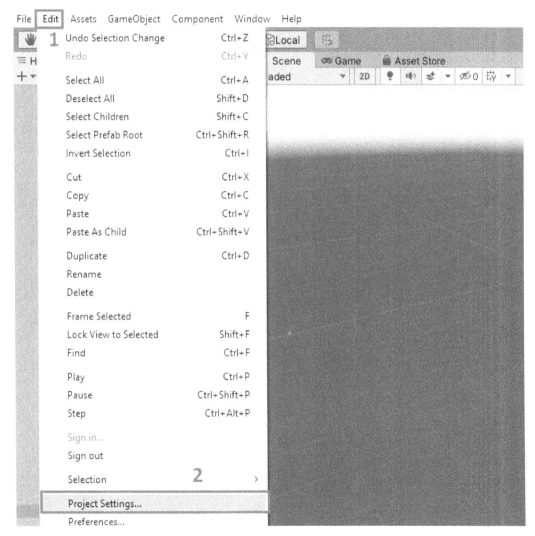

Figure 9.11 – Opening the Project Settings window

2.  Next, click the **Package Manager** item in the category list on the left to open the **Package Manager** settings panel.

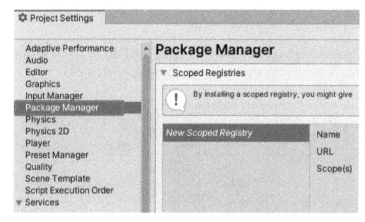

Figure 9.12 – Opening the Package Manager settings

3.  In the following screenshot, you can see that the **Enable Preview Packages** option is not selected by default. Let's check it to enable preview packages in the Unity Package Manager.

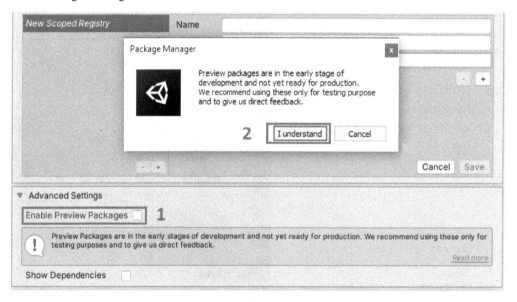

Figure 9.13 – Enable Preview Packages

Once done, you should be able to find the **Jobs** package and install it into your project.

Next, let's look at an example to understand how to use the Job System to improve the performance of a game.

## How to use the C# Job System

In this example, we will first use Unity's traditional way, that is, the GameObject+Components way, to create 10,000 cartoon cars in a game scene, with each car containing a Movement component to move it.

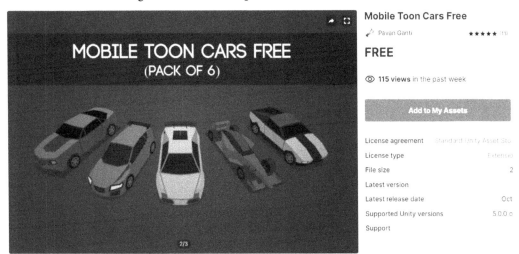

Figure 9.14 – The car models

The car models used in this example are from the Unity Asset Store, and you can download them here: `https://assetstore.unity.com/packages/3d/vehicles/land/mobile-toon-cars-free-99857`. Then, take the following steps:

1.  Let's create our first C# script, named `CarSpawner`, to generate the cars in the scene. In this script, we can create 10,000 new car instances from the car prefab by pressing the spacebar. As you can see in the following code, inside the `Update` method, we use the `Input.GetKeyDown(KeyCode.Space)` method to check whether the spacebar is pressed. If the spacebar is pressed, the `CreateCars` method is called to create new car instances:

```
using System.Collections;
using System.Collections.Generic;
using UnityEngine;

public class CarSpawner : MonoBehaviour
{
[SerializeField]
private List<GameObject> _carPrefabs;
[SerializeField]
```

```
private float _rightSide, _leftSide, _frontSide,
  _backSide;

  private void Update()
  {
      if(Input.GetKeyDown(KeyCode.Space))
      {
          CreateCars(10000);
      }
  }

  private void CreateCars(int count)
  {
      for(var i = 0; i < count; i++)
      {
          var posX = Random.Range(_rightSide,
            _leftSide);
          var posZ = Random.Range(_frontSide,
            _backSide);

          var pos = new Vector3(posX, 0f, posZ);
          var rot = Quaternion.Euler(0f, 0f, 0f);
          int index = Random.Range(0,
            _carPrefabs.Count);
          var carPrefab = _carPrefabs[index];
          var carInstance = Instantiate(carPrefab, pos,
            rot);
      }
  }
}
```

2.  Next, we also need another script that will be attached to each of the car objects to move them. As you can see, this Movement script is relatively simple; it moves the GameObject forward:

```
using UnityEngine;

public class Movement : MonoBehaviour
```

```
{
[SerializeField]
private float _speed;

    private void Update()
    {
        transform.position += transform.forward *
          _speed * Time.deltaTime;
    }
}
```

3.   Then, attach this Movement script to the car prefab.

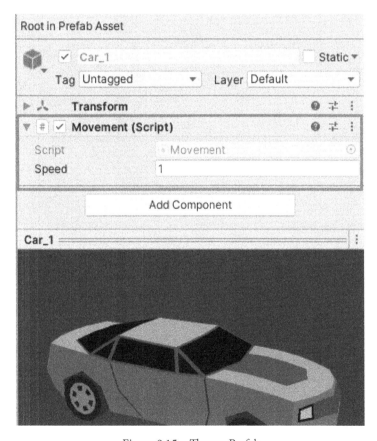

Figure 9.15 – The car Prefab

4.  Click the **Play** button in the Unity editor to run the example and press the spacebar to generate 10,000 cars in the scene. As shown in the following screenshot, when there are 10,000 cars in the scene, the value of **Frames per Second** (**FPS**) is around **12**:

Cars: 10000
FPS: 11.9

Figure 9.16 – The FPS

5.  We can view the CPU usage timeline of this example. Press *Ctrl + 7* or click the **Window | Analysis | Profiler** item in the Unity editor toolbar to open the **Profiler** window. Here, we can see that the Movement.Update for these 10,000 cars happens on the main thread while the job workers are idle.

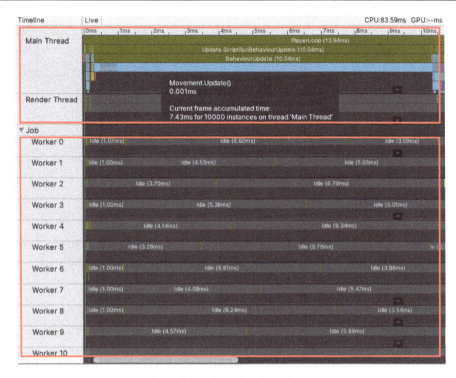

Figure 9.17 – The Timeline of CPU

Obviously, when we see all the logic being executed on the main thread, as game developers we definitely want to be able to have some operations running on other threads. However, before we start writing some real code, we should cover a little bit about how to write jobified code in Unity.

In Unity's Job System, each job can be seen as a method call. When writing a new job, you must follow these points:

- In order to ensure that the data is distributed contiguously in memory, and to reduce **Garbage Collection** (**GC**) pressure, a job must be a value type, which means it must be a struct, never a class.

- A new job struct needs to implement the IJob interface. There are many variants of the IJob interface, such as IJobParallelFor, IJobParallelForBatch, and IJobParallelForTransform. When implementing these interfaces, we need to implement the Execute method. It is worth noting that the parameters required by the Execute method are different when implementing different variants of the IJob interface, which allows us to handle different scenarios. For example, a new job implementing the IJobParallelForTransform interface can access transform data, such as position, rotation, and scale data, in parallel.

The following code is for a sample job that implements the `IJobParallelFor` interface:

```
using Unity.Jobs;

public struct SampleJob : IJobParallelFor
{
    public void Execute(int index)
    {
        throw new System.NotImplementedException();
    }
}
```

We have created a new job, but how do we make it work? Well, we have to schedule it. Usually, scheduling a job is very simple. The following code demonstrates how to schedule it:

```
SampleJob job = new SampleJob();
JobHandle handle = job.Schedule();
handle.Complete();
```

We've covered some basics on how to create a new job and how to make it work. Now, let's use the Job System to rewrite the `Movement` script to distribute the operation of moving the cars to different threads to run:

1.  First of all, let's create a job that moves cars. You can find the new `MotionJob` script below. `MotionJob` is a struct rather than a class and implements the `IJobParallelForTransform` interface, so this job can access the position data and modify it:

    ```
    using UnityEngine;
    using UnityEngine.Jobs;

    public struct MotionJob : IJobParallelForTransform
    {
        public float Speed, DeltaTime;
        public Vector3 Direction;

        public void Execute(int index, TransformAccess
            transform)
        {
    ```

```
transform.position += Direction * Speed *
    DeltaTime;
    }
}
```

2. Next, we need another script called `JobsManager` to create the job, provide it with transform data (specifically in the script, we use the `TransformAccessArray` struct to provide this data), and schedule it. Also, this script is similar to the previous `CarSpawner` script. It checks whether the spacebar is pressed and creates 10,000 cars in the game scene if the spacebar is pressed. First, let's see how to create and schedule a job on Unity's Job worker thread. In the `Update` method, we create a new `MotionJob` object and pass data to it, such as `deltaTime`, `speed`, and `direction` to create a new job, then we call `_motionJob.Schedule` to distribute the job to different threads:

```
using UnityEngine;
using UnityEngine.Jobs;
using Unity.Jobs;
using System.Collections.Generic;

public class JobsManager : MonoBehaviour
{
[SerializeField]
private List<GameObject> _carPrefabs;
[SerializeField]
private float _rightSide, _leftSide, _frontSide,
  _backSide, _speed;
    private TransformAccessArray _transArrays;
    private JobHandle _jobHandle;
    private MotionJob _motionJob;

    private void Start()
    {
        _transArrays = new
          TransformAccessArray(10000);
        _jobHandle = new JobHandle();
    }
```

```
    private void Update()
    {
        _jobHandle.Complete();


        if(Input.GetKeyDown(KeyCode.Space))
        {
            CreateCars(10000);
        }
        // Create the Job
        _motionJob = new MotionJob()
        {
            DeltaTime = Time.deltaTime,
            Speed = _speed,
            Direction = Vector3.forward
        };


        // Provide the transform data and schedule the
           Job.
        _jobHandle = _motionJob.Schedule(_transArrays);
    }
```

3. Next, let's see how to create cars in the code. Since we only need the position data for these cars this time, in the CreateCars method, we add the car's transform data to TransformAccessArray so that the job we just created can access TransformAccessArray to get that transform data. The CreateCars method is as follows:

```
    private void CreateCars(int count)
    {
        _jobHandle.Complete();
        _transArrays.capacity = _transArrays.length +
          count;


        for (var i = 0; i < count; i++)
        {
            var posX = Random.Range(_rightSide,
              _leftSide);
            var posZ = Random.Range(_frontSide,
```

```
      _backSide);

      var pos = new Vector3(posX, 0f, posZ);
      var rot = Quaternion.Euler(0f, 0f, 0f);
      int index = Random.Range(0,
        _carPrefabs.Count);
      var carPrefab = _carPrefabs[index];
      var carInstance = Instantiate(carPrefab,
        pos, rot);
      _transArrays.Add(carInstance.transform);
    }
  }
```

4.  This time, we no longer need to attach the Movement component to each car instance at runtime to move the car, so we need to remove the **Movement** component that was previously attached to the car prefab.

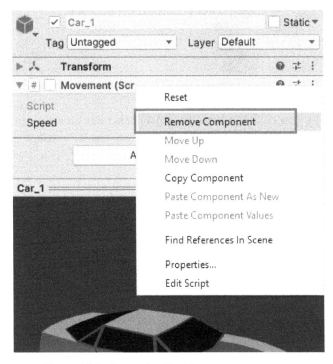

Figure 9.18 – Removing the Movement component

5.   Click the **Play** button in the Unity editor to run the example and press the spacebar to generate 10,000 cars in the scene. As shown in the following screenshot, when there are 10,000 cars in the scene, this time the value of FPS is around **19**. In a scene with 10,000 cars in motion, the game's frame rate nearly doubled:

Figure 9.19 – The FPS

6.   Let's press *Ctrl + 7* or click the **Window | Analysis | Profiler** item in the Unity editor toolbar to open the **Profiler** window to view the CPU usage timeline this time. Here, we can see that **MotionJob** is spread over multiple Job worker threads in Unity, instead of running on the main thread.

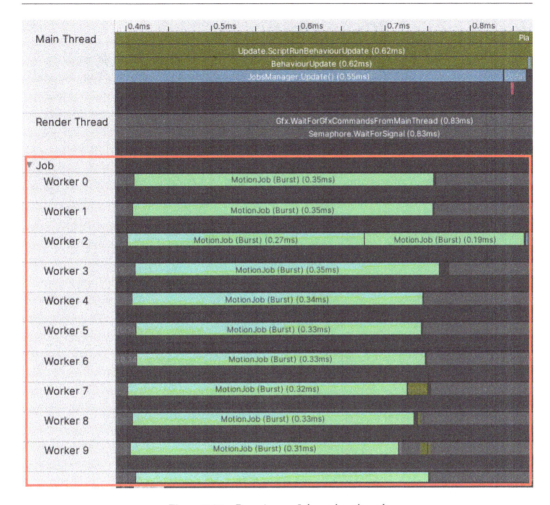

Figure 9.20 – Running on Job worker threads

Through this example, we saw how to use the Job System in Unity to improve the running performance of the game.

In this section, we discussed topics related to using asynchronous programming in Unity. Next, we will discuss another important topic in DOTS – namely, ECS.

# Working with ECS in Unity

Unity has always been centered around the concept of components; for example, we can add a Movement component to a GameObject so that the object can move. We can also add a Light component to the GameObject to make it emit light. We also add the AudioSource component, which can make the GameObject emit sound. In this case, the GameObject is a container to which game developers can attach different components to provide different behaviors. We can call this architecture a **GameObject-Components** relationship. In this architecture, we use the traditional OOP programming paradigm to write components, coupling data and behavior together. In the previous section, *Object-oriented design pattern versus DOTS*, we also discussed the impact of OOP on game performance.

So, to address these issues, Unity introduced ECS, which allows developers to write data-oriented code in Unity. In ECS, data and behavior are separated, which can greatly improve memory usage efficiency and thus improve performance.

> **Note**
> The so-called **behavior** here, specifically, is **methods**.

As its name suggests, ECS consists of three parts, namely the following:

- Entity
- Component
- System

We will introduce them respectively in the following sections.

## Entity

When using ECS, we talk more about entities, not GameObjects. You might think that there is not much difference between an entity and a GameObject, because you might think of an entity as a container for components, just like a GameObject. However, this is not the case. An entity is just an integer ID. It is neither an object nor a container. Its function is to associate the data of its components together.

### EntityManager and World

If you want to create new entities in your own C# code, Unity provides the EntityManager class to manage entities, which you can use to create entities, update entities, and destroy entities. ECS uses the World class to organize entities, and only one EntityManager instance can exist in a World.

When we click the **Play** button in the Unity Editor to run the game, Unity will create a `World` by default, so we can get the `EntityManager` that exists in the default `World` with the following code:

```
var entityManager =
    World.DefaultGameObjectInjectionWorld.EntityManager;
```

## Archetypes

ECS combines all entities with the same set of components in memory. ECS refers to this type of component set as an **Archetype**. Assuming that an entity has only two components, then those two components form a new Archetype. The following pseudocode demonstrates how to use `EntityManager` to create an Archetype that holds a set of components:

```
ComponentType[] types;
var archetype = entityManager.CreateArchetype(types);
```

## NativeArray

Undoubtedly, we also need an array to hold the newly created entities. But in ECS, we will use a different container than a traditional array in .NET programming, namely, `NativeArray`.

`NativeArray` provides a C# wrapper for accessing native memory so that game developers can share data directly between managed and native memory. Therefore, operations on `NativeArray` do not generate GC of managed memory like common arrays in .NET and require elements to be value types, that is, structs. The following pseudocode shows how to create a new `NativeArray` and create new entities:

```
var entityArray = new NativeArray<Entity>(count,
    Allocator.Temp);


entityManager.CreateEntity(entityArchetype, entityArray);
```

## Component

In ECS, there are also components, but the component in ECS is a different concept from the Movement "component" mentioned when talking about the GameObject-Components relationship previously. Before ECS was introduced, we usually thought of `MonoBehaviour` attached to GameObjects as components. `MonoBehaviour` contains data and behavior. ECS is different because entities and components do not have any behavioral logic; they only contain data, and logical operations will be handled by the system in ECS.

A component must be a struct rather than a class and needs to implement one of the following interfaces:

- IComponentData
- ISharedComponentData
- IBufferElementData
- ISystemStateComponentData
- ISharedSystemStateComponentData

The IComponentData interface is commonly used. The following uses it as an example to show how to create a new component in ECS:

```
using Unity.Entities;

public struct SampleComponent : IComponentData
{
    public int Value;
}
```

If you try to add this SampleComponent to a GameObject in the scene, you will find that you can't because it doesn't inherit from the MonoBehaviour class. But you can add the [GenerateAuthoringComponent] attribute to your component to mark it as an authoring component, as follows:

```
using Unity.Entities;

[GenerateAuthoringComponent]
public struct SampleComponent : IComponentData
{
    public int Value;
}
```

An authoring component can be added to a GameObject even if it does not inherit from MonoBehaviour.

## System

We already know that when using ECS, data and behavior are decoupled. In ECS, all logic is handled by **systems**, which takes a group of entities and performs the requested behavior based on the data contained in the grouped entities. As we already know, using ECS can make our code access memory efficiently, and in fact, systems in ECS can also be combined with the C# Job System to efficiently utilize multithreading and further improve game performance.

We can create a new system in ECS. The following code is an example:

```
using Unity.Entities;

public class SampleSystem : SystemBase
{
    protected override void OnUpdate()
    {
        Entities.ForEach((ref SampleComponent sample) =>
        {
            sample.Value = -1;
        }).
        ScheduleParallel();
    }
}
```

In this example, this new SampleSystem inherits the SystemBase class, and there is a ScheduleParallel Lambda function after the Entites.ForEach loop in OnUpdate for scheduling work to Unity's Job worker threads using the C# Job System.

Through these brief introductions, I believe you have a general understanding of ECS. Next, let's install ECS in our project.

## Installing the Entities and Hybrid Renderer packages

In order to install and enable ECS in your project, you need to install the **Entities** package first, as shown in the following screenshot:

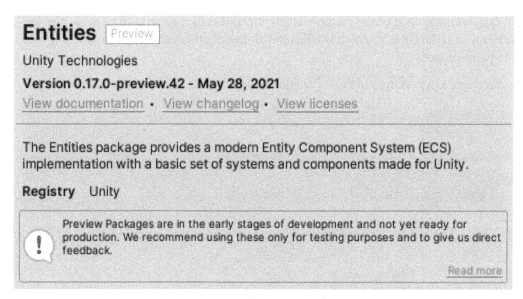

Figure 9.21 – The Entities package

As shown in the preceding screenshot, the **Entities** package is also in the preview state. Although we checked the **Enable Preview Packages** option in the previous subsection, the Package Manager still does not display this package. This is because starting from Unity 2020.1, this package is no longer hosted on Unity Registry, but hosted on GitHub, so we need to follow these steps to install it:

1.  The Package Manager window can be opened by clicking the **Window | Package Manager** item in the Unity editor toolbar, as shown in the following screenshot:

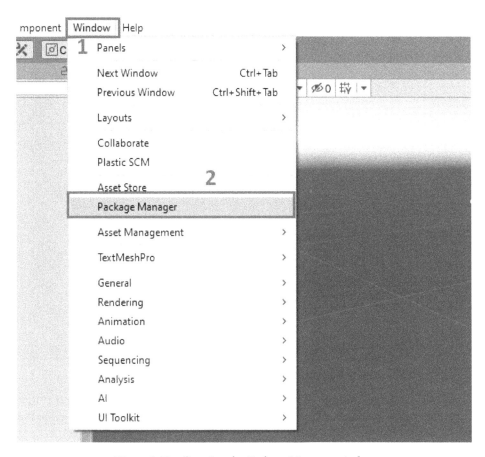

Figure 9.22 – Opening the Package Manager window

2.  Click the + button in the upper-left corner to add packages from other sources.

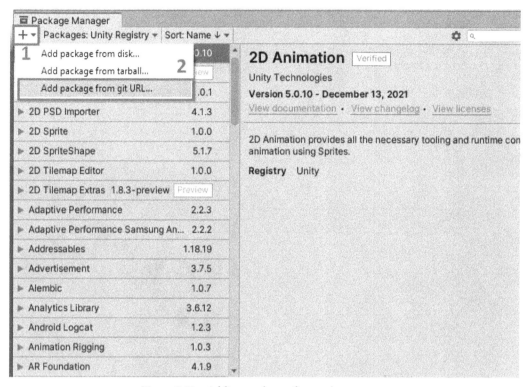

Figure 9.23 – Adding packages from other sources

3.  Click **Add package from git URL...** to install the **Entities** package in your project. The name of the package is `com.unity.entities`, so we enter it and click the **Add** button.

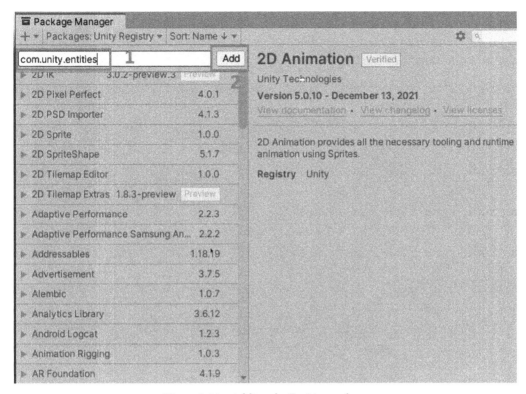

Figure 9.24 – Adding the Entities package

4.   Then, wait for the package installation to complete.

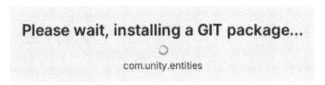

Figure 9.25 – Installing a Git package

Once done, you should be able to find the **Entities** package installed in your project.

Sometimes we also need another package, the **Hybrid Renderer** package. This package helps us render ECS entities.

The process of installing the **Hybrid Renderer** package is the same as the process of installing the **Entities** package, except that we need to enter the name of this package after clicking **Add package from git URL…**, which is `com.unity.rendering.hybrid`.

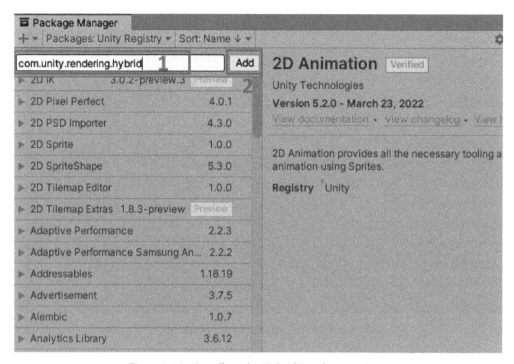

Figure 9.26 – Installing the Hybrid Renderer package

Wait for the package installation to complete, and then you'll find it's installed in your project.

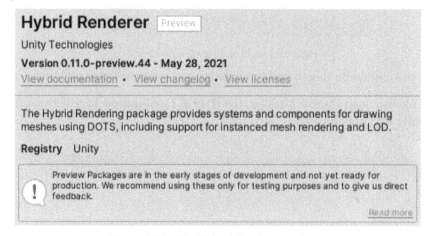

Figure 9.27 – The Hybrid Renderer package

Next, we will use the previous example to understand how to use ECS to further improve the performance of a game based on the use of the C# Job System.

## How to use ECS

In this example, we will create a new component, entities, and a new system and use the C# Job System to distribute work to Unity's Job worker threads. Let's get started!

1.  First, we will create a component script just for the data. In this case, it's the speed of cars:

    ```
    using Unity.Entities;


    public struct CarSpeed : IComponentData
    {
        public float SpeedValue;
    }
    ```

2.  Next, we also need a normal script called `CarsManager` to access the `EntityManager` object in `World` to create archetypes and entities. Here, we'll add some premade components from ECS to these entities, such as `Translation`, which contains only entity location data, and `RenderMesh`, which contains entity graphics attribute data:

    ```
    using UnityEngine;
    using Unity.Collections;
    using Unity.Mathematics;
    using Unity.Entities;
    using Unity.Rendering;
    using Unity.Transforms;
    using Random = UnityEngine.Random;


    public class CarsManager : MonoBehaviour
    {
    [SerializeField]
    private Mesh _mesh;


    [SerializeField]
    private Material _material;
    ```

```
[SerializeField]
private int _count = 10000;


[SerializeField]
private float _rightSide, _leftSide, _frontSide,
  _backSide, _speed;

    private void Start()
    {
        var entityManager =
          World.DefaultGameObjectInjectionWorld
          .EntityManager;


        // Create entity achetype
        var entityArchetype =
          entityManager.CreateArchetype(
            typeof(CarSpeed),
            typeof(Translation),
            typeof(LocalToWorld),
            typeof(RenderMesh),
            typeof(RenderBounds));
        var entityArray = new
          NativeArray<Entity>(_count, Allocator.Temp);


        // Create entities
        entityManager.CreateEntity(entityArchetype,
          entityArray);
        for (int i = 0; i < entityArray.Length; i++)
        {
          var entity = entityArray[i];
          entityManager.SetComponentData(entity, new
            CarSpeed { SpeedValue = 1f });
          entityManager.SetComponentData(entity, new
            Translation { Value = new
            float3(Random.Range(_rightSide,
```

```
            _leftSide),0,
            Random.Range(_frontSide, _backSide)) });

    entityManager.SetSharedComponentData(entity, new
      RenderMesh
            {
                mesh = _mesh,
                material = _material
            });
        }
    entityArray.Dispose();
    _information.CarCounts = _count;
    }
}
```

3.   Then, attach this **CarsManager** script to a GameObject in the scene and assign the appropriate properties, such as the car's mesh, and speed values.

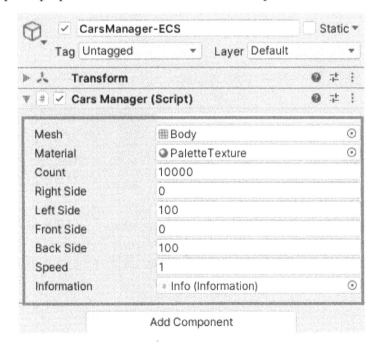

Figure 9.28 – The CarsManager object

4.  At this point, we have set up the components and entities. The next thing to do is to create the system. The system is also where the game logic is handled. In this example, we'll use the system to move these cars. As you can see in the following code, instead of searching for components in the traditional `Update` method and then operating on each instance at runtime, with ECS, we just statically declare that we need to process all entities with `Translation` and `CarSpeed` components attached. To find all of these entities, we just need to find the archetypes that match a specific "components set," which is done by the system:

```
using Unity.Entities;
using Unity.Transforms;


public class CarMotionSystem : SystemBase
{
    protected override void OnUpdate()
    {
        var deltaTime = Time.DeltaTime;

        Entities.ForEach((ref Translation translation,
            ref CarSpeed carSpeed) =>
        {
            translation.Value.z += carSpeed.SpeedValue
                * deltaTime;
        }).
        ScheduleParallel();
    }
}
```

5.  Click the **Play** button in the Unity editor to run the example. As shown in the following screenshot, when there are 10,000 cars in the scene, this time the value of FPS is around **260**! In this scene with 10,000 moving cars, using ECS increased the game's frame rate by nearly 30 times compared to the original traditional implementation:

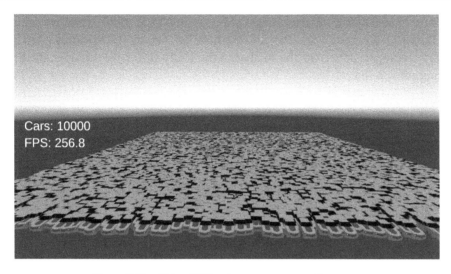

Figure 9.29 – Using ECS to improve game performance

6. If we look at the **Hierarchy** panel of this game scene, we won't see any car objects in the list. This is because when using ECS, traditional GameObjects and traditional components are not created, but entities and components from ECS are used to organize data.

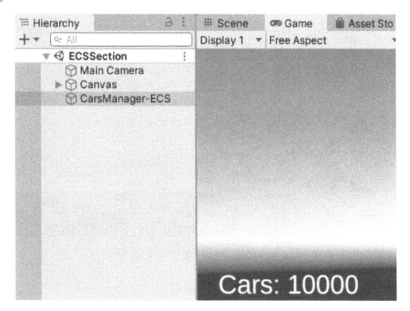

Figure 9.30 – No GameObjects are created

7.  In order to see the entities, components, and system used in the scene, we can use the **Entity Debugger** to view this information. By clicking the **Window | Analysis | Entity Debugger** item from the toolbar in the Unity editor, we can open the **Entity Debugger** window.

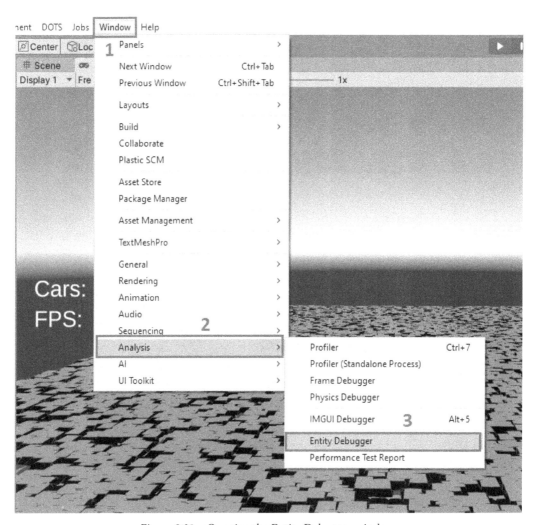

Figure 9.31 – Opening the Entity Debugger window

8.  We can see a list of entities as well as a list of systems in the Entity Debugger window. As shown in the following screenshot, there are 10,002 entities, including 10,000 car entities:

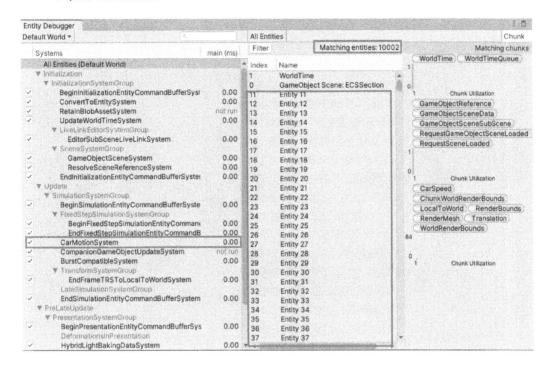

Figure 9.32 – The Entity Debugger

9.   If we select an entity in the entity list, the **Inspector** window for that entity will
     open, showing all components of this entity and data for those components.

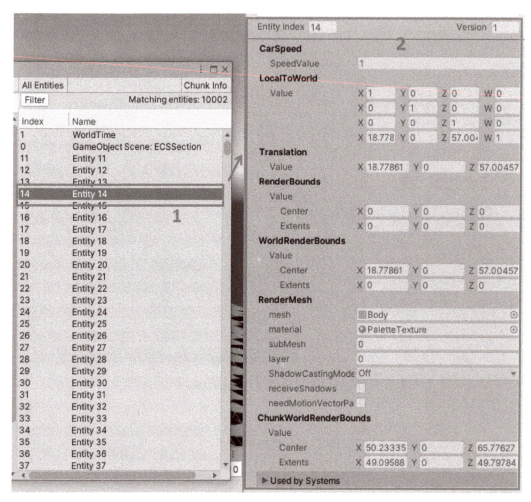

Figure 9.33 – Inspector window for an entity

10. Finally, let's view the CPU usage timeline in the **Profiler** window. If you forget how to open this window, just press *Ctrl + 7* or click **Window | Analysis | Profiler** in the Unity Editor toolbar. Here, we can see that the ECS work is distributed to multiple Job worker threads by the C# Job System as we expected:

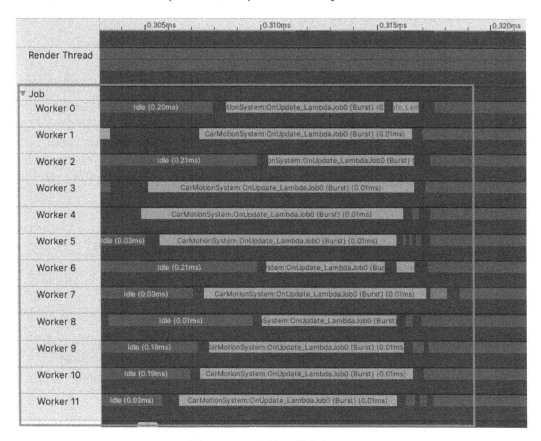

Figure 9.34 – ECS and Job System

Through the preceding steps, we changed the traditional GameObject-Components-style development method in Unity to the development method using ECS, adopting the data-oriented design method and using the C# Job System, making full use of multithreaded programming, and improving the running efficiency of the game.

Next, let's discuss another technology in DOTS, the Burst compiler.

# Using C# and the Burst compiler

The **Burst compiler** in Unity is an advanced compiler technology that can be used to convert a subset of .NET code into highly optimized native code for Unity games. It should be noted that it is not a general-purpose compiler, but a compiler designed for Unity to make Unity games run faster.

Burst works on a subset of C# called HPC#, so let's explore this subset of C# next.

## High-Performance C# (HPC#)

HPC# is a subset of C#. The standard C# language uses the concepts of "objects on the heap" and uses the garbage collector to reclaim unused memory automatically. So, as developers, we cannot control how the data is allocated in memory. On the other hand, HPC# doesn't support reference types, namely, classes, to avoid allocation in the heap and disable the garbage collector. In addition to these, some functions, such as `try-catch-finally`, are not supported in HPC# as well.

To summarize, we can use the following types in HPC#:

- Value types, such as int, float, bool, and char, enum types, and struct types
- `NativeArray` in Unity

## Enabling the Burst compiler

The Burst compiler is usually used with the C# Job System in Unity to optimize the code of a job. As we know, a job is a value type struct, so it is suitable for use with the Burst compiler. Enabling it in a job is very simple: just add the `[BurstCompile]` attribute to the job struct, as shown in the following code:

```
using Unity.Jobs;
using Unity.Burst;

[BurstCompile]
public struct SampleJobWithBurst : IJobParallelFor
{
    public void Execute(int index)
    {
        throw new System.NotImplementedException();
    }
}
```

If you also want to enable the Burst compiler in the Unity editor, you can find the settings for it at **Jobs | Burst** in the toolbar.

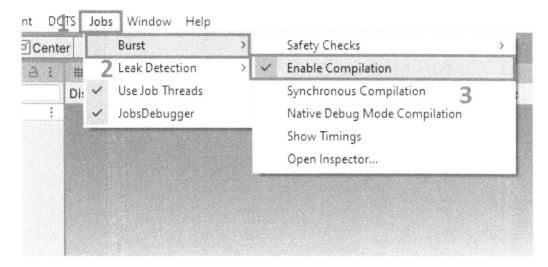

Figure 9.35 – Settings of Burst

By reading this section, you should know what the Burst compiler and HPC# are. You should also know that the Burst compiler is often used with the C# Job System in Unity and how to enable it in job code to generate more efficient native binary code.

# Summary

This chapter first introduced what data-oriented design is and the difference between data-oriented design and traditional object-oriented design. Then, we explored DOTS in Unity and the three technology modules that make it up, namely, the C# Job System, ECS, and the Burst compiler.

After that, we discussed in detail how to implement asynchronous programming in Unity and used an example to demonstrate how to use Unity's C# Job System to implement multithreading to improve game performance.

We also introduced the concept of ECS, discussed the difference between ECS and the traditional GameObject-Components architecture in Unity, and demonstrated how to use ECS and the C# Job System to further improve game performance.

Finally, we explored what the Burst compiler and HPC# are and how to enable them to generate highly optimized native code for your Unity games.

By reading this chapter, you should now understand how to work with DOTS correctly in Unity. In the next chapter, we will discuss topics related to assets management and serialization In Unity.

# 10
# Serialization System and Assets Management in Unity and Azure

In the last chapter, *Chapter 9, Using Data-Oriented Technology Stack in Unity*, we learned what the data-oriented technology stack is and how you can use this technology to take advantage of multicore processors to improve the performance of your game. In this chapter, we will cover some other important topics in Unity development, namely, **serialization** and **asset management** in Unity. Usually, a game not only has code but also consists of many different kinds of assets, such as models, textures, and audio. Hence, understanding what the serialization system in Unity is and what the assets workflow is can help you better develop games with Unity.

In the last section of this chapter, we will also explore an interesting topic – how to use the **Azure Cloud storage** services to host the content of a Unity game and load the content from the Azure Cloud to the Unity game by using Unity's **Addressable Assets system**.

The following key topics will be included in our learning path:

- Serialization system in Unity

- The Assets workflow in Unity

- Introducing the special folders in Unity

- Azure Blob storage with Unity's Addressable Assets system

By the end of this chapter, you will not only understand the serialization system and assets management in Unity, but you will also be familiar with Azure Cloud storage services.

Sounds exciting!

# Technical requirements

Since this chapter will be covering Azure's Storage account service, if you don't have an Azure account available, I recommend you set up a free Azure trial account first before starting this chapter. You can click the following link to create a free Azure trial account with $200 credit:

```
https://azure.microsoft.com/en-us/free/
```

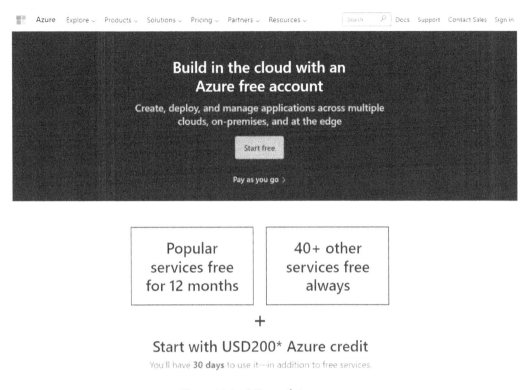

Figure 10.1 – Microsoft Azure page

Now, let's get started!

# Serialization system in Unity

When developing a game, adding a reliable content saving and loading feature is a critical part of the development process. If you're using a game engine editor, such as the Unity engine editor, you'll also need some common editor features, such as undo, saving editor settings, and more. All of this, whether the game saves or loads content at runtime, or whether the developer uses the editor to develop the game, is built on **serialization**.

## What is Unity's serialization system?

So, what is **serialization**? According to Wikipedia, the definition of serialization is *the process of translating a data structure or object state into a format that can be stored or transmitted and reconstructed later*. The opposite operation is deserialization.

In Unity, there are three serialization formats, namely the following:

- Binary serialization
- YAML serialization
- JSON serialization

## YAML and binary serialization in Unity

Assets created by Unity, such as Scenes and Prefabs, will be saved in YAML format by default. For example, if we open the Scene of this chapter, namely, Chapter10.unity, in a text editor such as **Sublime Text**, you can see that this Scene is serialized in YAML format, and you will see that there are options including OcclusionCullingSettings and RenderSettings. If you scroll down, you can also find the GameObjects and components contained in this Scene.

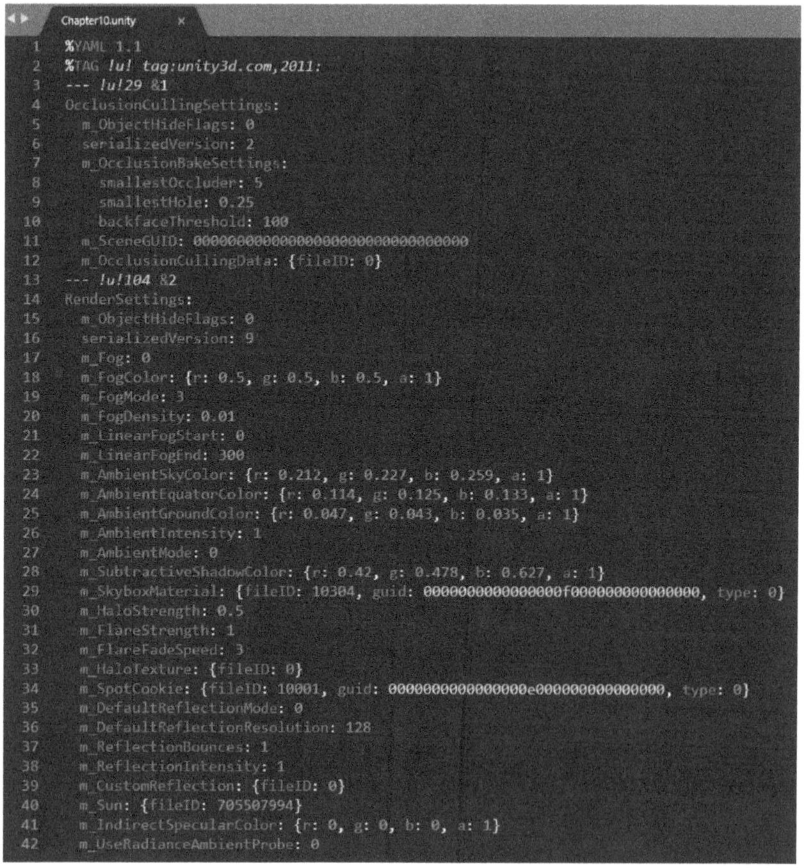

Figure 10.2 – The Scene in YAML format

As shown in *Figure 10.2*, there is no doubt that the YAML format is human-readable and makes it easy for the version control tools to work with. However, YAML is a text-based format, so you can also choose to use binary serialization for the more efficient use of space and increased security. Let's perform the following steps to set Unity's serialization mode:

1.  Open the **Project Settings** window by clicking the **Edit | Project Settings...** item in the Unity Editor toolbar, as shown in the following screenshot:

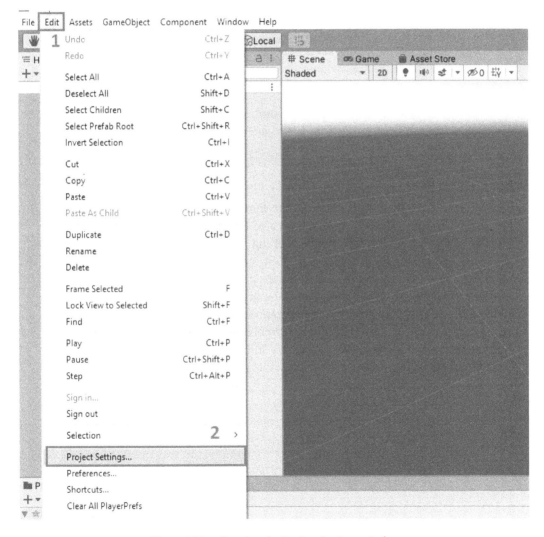

Figure 10.3 – Opening the Project Settings window

2.  Next, click the **Editor** item in the category list on the left to open the **Editor** settings panel, as shown in *Figure 10.4*:

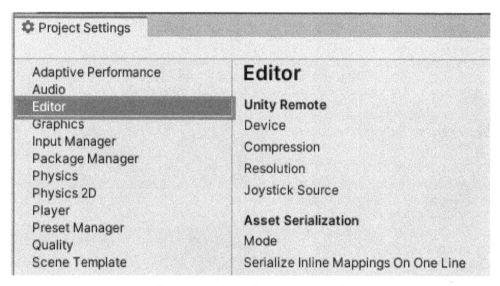

Figure 10.4 – The Editor settings panel

3.  In the **Asset Serialization** section, we can find that the **Mode** option is **Force Text** by default. In this mode, all the assets created by Unity will be serialized in YAML format. This is also the recommended setting if you use a version management tool such as Git, as using plain text serialization can often avoid unresolvable merge conflicts. As shown in *Figure 10.5*, in the drop-down window, we can select **Force Binary** mode to convert all the assets to binary format, and we can also choose the **Mixed** mode option to retain the serialization format of the current assets; that is, the assets that are serialized in binary format are still in binary format, and assets that are serialized using YAML format are still in YAML format. However, newly created assets will be serialized in binary format.

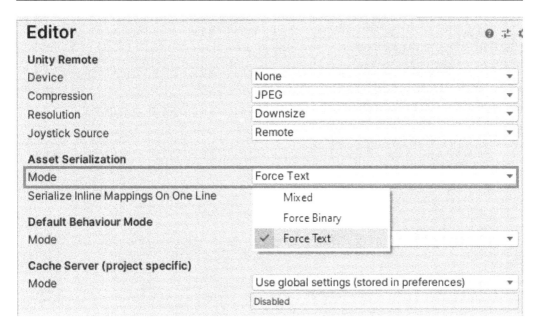

Figure 10.5 – Asset Serialization mode

4. Here, we can select **Force Binary** mode and check the same Scene file in our text editor again. The Scene file is converted to binary format, as shown in the following screenshot:

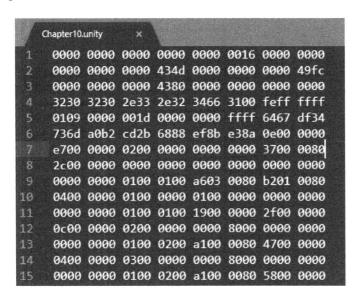

Figure 10.6 – The Scene file in binary format

As we mentioned earlier, serialization is also an important part of implementing the Unity Editor. Not only are the assets created by Unity as used in the game, such as game Scenes, serialized by Unity, but the various settings in the Unity Editor are also serialized by Unity.

In the project root directory, we can find the `ProjectSettings` folder, which is automatically created by the Unity Editor when the project is created, as shown in *Figure 10.7*:

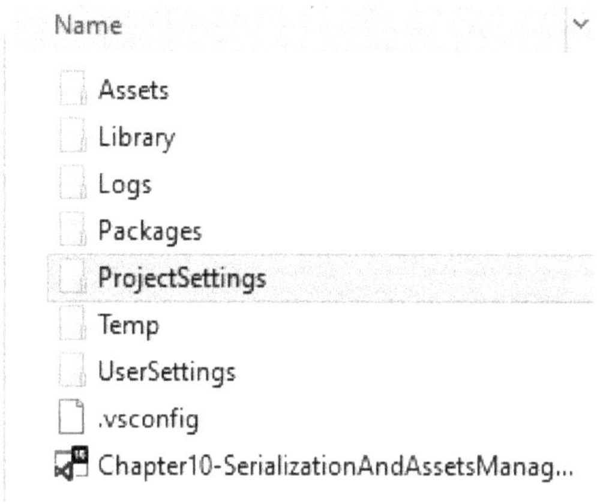

Figure 10.7 – The ProjectSettings folder

Double-click this folder to open it. We can find all the settings files of the current project here.

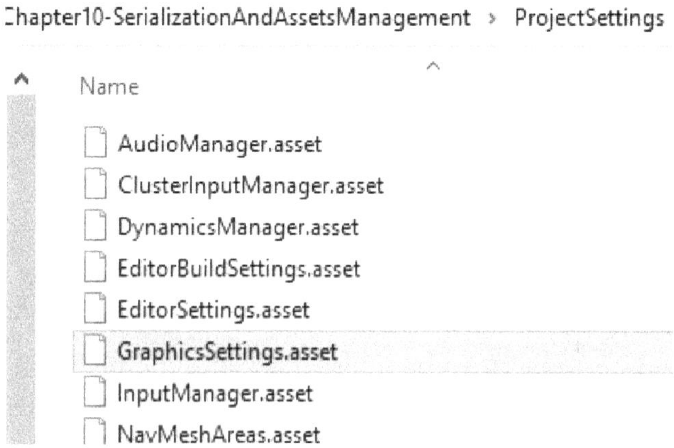

Figure 10.8 – The settings files in the ProjectSettings folder

Next, we still use the text editor to open a settings file, such as `GraphicsSettings.asset`, and serialize this file using Unity's binary serialization mode and text serialization mode, respectively. *Figure 10.9* shows the settings file serialized in binary format:

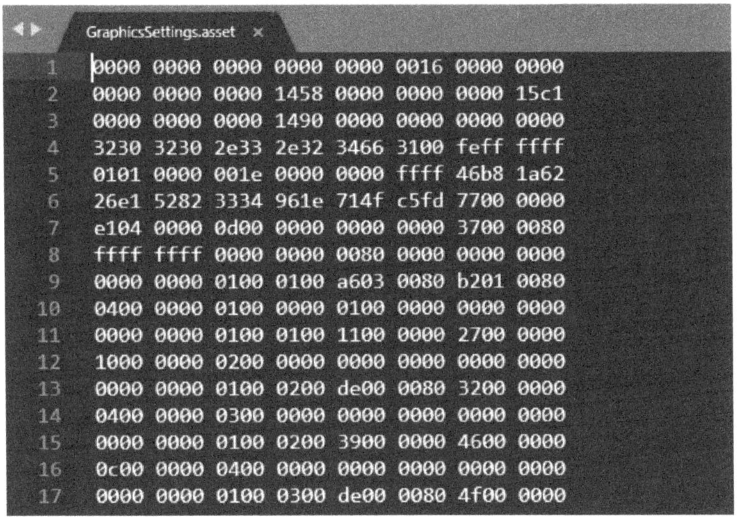

Figure 10.9 – The settings file in binary format

On the other hand, you can see the settings file serialized in YAML format in *Figure 10.10*:

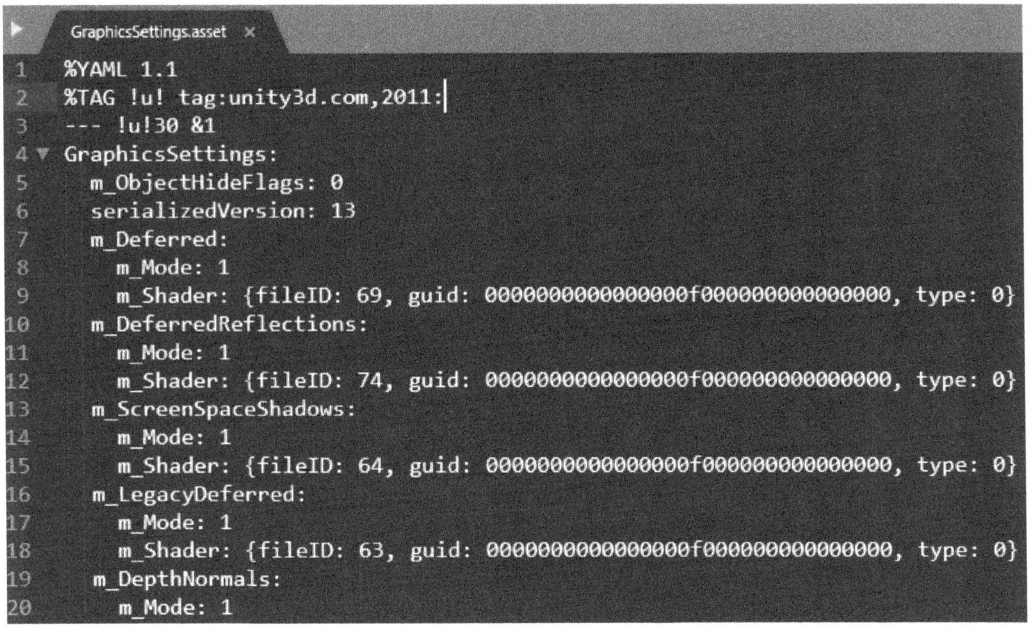

Figure 10.10 – The settings file in YAML format

So far, we've discussed Unity's binary serialization and text-based YAML serialization, but we haven't covered the JSON serialization provided by Unity yet. Next, let's take a look at JSON serialization in Unity.

## JsonUtility class and JSON serialization in Unity

If you have previous experience of developing .NET projects, you are probably familiar with JSON serialization. You can choose the solutions provided by .NET, such as using the `DataContractJsonSerializer` class defined in the `System.Runtime.Serialization.Json` namespace or using the `JsonSerializer` class defined in the `System.Text.Json` namespace, and there are also solutions from the open source community, such as `Newtonsoft.Json`, which is a very popular JSON framework for .NET. Unity also provides game developers with JSON serialization capabilities in Unity development, namely, the `JsonUtility` class. We can call `JsonUtility`'s `ToJson` method to serialize an object into a JSON string, and conversely, `JsonUtility`'s `FromJson` method can deserialize a JSON string into an object. Next, let's look at an example of how to use the `JsonUtility` class in Unity:

1.  Create a new folder named `Scripts` by clicking the **Create | Folder** item in the **Project** window.

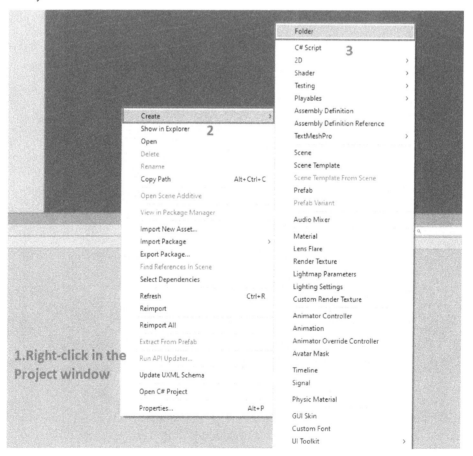

Figure 10.11 – Creating the Scripts folder

2.   Double-click on the `Scripts` folder to enter it, and then create a new C# script in this folder, name it `PlayerData`, and add the following to this script. The `PlayerData` struct is used to store the data of a player, and an object of it will be serialized to a JSON string later. And you should note that fields of the structs or classes should be `public`; otherwise, the Unity serializer will ignore these fields:

```
public struct PlayerData
{
    public string Name;
    public int Age;
    public float HP;
    public float Attack;

    public PlayerData(string name, int age, float hp,
        float attack)
    {
        Name = name;
        Age = age;
        HP = hp;
        Attack = attack;
    }
}
```

3.   Next, we also need to create another C# script in the same folder and name it `JSONSerializationSample`. The code in `JSONSerializationSample` is as follows. In the `Start` method, we create a new `PlayerData` object and assign values to its fields, and then call the `JsonUtility.ToJson` method to serialize this object into a JSON string and print the string to the `Console` window:

```
using UnityEngine;

public class JSONSerializationSample : MonoBehaviour
{
    private void Start()
    {
        var playerData = new PlayerData("player1", 50,
            100, 100);
```

```
        var jsonString =
            JsonUtility.ToJson(playerData);
        Debug.Log(jsonString);
    }
}
```

4. Create a new GameObject in the Scene, attach the `JSONSerializationSample` script to it, and run the game in the editor. The JSON string, as shown in the following screenshot, will be printed:

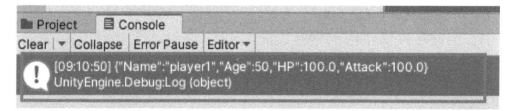

Figure 10.12 – The JSON string

5. Deserializing a JSON string to an object is fairly straightforward; you just need to call `JsonUtility.FromJson<T>`, which is a generic method. If you don't know about generic methods in C#, generic methods are methods declared with type parameters. So, let's go back to `JSONSerializationSample` and update the code in the `Start` method. This code will deserialize the JSON string into a new object, and the object's `Name` field will be printed in the **Console** window:

```
using UnityEngine;

public class JSONSerializationSample : MonoBehaviour
{
    private void Start()
    {
        var playerData = new PlayerData("player1", 50,
            100, 100);
        var jsonString =
            JsonUtility.ToJson(playerData);
        Debug.Log(jsonString);
```

```
        var deserializedObject =
        JsonUtility.FromJson<PlayerData>(jsonString);
        Debug.Log(deserializedObject.Name);
    }
}
```

6.  Run the game in the editor. The name of this player is printed as shown in the following screenshot:

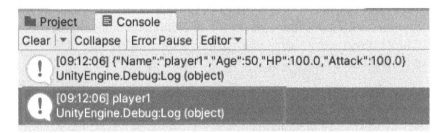

Figure 10.13 – Deserializing the JSON string

7.  If you want PlayerData as a field of another class and you want to serialize this class, PlayerData needs to be marked with the [System.Serializable] attribute, otherwise, PlayerData as a field won't be serialized correctly. So, let's go back to PlayerData and update the code to add the [System.Serializable] attribute:

```
[System.Serializable]
public struct PlayerData
{
    //No Change
}
```

Now that you know how to use the JsonUtility class to serialize an object to a JSON string and deserialize a JSON string to an object in Unity, it's time to discuss the advantages and limitations of Unity's JsonUtility class.

## Advantages and limitations of Unity's JsonUtility class

Let's start with the advantages of Unity's `JsonUtility` class. Using the `JsonUtility` class in Unity can achieve relatively high performance in terms of serializing and deserializing JSON. The `ToJson` method and the `FromJson` method of `JsonUtility` use the Unity serializer internally, and it has better support for some built-in types of Unity, such as `Vector2` and `Vector3`. In addition, since it is provided by the Unity game engine, there is no need to install additional packages.

However, `JsonUtility` has limited functionality compared to other popular JSON frameworks such as `Newtonsoft.Json`. The two most obvious limitations are that `JsonUtility` does not support the serialization of dictionaries and that the root element must be an object, not an array or a list. Let's look at an example of the limitations of the `JsonUtility` class:

1.  Create a new C# script in the `Scripts` folder, name it `TeamData`, and add the following to this script. As shown in the following code, this class has two fields, a `PlayerData` list and a dictionary:

```csharp
using System.Collections.Generic;

public class TeamData
{
    public List<PlayerData> Players;
    public Dictionary<string, PlayerData> Roles;

    public TeamData()
    {
        Players = new List<PlayerData>();
        Roles = new Dictionary<string, PlayerData>();
    }
}
```

2.  Next, we also need to create another C# script in the same folder
    and name it `JsonUtilityLimitationsSample`. The code in
    `JsonUtilityLimitationsSample` is as follows. In the `Start` method, we
    create a new `TeamData` object, add an element to the `Players` list, and add a key
    and value to the `Roles` dictionary. Then, call the `JsonUtility.ToJson` method
    to serialize this object into a JSON string and print the string to the **Console** window:

```
using UnityEngine;

public class JsonUtilityLimitationsSample :
  MonoBehaviour
{
    private void Start()
    {
        var playerData = new PlayerData("player1", 50,
          100, 100);
        var teamData = new TeamData();
        teamData.Players.Add(playerData);
        teamData.Roles.Add("leader", playerData);

        var jsonStringFromTeamData =
          JsonUtility.ToJson(teamData);
        Debug.Log(jsonStringFromTeamData);
    }
}
```

3.  Run the game in the editor; you can find that only the `Players` list is serialized,
    but the `Roles` dictionary is not serialized as expected, as shown in the following
    screenshot. This is because `JsonUtility` does not support serializing dictionaries
    in Unity.

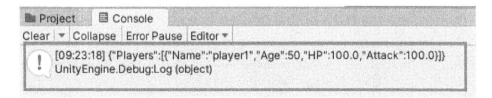

Figure 10.14 – The Roles dictionary is not serialized

4.  Then, let's go back to JsonUtilityLimitationsSample and update the code
    in the Start method to try to serialize the Players list individually:

```
public class JsonUtilityLimitationsSample :
  MonoBehaviour
{

    private void Start()
    {
        // No Change
        var jsonStringFromList =
          JsonUtility.ToJson(teamData.Players);
        Debug.Log(jsonStringFromList);

    }

}
```

5.  Run the game in the editor again and you will find that the Players list is not
    serialized this time, as shown in the following screenshot. This is because if using
    JsonUtility for serialization, the root element must be an object, not an array
    or list.

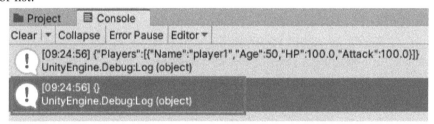

Figure 10.15 – The Players list is not serialized

## Newtonsoft.Json framework

It is a real headache to encounter the problems mentioned in the preceding example
during development, so some other JSON frameworks may also be worth trying. Next,
we will use Newtonsoft.Json to modify the preceding example so that the Roles
dictionary in the TeamData class and the individual Players list can be serialized
into JSON strings correctly:

1.  First, if the Newtonsoft.Json package is not installed in your project, you can
    install it through Unity's Package Manager. You can open it by clicking the **Window
    | Package Manager** item in the toolbar.

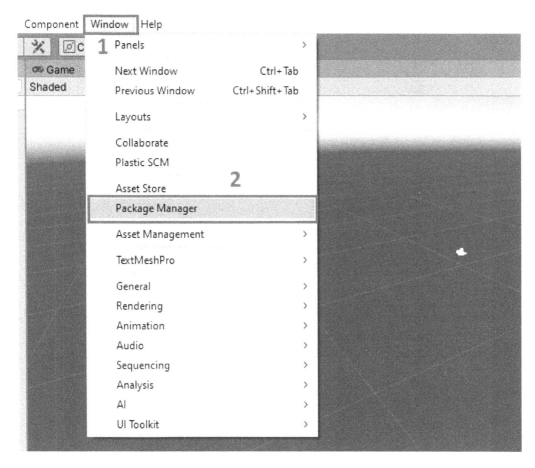

Figure 10.16 – Opening Package Manager

2.  Then, click the + in the upper-left corner to open the drop-down menu, and select the **Add package from git URL…** item in the drop-down menu.

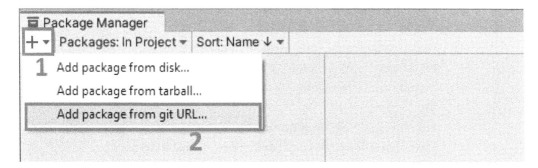

Figure 10.17 – Add package from git URL

3.  Enter `com.unity.nuget.newtonsoft-json` in the input box that appears, click the **Add** button, and wait for Package Manager to install this package.

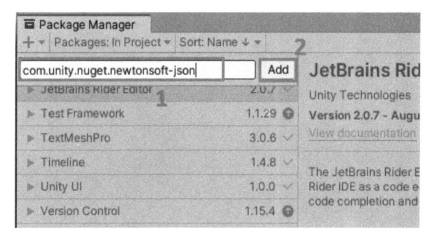

Figure 10.18 – Adding Newtonsoft.Json

4.  After installing the package in the project, we can use the `Newtonsoft.Json` framework in our C# script, so let's go back to `JsonUtilityLimitationsSample.cs` and update the code:

```
using UnityEngine;
using Newtonsoft.Json;

public class JsonUtilityLimitationsSample :
  MonoBehaviour
{
    private void Start()
    {
      var playerData = new PlayerData("player1", 50,
        100, 100);
      var teamData = new TeamData();
      teamData.Players.Add(playerData);
      teamData.Roles.Add("leader", playerData);

      var jsonStringFromTeamData =
        JsonConvert.SerializeObject(teamData);
      Debug.Log(jsonStringFromTeamData);
```

```
        var jsonStringFromList =
            JsonConvert.SerializeObject(teamData.Players);
        Debug.Log(jsonStringFromList);
        }
    }
```

Let's break down the code as follows:

- We add the `Newtonsoft.Json` namespace with the `using` keyword, which provides classes and methods for JSON serialization and deserialization.

- In the `Start` method, we replace the `JsonUtility.ToJson` method with the `JsonConvert.SerializeObject` method that is defined in the `Newtonsoft.Json` namespace.

5. Run the game. You will find that the `Roles` dictionary field of the `TeamData` object is serialized as expected, while the `Players` list as the root element is also serialized correctly.

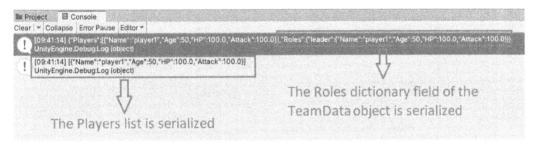

Figure 10.19 – Newtonsoft.Json works correctly

In this section, we have explained what Unity's serialization system is and how to use JSON serialization in your Unity project. Now I think you're ready to continue exploring how assets in your game project are managed by the Unity engine!

# The assets workflow in Unity

Unity's assets workflow is another very interesting topic that is also very closely related to serialization. So, what is an `asset` in Unity? If you look at a Unity project, you will find that there is a folder called `Assets` in the root directory of this project, and an asset is a file stored in this folder.

In Unity development, assets can be divided into the following two categories according to their sources:

- External assets that are imported into Unity; the most common in this case are models, textures, and audio. They are often created by third-party tools, such as Maya, 3Ds Max, and Photoshop, and then imported into Unity for use.
- Assets created by Unity itself, such as Prefab and Scene files.

Whether it's an imported asset or an asset created by Unity, Unity does the following three things with them:

1. Unity will assign a GUID to this asset.
2. Then, a meta file will be created automatically by Unity to store some additional information about the asset, such as the GUID and the import settings of this asset. *Figure 10.20* shows an example of an automatically created meta file. When a PNG file named SampleTexture is imported into the Unity project, Unity automatically creates a meta file and names it SampleTexture.PNG.meta.

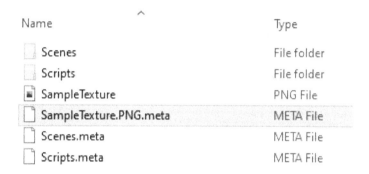

Figure 10.20 – A meta file

3. Finally, Unity will process the asset file, convert its content into an internal representation in Unity, and store the internal representation in the Library folder in the project root. We will cover this in detail when we introduce the Library folder later.

pter10-SerializationAndAssetsManagement  ›  Library  ›  Artifacts  ›  00

| Name | Type |
| --- | --- |
| 00a3fb15f588b7cb0b6eae2df697c888 | File |
| 00d5974b5138b9f6567d126ce3bca64f | File |
| 00e1d866828f41c19cab319ce64a405d | File |
| 00e4d91b4edb06f4779f920ec758e587 | File |
| 004d04e9ca86f273965b8bf7613e9ea9 | File |
| 005a83b308abcac14a36f25acac9caec | File |
| 008aa1ccf2e6d32c0afc90af59be1dd2 | File |
| 00342a1f96d4b2d1af4890865196315e | File |
| 004779a5e5d4b92873e9dadcc338dadc | File |
| 000771635a267e72a29a00000acbbf5c | File |

Figure 10.21 – The Library folder

Armed with an understanding of Unity's assets workflow, let's introduce the three things involved in this workflow in more detail: GUID and File ID, meta files, and the `Libary` folder.

# GUID and File ID

GUID and File ID are obviously an important topic when we discuss Unity's asset workflow. This is because no matter whether we use Unity to create an asset or import an external asset, Unity has to uniquely identify this asset, and this unique value is the GUID. File ID is often used together with GUID; it is not used to identify an asset like GUID, but is used to identify a reference to another object within an object.

Now that we have a brief understanding of GUID and File ID, it's time to move on to exploring GUID and File ID in more detail!

## GUID

As we just mentioned, Unity assigns a GUID to each asset in the `Assets` folder as the asset's identifier. We can use a text editor to open the meta file associated with this asset to find the GUID of this asset within the Unity engine.

Let's now perform the following steps to create a new C# script as an asset and check the GUID of this C# script in Unity:

1.  Create a new C# script in the Scripts folder, name it AssetSample, and add the following to this script. As shown in the following code, this class has a Texture field:

    ```
    public class AssetSample : MonoBehaviour
    {
        [SerializeField]
        private Texture _texture;
    }
    ```

2.  A meta file called AssetSample.cs.meta is created next to the C# script file in the file explorer, as shown in the following screenshot:

Figure 10.22 – The AssetSample.cs.meta file

3.  Open the AssetSample.cs.meta file in a text editor, and you will discover that the GUID of this C# script asset in Unity is e35f96b75211edd4bad6451a26675090, as shown in the following screenshot:

```
AssetSample.cs.meta    ×
fileFormatVersion: 2
guid: e35f96b75211edd4bad6451a26675090
MonoImporter:
    externalObjects: {}
    serializedVersion: 2
    defaultReferences: []
    executionOrder: 0
    icon: {instanceID: 0}
    userData:
    assetBundleName:
    assetBundleVariant:
```

Figure 10.23 – The GUID of this C# script

After reading this, you should know how to find the GUID of an asset in Unity; however, where is the File ID stored, and how does Unity use it to create and maintain references between objects? So, let's continue our journey with another example.

## File ID

We mentioned earlier that Unity uses a `File ID` to refer to another object within an object, which is the unique ID of the object referenced within that object.

Now, let's take a look at an example to learn how to find the `File IDs` and how Unity uses the `File IDs` to maintain the reference relationship between objects. In this example, we will still use the `AssetSample` script we just created, so now let's get started!

1.  First, create a new GameObject in the Scene and name it `AssetSampleGameObject`. You already know that a `Transform` component is automatically created and attached to this GameObject, as shown in *Figure 10.24*:

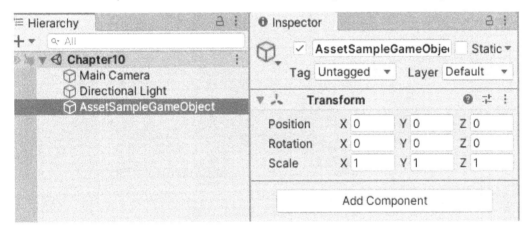

Figure 10.24 – Creating an AssetSampleGameObject

2. Attach an `AssetSample` component to `AssetSampleGameObject`, and then assign a texture from the **Project** window to the `Texture` field of `AssetSample`. Then, attach another `AssetSample` component to the same GameObject; however, this time, we set the **Texture** field of `AssetSample` to **None** and save the Scene.

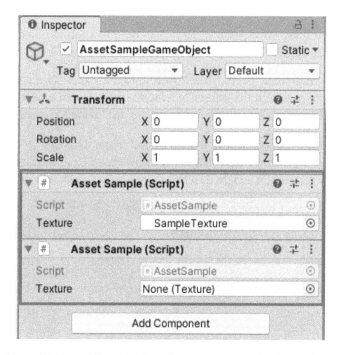

Figure 10.25 – Adding AssetSample components to the GameObject

3. Make sure your project's **Asset Serialization** mode is now **Force Text** (we covered this topic in the *YAML and binary serialization in Unity* section), and then use a text editor to open the Scene file from **File Explorer**. You will see a lot of content in the Scene file, as shown in the following screenshot:

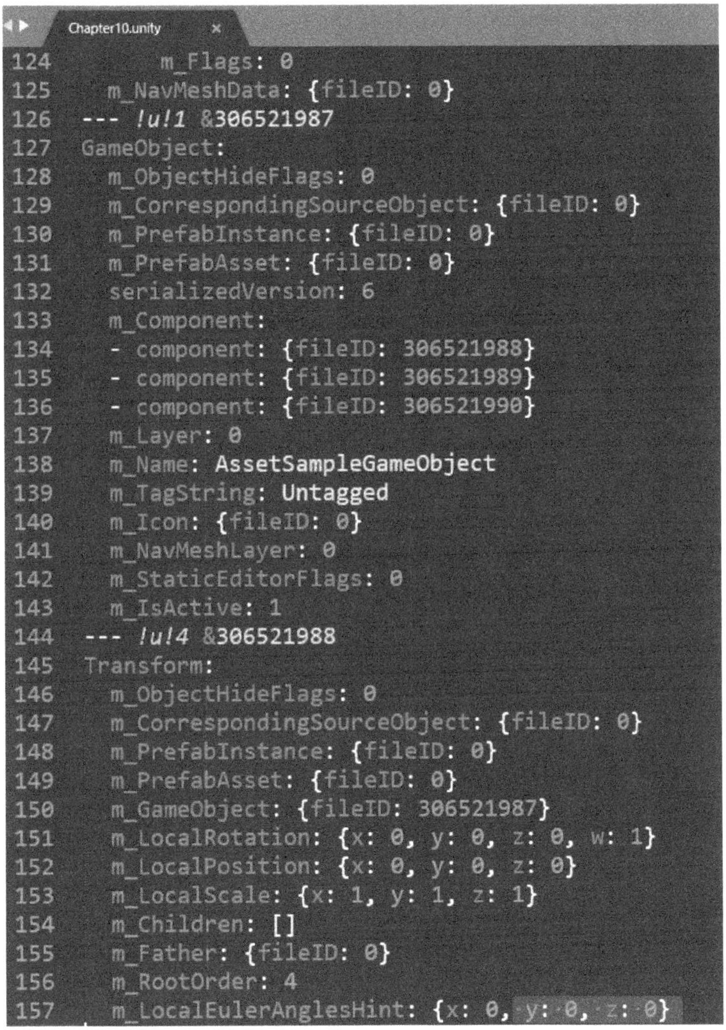

```
     Chapter10.unity      ✕
124         m_Flags: 0
125     m_NavMeshData: {fileID: 0}
126   --- !u!1 &306521987
127   GameObject:
128     m_ObjectHideFlags: 0
129     m_CorrespondingSourceObject: {fileID: 0}
130     m_PrefabInstance: {fileID: 0}
131     m_PrefabAsset: {fileID: 0}
132     serializedVersion: 6
133     m_Component:
134     - component: {fileID: 306521988}
135     - component: {fileID: 306521989}
136     - component: {fileID: 306521990}
137     m_Layer: 0
138     m_Name: AssetSampleGameObject
139     m_TagString: Untagged
140     m_Icon: {fileID: 0}
141     m_NavMeshLayer: 0
142     m_StaticEditorFlags: 0
143     m_IsActive: 1
144   --- !u!4 &306521988
145   Transform:
146     m_ObjectHideFlags: 0
147     m_CorrespondingSourceObject: {fileID: 0}
148     m_PrefabInstance: {fileID: 0}
149     m_PrefabAsset: {fileID: 0}
150     m_GameObject: {fileID: 306521987}
151     m_LocalRotation: {x: 0, y: 0, z: 0, w: 1}
152     m_LocalPosition: {x: 0, y: 0, z: 0}
153     m_LocalScale: {x: 1, y: 1, z: 1}
154     m_Children: []
155     m_Father: {fileID: 0}
156     m_RootOrder: 4
157     m_LocalEulerAnglesHint: {x: 0, y: 0, z: 0}
```

Figure 10.26 – Opening the Scene file in a text editor

This file gives us a lot of information, recording the GameObjects, components, and referenced assets in the Scene. So let's break it down:

- First of all, we can find the record of the GameObject called AssetSampleGameObject in the file. In the following screenshot, you can see that there are three components attached to this GameObject, with File IDs of 306521988, 306521989, and 306521990, respectively:

```
126   --- !u!1 &306521987
127   GameObject:
128     m_ObjectHideFlags: 0
129     m_CorrespondingSourceObject: {fileID: 0}
130     m_PrefabInstance: {fileID: 0}
131     m_PrefabAsset: {fileID: 0}
132     serializedVersion: 6
133     m_Component:
134     - component: {fileID: 306521988}
135     - component: {fileID: 306521989}
136     - component: {fileID: 306521990}
137     m_Layer: 0
138     m_Name: AssetSampleGameObject
139     m_TagString: Untagged
140     m_Icon: {fileID: 0}
141     m_NavMeshLayer: 0
142     m_StaticEditorFlags: 0
143     m_IsActive: 1
```

Figure 10.27 – The AssetSampleGameObject record

- If we search these three File IDs, we can find records for three components in this file – a Transform component, which is created and attached to this GameObject when the GameObject is created, and two MonoBehaviour components, which represent C# script components.

```
144   --- !u!4 &306521988
145 ▼ Transform:
146     m_ObjectHideFlags: 0
147     m_CorrespondingSourceObject: {fileID: 0}
148     m_PrefabInstance: {fileID: 0}
149     m_PrefabAsset: {fileID: 0}
150     m_GameObject: {fileID: 306521987}
151     m_LocalRotation: {x: 0, y: 0, z: 0, w: 1}
152     m_LocalPosition: {x: 0, y: 0, z: 0}
153     m_LocalScale: {x: 1, y: 1, z: 1}
154     m_Children: []
155     m_Father: {fileID: 0}
156     m_RootOrder: 4
157     m_LocalEulerAnglesHint: {x: 0, y: 0, z: 0}
158   --- !u!114 &306521989
159 ▼ MonoBehaviour:
160     m_ObjectHideFlags: 0
161     m_CorrespondingSourceObject: {fileID: 0}
162     m_PrefabInstance: {fileID: 0}
163     m_PrefabAsset: {fileID: 0}
164     m_GameObject: {fileID: 306521987}
165     m_Enabled: 1
166     m_EditorHideFlags: 0
167     m_Script: {fileID: 11500000, guid: e35f96b75211edd4bad6451a26675090, type: 3}
168     m_Name:
169     m_EditorClassIdentifier:
170       texture: {fileID: 2800000, guid: 908f698325672e74ebdabdc3c2c05477, type: 3}
171   --- !u!114 &306521990
172 ▼ MonoBehaviour:
173     m_ObjectHideFlags: 0
174     m_CorrespondingSourceObject: {fileID: 0}
```

Figure 10.28 – File IDs

- So, what is the difference between File ID and GUID? If we focus on these two `MonoBehaviour` components, we can see that the `m_Script` field of both components references the same C# script with a GUID of `e35f96b75211edd4bad6451a26675090`.

```
158  --- !u!114 &306521989
159  MonoBehaviour:
160    m_ObjectHideFlags: 0
161    m_CorrespondingSourceObject: {fileID: 0}
162    m_PrefabInstance: {fileID: 0}
163    m_PrefabAsset: {fileID: 0}
164    m_GameObject: {fileID: 306521987}
165    m_Enabled: 1
166    m_EditorHideFlags: 0
167    m_Script: {fileID: 11500000, guid: e35f96b75211edd4bad6451a26675090, type: 3}
168    m_Name:
169    m_EditorClassIdentifier:
170    _texture: {fileID: 2800000, guid: 908f698325672e74ebdabdc3c2c05477, type: 3}
171  --- !u!114 &306521990
172  MonoBehaviour:
173    m_ObjectHideFlags: 0
174    m_CorrespondingSourceObject: {fileID: 0}
175    m_PrefabInstance: {fileID: 0}
176    m_PrefabAsset: {fileID: 0}
177    m_GameObject: {fileID: 306521987}
178    m_Enabled: 1
179    m_EditorHideFlags: 0
180    m_Script: {fileID: 11500000, guid: e35f96b75211edd4bad6451a26675090, type: 3}
181    m_Name:
182    m_EditorClassIdentifier:
183    _texture: {fileID: 0}
```

Figure 10.29 – The MonoBehaviour components

Therefore, we can find that although these two component objects refer to the same C# script, namely, `AssetSample`, they are two different instances of `AssetSample`; the file ID of the first `MonoBehaviour` component object is `306521989`, and the file ID of the second `MonoBehaviour` component object is `306521990`.

Moreover, the `_texture` field of one instance refers to a texture asset, and the `_texture` field of the other instance does not refer to any texture asset.

By reading this section, we learned that Unity uses GUID to identify an asset and File ID to identify a referenced object.

# Meta files

We already know that a meta file records the GUID of its associated asset in a Unity project, and that a meta file also records the import settings of this asset. In this section, we will talk about meta files that look inconspicuous but are actually very important.

## Meta files and version management

A common mistake developers new to Unity make is not paying attention to these autogenerated meta files. One such example is ignoring meta files when using Git or other version control systems to manage the version of the Unity project.

If you remember from the previous section, Unity assigns each asset a GUID, uses this GUID to identify the asset, and records this GUID in the meta file.

So, if your version management system does not include meta files, your Unity development progress may be disrupted.

To illustrate this, let's imagine a scenario where, when a Unity project that does not contain meta files is cloned from a remote repository to your colleague's local machine, the Unity Editor will reimport those assets and assign them new GUIDs and create meta files to store this information. As a result, references that previously existed between objects in your Unity project will no longer be valid.

As an example, assuming that the `AssetSample.cs.meta` meta file of the `AssetSample` C# script we created earlier is not managed by the version management system, then you will encounter the `Script Missing` error, as shown in *Figure 10.30*, after cloning and opening the project on another computer:

Figure 10.30 – The Script Missing error

At this point, the script actually exists, but since its GUID has been regenerated, the previous reference relationship is invalid.

Therefore, when developing a Unity project, please make sure that the meta files are managed by your version management tool.

## Import settings in meta files

In addition to storing the GUID of an asset, a meta file also stores the import settings of this asset. Of course, the meta files that will be discussed in this subsection mainly refer to the meta files of assets created in third-party software and that are then imported into the Unity Editor, such as models, textures, and audio.

Let's use a meta file of an audio asset as an example to see how the import settings of the asset are saved.

The audio asset we are using here is from Unity's Asset Store and you can download it from here: `https://assetstore.unity.com/packages/audio/sound-fx/weapons/ultra-sci-fi-game-audio-weapons-pack-vol-1-113047`.

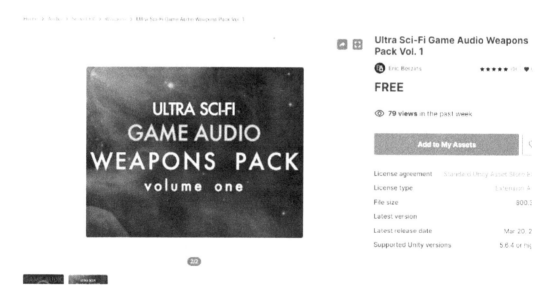

Figure 10.31 – Audio pack

After importing the audio into the Unity project, we can select the first audio file in the Ultra SF Game Audio Weapons Pack v.1 folder to open the audio's Inspector window in the Unity Editor, which shows the asset's import settings. Then we use a text editor to open the meta file of the same audio asset in the folder explorer and, as shown in *Figure 10.32*, we can see that AudioImporter in the meta file corresponds to the import settings in the editor:

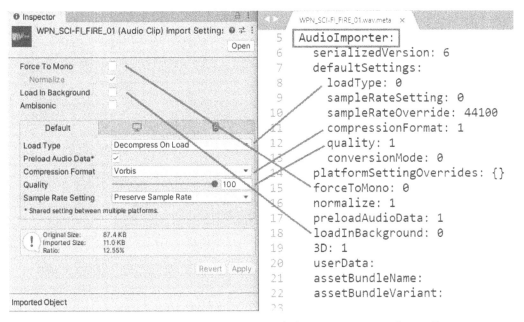

Figure 10.32 – WPN_SCI-FI_FIRE_01 audio's import settings and meta file

The import settings of a texture asset and a model asset are also stored in their meta files. The following screenshot shows the import settings for a texture and a model:

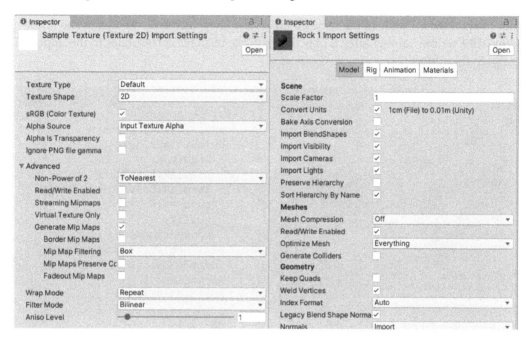

Figure 10.33 – Import settings of a texture (left) and the import settings of a model (right)

Since the meta file stores the import settings of the asset, once we modify the import settings of the asset in the Unity Editor, the corresponding meta file will be updated.

The import settings often affect how Unity processes these assets, so it is important to ensure that the import settings can be managed according to the requirements of the project. For example, in many mobile game projects, we should check the **Force To Mono** option on the audio import settings to reduce the memory usage of this audio file.

Next, let's take a look at how to manage import settings through a C# script in Unity.

## The AssetPostprocessor class and the import pipeline

Unity provides the AssetPostprocessor class for game developers to hook into the assets import pipeline in Unity. When importing an asset, we can manage the import pipeline according to the asset type.

In the following example, we will create a new C# script to set the **Force To Mono** option enabled in the import settings of all audio files in the Unity projec:

1.  Create a subfolder in the `Scripts` folder and name it `Editor`. This is because the C# class that we will create inherits from the `AssetPostprocessor` class, which is a class for the editor, so it needs to be placed in an `Editor` folder.

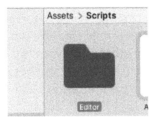

Figure 10.34 – Creating an Editor folder

2.  Double-click on the `Editor` folder to enter it, create a new C# script in this folder, name it `AssetImporterSample`, and then add the following to this script:

```csharp
using UnityEditor;

public class AssetImporterSample : AssetPostprocessor
{
    private void OnPreprocessAudio()
    {
        var audioImporter =
            (AudioImporter)assetImporter;
        if(audioImporter == null)
        {
            return;
        }
        audioImporter.forceToMono = true;
        audioImporter.SaveAndReimport();
    }
}
```

Let's break down how this works:

- First, the code is using the `UnityEditor` namespace. This is because the `AssetPostprocessor` class is defined in this namespace, which also means that the `AssetImporterSample` C# script is used in the Unity Editor and not at runtime.

- The `AssetImporterSample` class inherits the `AssetPostprocessor` class and implements the `OnPreprocessAudio` method, which will be called before the audio asset is imported. We can also implement other similar methods to be called when other asset types will be imported. For example, the `OnPreprocessTexture` method will be called before the texture asset is imported, and the `OnPreprocessModel` method will be called before the model asset is imported.

- In the `OnPreprocessAudio` method, we can get an instance of `AudioImporter`, set the `forceToMono` option to `true`, and then save and re-import the asset to ensure that the new import settings for the asset take effect.

3.  Save the C# script and the Unity Editor should modify the import settings of these audio assets in the project and then re-import them, as shown in *Figure 10.35*:

Figure 10.35 – Importing audio assets

4.  Let's now select an audio file to check its import settings. As shown in *Figure 10.36*, the new import settings work as expected:

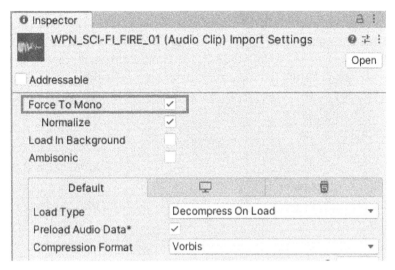

Figure 10.36 – New import settings

In this subsection, we introduced how to use C# code to manage the asset import pipeline. Next, let's explore another assets workflow topic in Unity – the Library folder.

## The Library folder

In a Unity project, Unity will process and convert the external assets into Unity internal format assets and save them in the Library folder. Because the data stored in the Library folder is cached data that can always be regenerated from the source asset files based on the import settings, the Library folder should generally not be included in a version management system.

> **Note**
>
> In addition to the Library folder, there are some other Unity folders that need to be excluded from version management, including Temp, Obj, and Logs. If you are using Git as your version management tool, you can find the .gitignore file for Unity projects at this link: https://github.com/github/gitignore/blob/main/Unity.gitignore.

You can find the Library folder in the root directory of your Unity project, as shown in *Figure 10.37*. If there is no Libary folder in the root directory of your Unity project, you need to open the project with the Unity Editor. The Unity Editor will import the assets in the Assets folder and generate the Library folder automatically.

| Name | Type |
| --- | --- |
| .git | File folder |
| .vs | File folder |
| Assets | File folder |
| Library | File folder |
| Logs | File folder |

Figure 10.37 – The Library folder

Double-click the Library folder to enter it and you will see the ScriptAssemblies subfolder, which saves the assemblies of the C# code in the project, and you can also see the PackageCache subfolder, which saves the cache of Unity packages used by the project. In addition to these, you also can see the Artifacts subfolder, where the assets processed by Unity are saved.

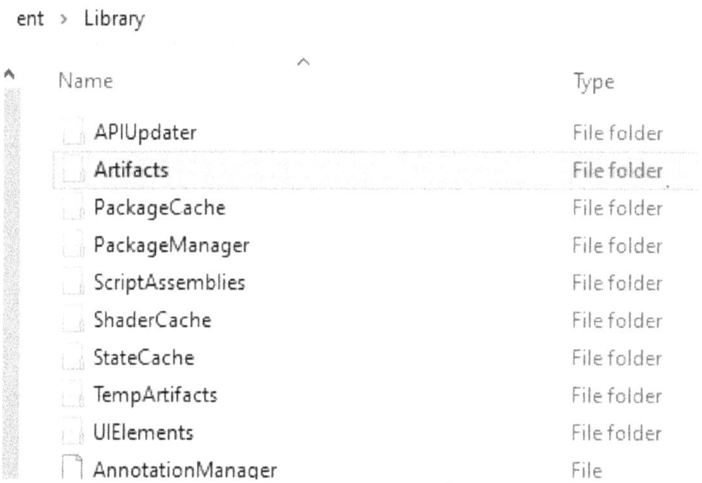

ent  >  Library

| Name | Type |
| --- | --- |
| APIUpdater | File folder |
| Artifacts | File folder |
| PackageCache | File folder |
| PackageManager | File folder |
| ScriptAssemblies | File folder |
| ShaderCache | File folder |
| StateCache | File folder |
| TempArtifacts | File folder |
| UIElements | File folder |
| AnnotationManager | File |

Figure 10.38 – The Artifacts folder

In this section, we introduced Unity's assets workflow, covering topics such as GUIDs, File IDs, meta files, and the Library folder. Next, let's take a look at the special folders created and managed by developers related to assets management in Unity.

# Introducing the special folders in Unity

We already covered some of these special folders related to scripting in Unity in *Chapter 2, Scripting Concepts in Unity*. In this section, we will introduce the remaining special folders, which are related to asset management in Unity.

## Resources folder

First, let's take a look at the `Resources` folder in Unity. `Resources` is a special folder name in Unity, but Unity does not automatically create a `Resources` folder for you. If you want to use a `Resources` folder to manage assets, you need to create it yourself. It should be noted that there can be multiple `Resources` folders in the `Assets` directory in a Unity project.

Unity provides the `Resources.Load` method to load assets in `Resources` folders. Next, we will use an example to learn how to use `Resources` folders to manage assets:

1. Create a new folder named `Resources` by clicking the **Create | Folder** item in the **Project** window.

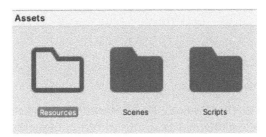

Figure 10.39 – Creating a Resources folder

2. Create an empty GameObject, name it `SamplePrefab`, and drag it into the `Resources` folder to create a new prefab, as shown in *Figure 10.40*:

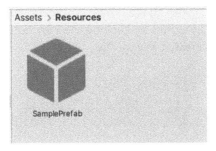

Figure 10.40 – SamplePrefab

3.  Create a new C# script in the **Scripts** folder, name it `ResourcesLoadExample`, and add the following to this script:

```csharp
using UnityEngine;

public class ResourcesLoadExample : MonoBehaviour
{
    private GameObject _gameObjectInstance;

    private void Start()
    {
        var samplePrefab =
            Resources.Load<GameObject>("SamplePrefab");
        if (samplePrefab != null)
        {
            _gameObjectInstance =
                Instantiate(samplePrefab);
        }
    }
}
```

Let's break down how this works:

I.    In the `Start` method, we call the `Resources.Load` method and pass the path to the asset to load as an argument to this method, which is `SamplePrefab`.

II.   Then, if the prefab asset is loaded, we instantiate it to create a new GameObject in the game Scene.

4.  Create a new GameObject and attach the `ResourcesLoadExample` script to it. Run the game in the Unity Editor by clicking the **Play** button. We can see that a new instance of the prefab is created as expected.

Figure 10.41 – Loading assets from the Resources folders

Through this example, we see that using `Resources` folders to manage assets is very convenient, especially when you need to develop a prototype quickly, but managing assets in a Unity project by using `Resources` folders is not recommended for the following reasons:

- When the Unity Editor builds the game, the assets in `Resources` folders will be included in the build, even if the assets are not used, so improper use of the `Resources` folder may cause the build game to be too large. In addition, it will also affect the game's startup speed.

- Using `Resources` folders will make incremental content upgrades to the game very difficult or impossible.

Now that we have an understanding of the `Resources` folders, we know the situations in which they will be suitable, such as developing a rapid prototype, as well as its limitations.

Next, we will continue to introduce another special folder in Unity, namely, `StreamingAssets`

## StreamingAssets folder

In Unity, `StreamingAssets` is also a special folder name. We actually already covered this in *Chapter 6, Integrating Audio and Video in a Unity Project*. In this subsection, we will discuss it in more detail.

We mentioned earlier that Unity will process assets in the `Assets` folder in a format that the Unity engine understands, but there is an exception.

The assets in the `StreamingAssets` folder in the Unity project will still be in the original format and these assets will not be built into the game along with the other assets when Unity builds the game. Instead, all assets in the folder will be copied to a specific folder on the target device.

Since the location of this special folder is different on different platforms, Unity provides the `Application.streamingAssetsPath` property so that we can access the correct path to this folder from C# code.

The following code snippet is from the example used in *Chapter 6, Integrating Audio and Video in a Unity Project*. We can see how to use `Application.streamingAssetsPath` in C# code:

```
public void OnClickSetVideoURL()
{
    _videoPlayer.url =
```

```
            Path.Combine(Application.streamingAssetsPath,
            _videoFileName);
    }
```

Similar to the `Resources` folder, Unity does not automatically create the `StreamingAssets` folder for you. If you wish to use it, you need to create it yourself, as shown in *Figure 10.42*:

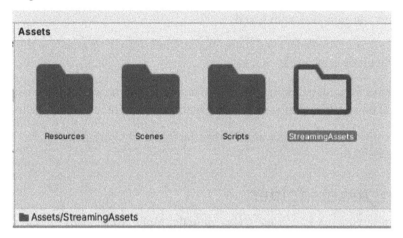

Figure 10.42 – Creating a StreamingAssets folder

Then we can place an audio WAV file in the `StreamingAssets` folder. As you can see from the following screenshot, the icon of this WAV file is not the same as the icon of an audio clip in Unity that we are already familiar with. This is because Unity does not process the WAV file; it still maintains its original format.

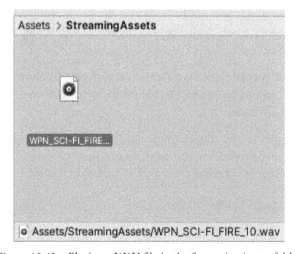

Figure 10.43 – Placing a WAV file in the StreamingAssets folder

In this section, we explored the `Resources` folder and the `StreamingAssets` folder, which are special folders in Unity, and understood that what they do can help you better develop games with Unity.

Next, we'll cover another interesting topic; how to use Azure Blob storage in the Azure Cloud with Unity's Addressable Asset system.

# Azure Blob storage with Unity's Addressable Asset system

In this section, we'll cover the Azure Blob storage service in Microsoft's Azure Cloud and how to use it with Unity's Addressable Asset system.

**Azure Blob storage** is a type of Azure Storage account in Azure. Other types of Azure Storage accounts include **queues**, **file shares**, and **tables**. Among them, Blob storage is very suitable for storing large amounts of unstructured data such as binary data.

> **Note**
>
> You can find additional information and resources about the Azure Storage account in Microsoft's Azure Cloud at `https://docs.microsoft.com/en-us/azure/storage/common/storage-introduction`.

Unity's Addressable Asset system, as the name suggests, provides a convenient method for loading specific assets, whether on the local or remote server, according to a specific address. When discussing the `Resources` folder in the previous section, we discussed various limitations when using it in terms of managing assets, and the Addressable Asset system can solve these problems very well; for example, the size of the game package can be well controlled, there is no need to include unnecessary assets in game builds, and assets can be hosted on remote servers, such as the Azure Cloud, to incrementally update assets within the game.

> **Note**
>
> Before the Addressable Asset system was introduced, developers could also use `AssetBundles` to manage assets; `AssetBundles` is beyond the scope of what we need here, but if you're interested, you can find out more at `https://docs.unity3d.com/Manual/AssetBundlesIntro.html`.

Well, now we have an understanding of Azure Blob storage and the Addressable Asset system. Next, we will explore how to use Azure Blob storage to host assets and use the Addressable Asset system to manage them.

Let's start!

# Setting up an Azure Blob storage service

First, make sure you have an available Azure subscription. You can apply for a free Azure trial account on the page introduced at the beginning of this chapter.

If everything is ready, we can create our first resource in Azure, namely, an **Azure resource group**.

## Creating a new resource group

Usually, a resource group is our first resource in the Azure Cloud. This is because a resource group is a container for holding other Azure resources.

We can create a resource group in just a few steps:

1.  Sign in to the Azure portal page with your account at `https://portal.azure.com/`.

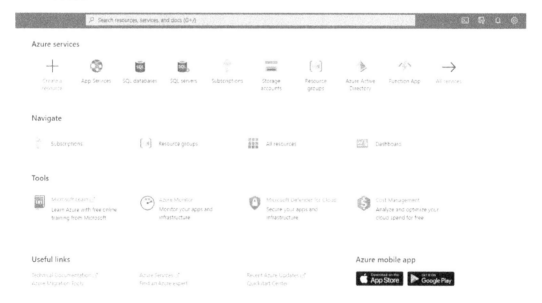

Figure 10.44 – Azure portal page

2.  The Azure portal page does not display the portal menu by default. We can click the **Show portal menu** button in the upper-left corner of the page to open the portal menu.

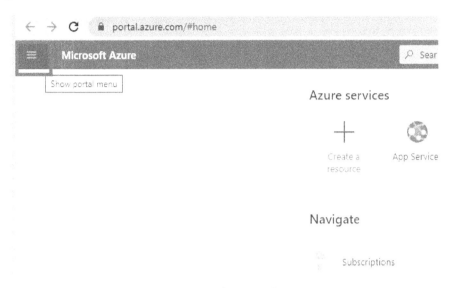

Figure 10.45 – Show portal menu

3.  Select **Resource groups** from the portal menu.

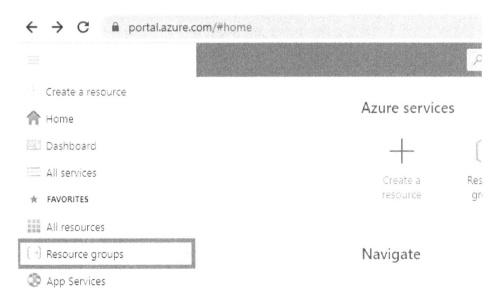

Figure 10.46 – Selecting the Resource groups service

4.  The **Resource groups** page will then open. Click the **Create** button on this page, as shown in *Figure 10.47*:

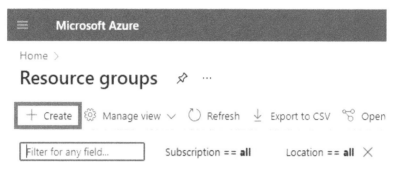

Figure 10.47 – Creating a resource group

5.  Then, you will see the **Create a resource group** page. Select your Azure subscription and enter the name of the resource group. Here, it is `rg-unitybook-dev-001`. Select the region of the resource group as `(Asia Pacific) Australia East` and then click on **Review + create** to verify the settings of this resource group and create it, as shown in *Figure 10.48*:

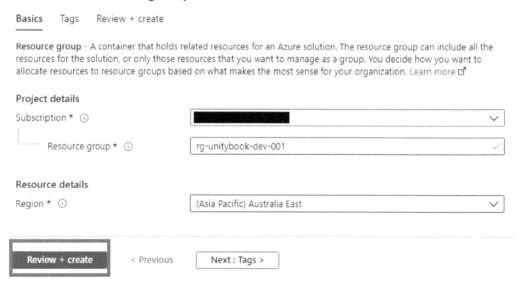

Figure 10.48 – Creating a resource group

We've created a resource group in Azure. Next, let's create an Azure Storage account resource.

> **Note**
>
> You can find additional information about the naming convention in Microsoft's Azure Cloud at `https://docs.microsoft.com/en-us/azure/cloud-adoption-framework/ready/azure-best-practices/resource-naming`.

## Creating a new Azure Storage account resource

In order to set up an Azure Blob storage service, we will need to create an Azure Storage account to provide a unique namespace in Azure for the assets that will be hosted first.

We will perform the following steps:

1. Go back to the Azure portal page, repeat the steps introduced previously to open the portal menu, and then click **Storage accounts** this time, as shown in *Figure 10.49*:

Figure 10.49 – Clicking Storage accounts

2. The Storage accounts page will then open. Click the **Create** button on this page, as shown in *Figure 10.50*:

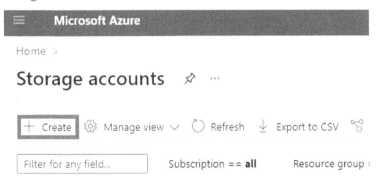

Figure 10.50 – Creating a storage account

3.  Similar to creating a resource group, on the **Create a storage account** page, we also need to select the Azure subscription first and then select the resource group we just created. Then, in the **Instance details** section, enter the name of the storage account and the location of the resource, `unitybookchapter10` and `(Asia Pacific) Australia East`, respectively. The other settings can be left as their defaults, and then click the **Review + create** button to create the resource.

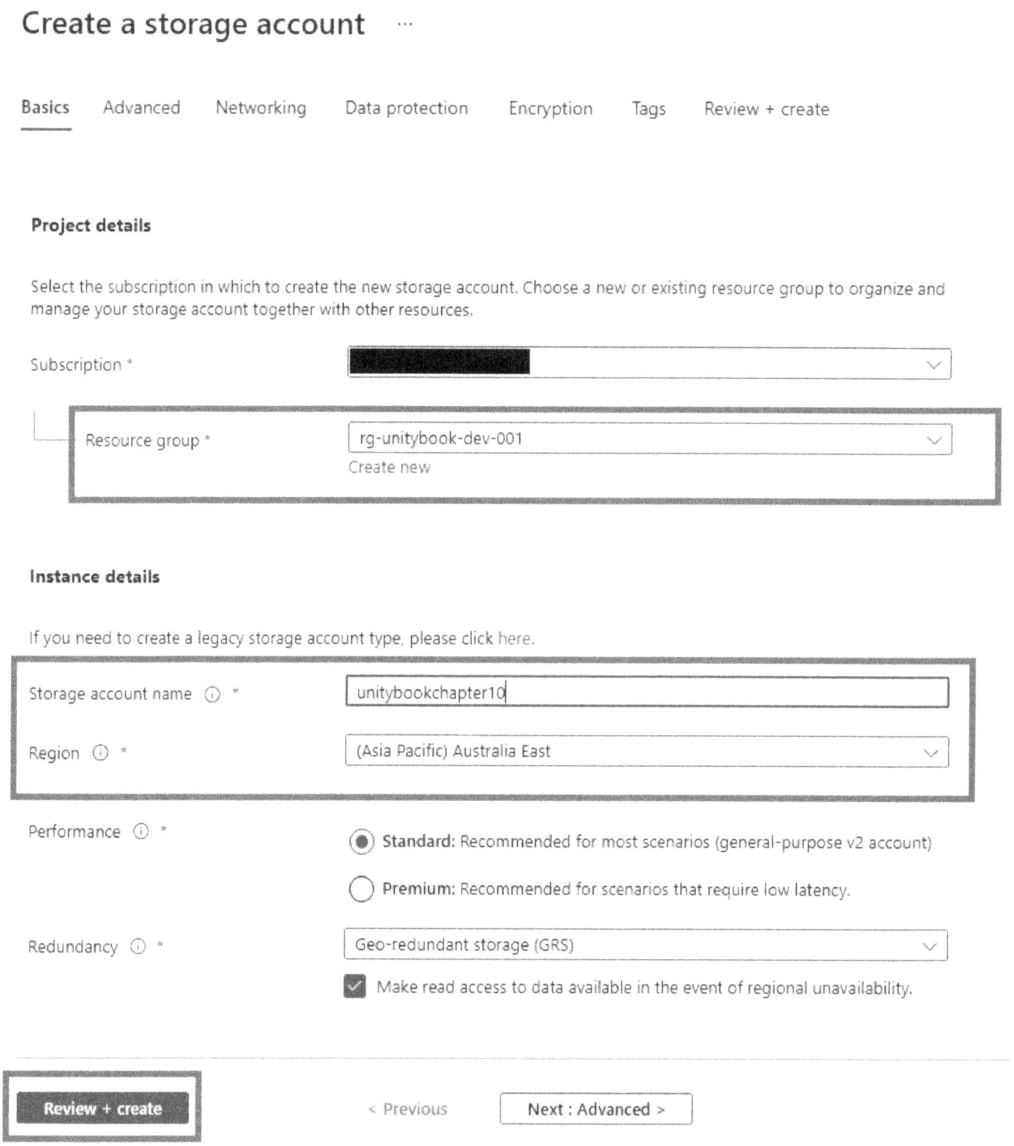

## Create a storage account  ···

Basics    Advanced    Networking    Data protection    Encryption    Tags    Review + create

### Project details

Select the subscription in which to create the new storage account. Choose a new or existing resource group to organize and manage your storage account together with other resources.

Subscription *                              ▮▮▮▮▮▮▮▮▮▮▮▮                                        ∨

Resource group *                            rg-unitybook-dev-001                                ∨
                                            Create new

### Instance details

If you need to create a legacy storage account type, please click here.

Storage account name ⓘ *                    unitybookchapter10|

Region ⓘ *                                  (Asia Pacific) Australia East                        ∨

Performance ⓘ *                             ◉ **Standard:** Recommended for most scenarios (general-purpose v2 account)

                                            ○ **Premium:** Recommended for scenarios that require low latency.

Redundancy ⓘ *                              Geo-redundant storage (GRS)                          ∨

                                            ☑ Make read access to data available in the event of regional unavailability.

[ Review + create ]              < Previous        [ Next : Advanced > ]

Figure 10.51 – Creating a storage account

4. We can click the notifications button in the upper-right corner of the page to view the progress of the resource deployment. When the resource is deployed, we can click **Go to resource** to go to the resource page.

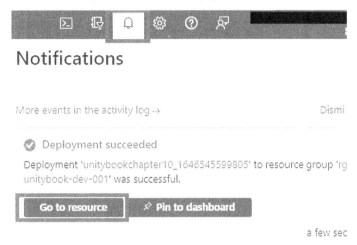

Figure 10.52 – Notifications

5. As shown in *Figure 10.53*, a Storage account named `unitybookchapter10` is created:

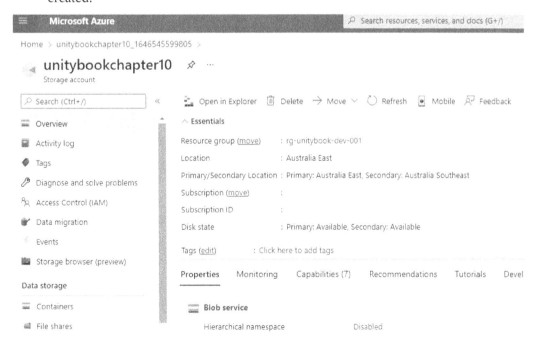

Figure 10.53 – The Storage account page

At this point, we have set up a Storage account resource in Azure. Next, let's set up Blob Storage.

## Creating a container

As we mentioned at the beginning of this section, Blob Storage is a type of Azure Storage account, so we can find the settings for Blob Storage on the Storage account page we just opened. We can perform the following steps to set up Blob Storage:

1.  First, we need to create a container, similar to a directory in the filesystem on our computers, to organize a group of files, and a container to organize a group of blobs on the Azure Cloud. Scroll down the menu on the left side of the Storage account page and, in the **Data storage** section, we can see four different storage types. Then, select **Containers**.

Figure 10.54 – Selecting containers

2.  Then, click the **+ Container** button, as shown in *Figure 10.55*:

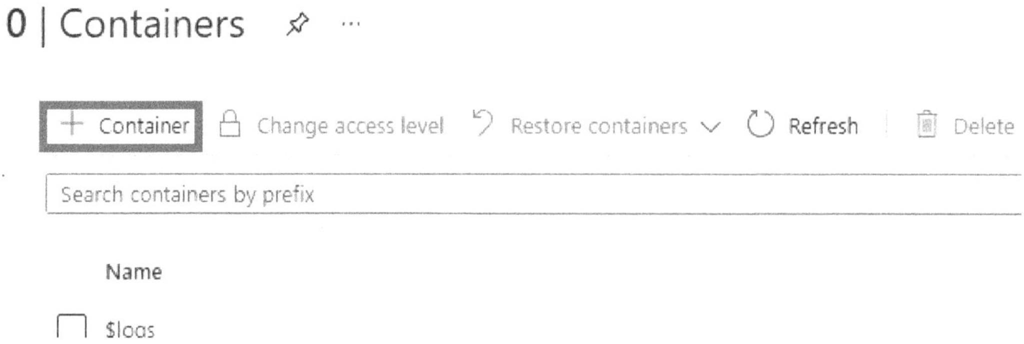

0 | Containers  📌  ⋯

    ＋ Container    🔒 Change access level    ↺ Restore containers ∨    ⟳ Refresh    🗑 Delete

Search containers by prefix

Name

☐ $loɑs

Figure 10.55 – Clicking the + Container button

3.  In the **New container** panel, we enter `remotedata` as the name of the container, and for simplicity, we set **Public access level** to **Blob** to allow anonymous access to blobs inside the container.

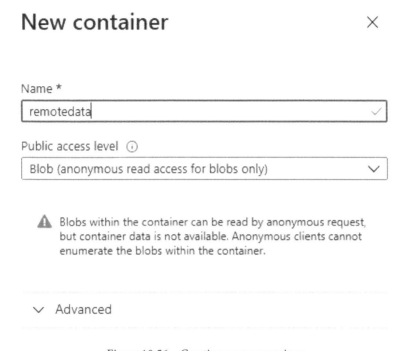

Figure 10.56 – Creating a new container

> **Note**
>
> For security purposes, you should try to manage access to blobs in a more secure way, for example, by using an **access key** for authorization, or by using a **Shared Access Signature (SAS)** to delegate access. If you're interested, you can find out more at `https://docs.microsoft.com/en-us/azure/storage/blobs/authorize-data-operations-portal`.

Now that we have set up Azure Blob storage, we also need to use the Addressable Assets system in Unity to create asset packages and deploy them to Azure.

# Installing the Addressable Assets system package

By default, the Addressable Asset system is not available in a Unity project. So, we need to install the `Addressables` package first.

As shown in *Figure 10.57*, we can find this package in Unity's Package Manager and install it in our project.

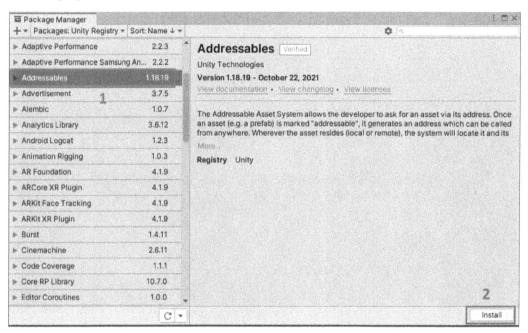

Figure 10.57 – Installing the Addressables package

Once installed, you can find the **Addressables** item in the **Window** menu of the Unity Editor.

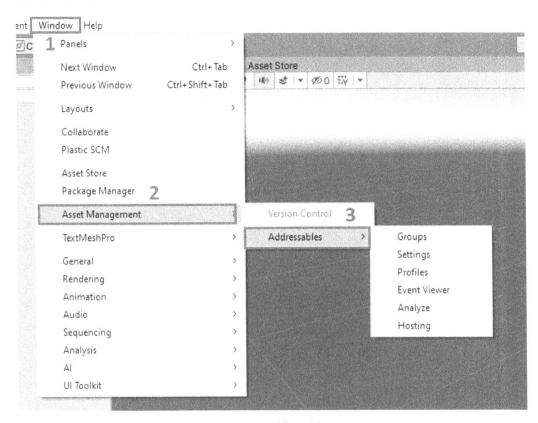

Figure 10.58 – Addressables item

Next, let's build addressable content by using the Addressable Assets system.

## Building addressable content

Building addressable content that can be hosted on the Azure Cloud sounds complicated, but we can break this task down further into the following tasks:

1.  First, mark an asset as addressable.

2.  Then, enable the remote catalog.

3.  And finally, build the content.

Now, let's move on to explore the first task.

## Marking addressable assets

In the Unity Editor, we can easily mark an asset as addressable. Before we mark an addressable asset, let's create a new asset first. We can create a new cube in the Scene, name it `SampleContentOnAzure`, and drag it into the **Project** window to create a new prefab asset.

Then, select this new prefab to open its **Inspector** window and you can see the **Addressable** checkbox in the window, as shown in *Figure 10.59*:

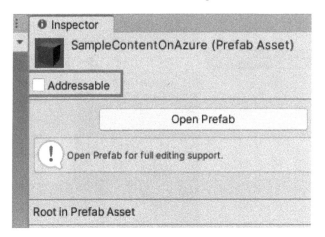

Figure 10.59 – Marking an addressable asset

By checking this checkbox, we will mark the prefab asset as addressable.

## Enabling the remote catalog

Before enabling the remote catalog in the Addressable Asset settings, we can first create a new profile that defines variables such as `RemoteLoadPath`.

### Creating a profile

So, let's start by creating a profile using the following steps:

1.  In the toolbar, click on **Window | Asset Management | Addressables | Profiles**.

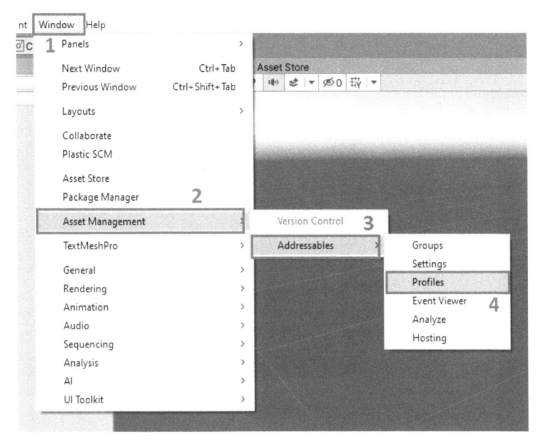

Figure 10.60 – Opening the Profiles window

2.  In the **Addressables Profiles** window, click the **Create** button and select **Profile** in the drop-down menu to create a new profile.

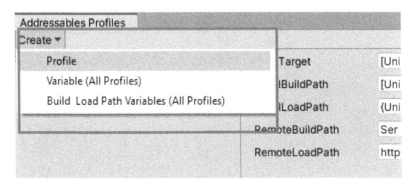

Figure 10.61 – Creating a new profile

3.  Then, rename this new profile to **AzureCloud** and enter the URL of the Azure Blob container in relation to the **RemoteLoadPath** variable.

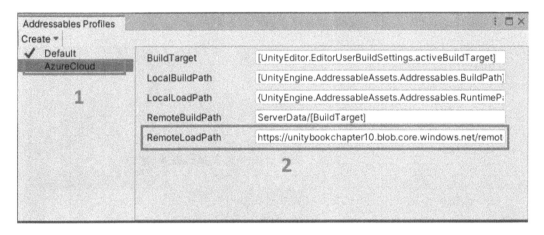

Figure 10.62 – Setting up the new profile

If you don't know the URL of the Azure Blob container, you can find it on the container's **Properties** page in Azure, as shown in the following screenshot:

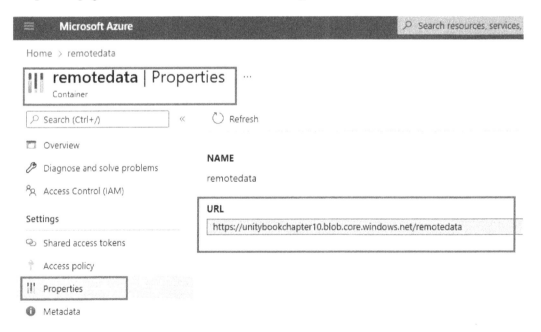

Figure 10.63 – Container Properties

## Creating a new addressables group

Next, we also need to create a new **addressables group**, which is a container for addressable assets and their data, and can determine whether the assets within the group will be hosted on a remote server or stored locally. We can then place assets that need to be hosted on a remote server in this new group without changing the local location configured in the default group.

Let's perform the following steps:

1.  In the toolbar, click on **Window | Asset Management | Addressables | Groups**.

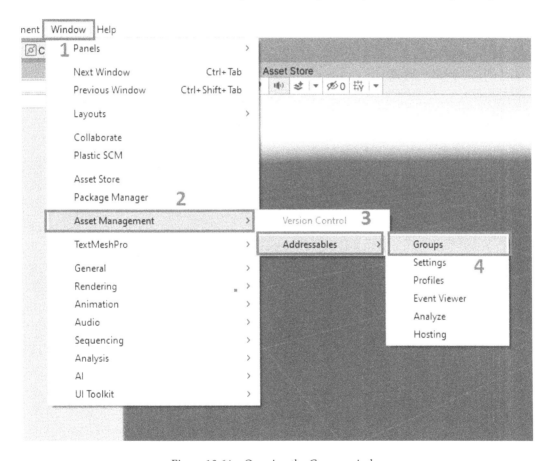

Figure 10.64 – Opening the Groups window

2. In the **Addressables Groups** window, click the **Create** button, then select **Group >
   Packed Assets** to create a new group.

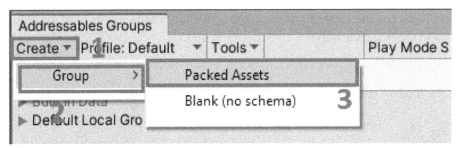

Figure 10.65 – Creating a new group

Rename it to **Azure Remote Group**.

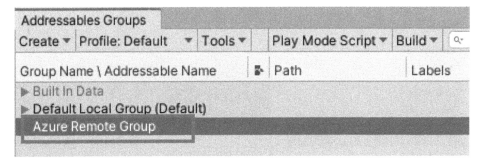

Figure 10.66 – Azure Remote Group

3. Change the active profile from **Default** to **AzureCloud** so that the Addressable
   Assets system can access the variables in **AzureCloud**.

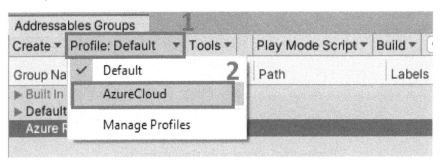

Figure 10.67 – Activating the AzureCloud profile

4.  Select the Azure remote group in the **Addressables Groups** window to open its **Inspector** window and set **Content Packing & Loading** using the remote path defined in the **AzureCloud** profile.

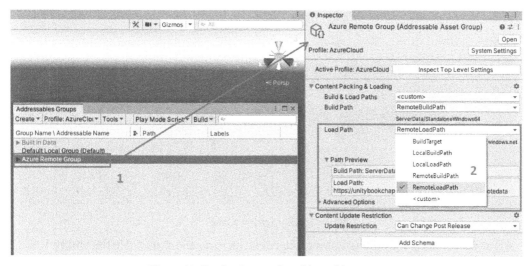

Figure 10.68 – Setting up the addressables group

5.  By default, the marked addressable asset will be under **Default Local Group**; we need to move it to the Azure remote group we just created.

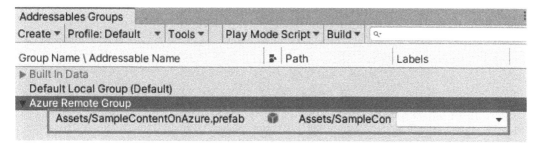

Figure 10.69 – Moving the asset to Azure Remote Group

6.   Finally, we also need to set a label, **Azure**, for this asset. You can think of it as a key so that we can then load this specific asset through the `Addressables.LoadResourceLocationsAsync` method with this key in C# code.

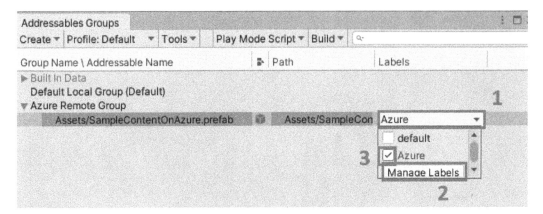

Figure 10.70 – Setting a label

Now that we've set up the addressables group, next, let's move on to enable the ability to build remote content.

## Enabling the Build Remote Catalog checkbox

1.   Back to the toolbar, click on **Window | Asset Management | Addressables | Settings** to open the **Addressable Asset Settings** window.

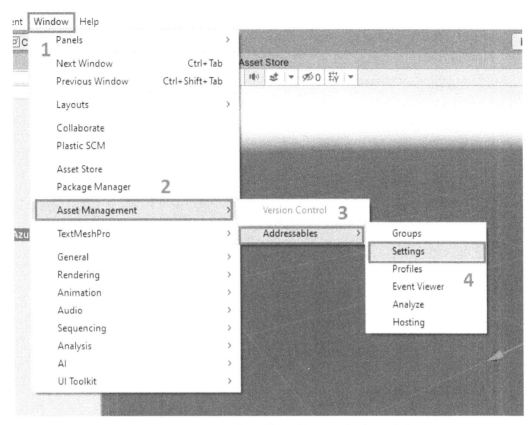

Figure 10.71 – Opening the Addressable Asset Settings window

2.  Scroll down the window and you will find the **Content Update** section. Then, check the **Build Remote Catalog** checkbox and set the **Build Path** and **Load Path** fields, respectively.

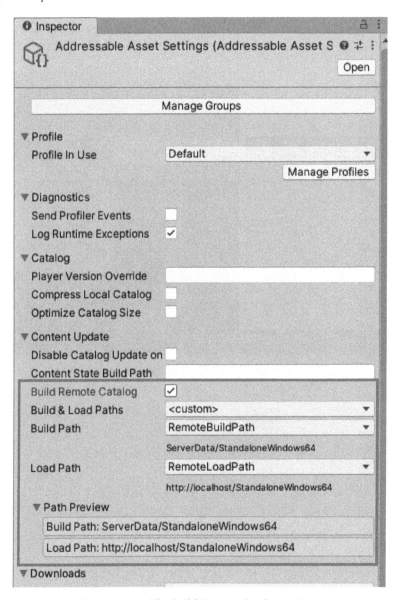

Figure 10.72 – The Build Remote Catalog settings

Now that you know how to enable the remote catalog in the Addressable Asset system, I think you're ready and can't wait to learn how to build the content. Let's go!

## Building the content

It's finally time to build the content with the help of the following steps:

1. Go back to the **Addressables Groups** window, click **Play Mode Script**, and select **Use Existing Build (requires built groups)** in the drop-down menu, as shown in *Figure 10.73*:

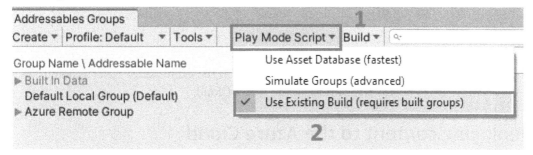

Figure 10.73 – Setting up the Play mode scripts

> **Note**
> Unity provides developers with three build scripts to create play mode data. Here, we are using **Use Existing Build** mode, which best matches the game build deployed. You can find more information about build scripts in the Addressable Asset system at `https://docs.unity3d.com/Packages/com.unity.addressables@1.9/manual/AddressableAssetsDevelopmentCycle.html`.

2. Then, click **Build | New Build | Default Build Script** to build the content.

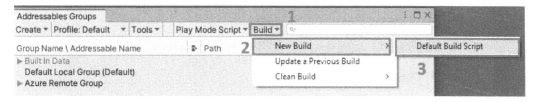

Figure 10.74 – Building the content

3.  Wait for the build to complete and then you can find the build in the `ServerData` folder in your project root.

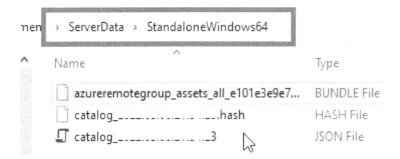

Figure 10.75 – ServerData

Now that you know how to build addressable content in the Addressable Asset system, next, let's move on to deploy the content to the Azure Cloud.

## Deploying content to the Azure Cloud

To deploy the addressable content we just built to the Azure Cloud, follow these steps:

1.  Navigate to the **remotedata** container we created in Azure and then click the **Upload** button.

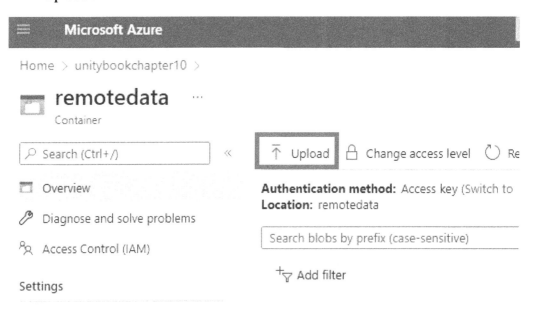

Figure 10.76 – The remotedata container page

2.  An **Upload blob** panel will then appear. Select the files you want to upload and click the **Upload** button.

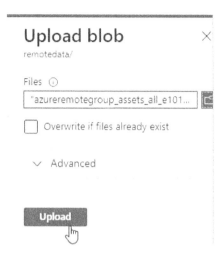

Figure 10.77 – Uploading the content

3.  Wait for the upload to finish and then we can see our addressable content in the blobs list in the **remotedata** container.

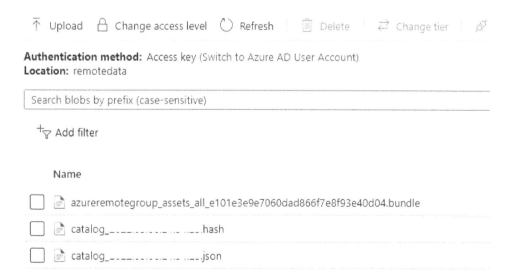

Figure 10.78 – The addressable content in Azure

Now that you know how to deploy addressable content to the Azure Cloud, next, let's move on to exploring how to load content into your game from Azure.

# Loading addressable content from the Azure Cloud

Since we are using the Addressable Asset system to manage assets, loading content from the Azure Cloud into the game also needs to use the methods provided by the Addressable Asset system.

Let's get started!

1.  Create a new C# script in the `Scripts` folder, name it `LoadAddressableContentFromAzureCloud`, and add the following to this script:

    ```
    using UnityEngine;
    using UnityEngine.AddressableAssets;

    public class LoadAddressableContentFromAzureCloud :
      MonoBehaviour
    {
        [SerializeField]
        private string _assetKey;

        private void Start()
        {
            GetContentFromAzureCloud();
        }

        private async void GetContentFromAzureCloud()
        {
            var resourceLocations = await
              Addressables.LoadResourceLocationsAsync
              (_assetKey).Task;

            foreach (var resourceLocation in
              resourceLocations)
            {
                await Addressables.InstantiateAsync
    ```

```
                    (resourceLocation).Task;
        }
    }
}
```

As you can see in the code, we first provide `_assetKey`, whose value is the label of the asset we set in the previous section. Then, we call the `Addressables.LoadResourceLocationsAsync` method to load content and `Addressables.InstantiateAsync` to instantiate a GameObject.

2.  Create a new GameObject, attach the `LoadAddressableContentFromAzureCloud` script to it, set the value of `Asset Key` to `Azure`, and then run the game in the Unity Editor by clicking the **Play** button. We can see that a new instance of the prefab is created as expected.

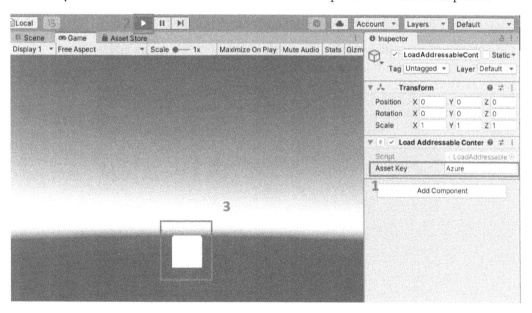

Figure 10.79 – Loading the addressable content from the Azure Cloud

By reading this section, you learned what the Azure Blob storage service in the Microsoft Azure Cloud is and how to use it with Unity's Addressable Asset system to host and update game content. This section also brings us to the end of the chapter!

# Summary

We've come a long way in this chapter. We started by introducing Unity's serialization system, discussing binary serialization, YAML serialization, and JSON serialization in Unity. Then we explored the assets workflow in Unity, covering important concepts such as GUIDs, File IDs, meta files, the `Library` folder, and how to manage the assets import pipeline from C# code. Next, we discussed the `Resources` folder and the `StreamingAssets` folder in detail, which are special folders in Unity, and understood that what they do can help you better develop games with Unity. Finally, we covered quite a bit about Azure Blob storage and Unity's Addressable Asset system, from how to create an Azure Blob storage service in the Azure Cloud to how to load the addressable content from Azure into a Unity project. It's been an amazing journey!

The knowledge you have acquired in this chapter will help you choose the appropriate serialization mode in Unity according to your needs, manage assets reasonably, and use the Azure Cloud to achieve incremental updates of game content.

In the next chapter, we will continue this wonderful journey to explore how to create games with Unity, Microsoft Game Dev, and the Azure Cloud.

# 11
# Working with Microsoft Game Dev, Azure Cloud, PlayFab, and Unity

This is the last chapter of this book. In the previous chapters, we have learned about the various modules that can be used to develop games with the Unity engine, such as the UI module, the physics module, and the animation module, and also covered some advanced topics – for example, Unity's rendering pipelines and the new Data-Oriented Technology Stack. Also, in *Chapter 10, Serialization System and Assets Management in Unity and Azure*, we not only discussed Unity's serialization system and assets management but also covered some knowledge related to Microsoft Azure Cloud.

This chapter will continue to explore **Microsoft Game Dev** (previously known as Microsoft Game Stack), the **Microsoft Azure cloud**, and **Microsoft Azure PlayFab** because the tools needed in modern game development are not limited to game engines; other tools and services such as the cloud are increasingly used in game development.

The following key topics will be included in our learning path:

- Introducing Microsoft Game Dev, Microsoft Azure Cloud, and Azure PlayFab

- Setting up Azure PlayFab for a Unity project

- Signing up and logging in players using Azure PlayFab in Unity

- Implementing a leaderboard using Azure PlayFab in Unity

By the end of this chapter, you will understand what Microsoft Game Dev, Microsoft Azure Cloud, and Microsoft Azure PlayFab are and how to set up Azure PlayFab in a Unity project and use Azure PlayFab's API to implement registration, login, and leaderboard functions in Unity.

Sounds exciting! Now, let's get started!

# Technical requirements

You can find the example project that will be used in this chapter, namely `Chapter11-AzurePlayFabAndUnity`, in the following GitHub repository: `https://github.com/PacktPublishing/Game-Development-with-Unity-for-.NET-Developers`.

# Introducing Microsoft Game Dev, Microsoft Azure Cloud, and Azure PlayFab

We have learned how to use the Unity engine to develop games. However, modern game development requires not only game engines but also other tools, such as cloud services.

## Microsoft Game Dev

In 2019, Microsoft announced Microsoft Game Stack, now known as Microsoft Game Dev, which aims to provide game developers with the tools and services they need to easily create and operate games:

Figure 11.1 – Microsoft Game Dev products (from the Game Dev website)

These tools and services in Microsoft Game Dev include not only DirectX, Visual Studio, Xbox Services, App Center, and Havok, which are commonly used by game developers to complete game development and content creation, but also cloud-based services such as Microsoft Azure Cloud and Azure PlayFab, which all come together to form a powerful ecosystem that every game developer can use, as shown in *Figure 11.1*.

The Microsoft Azure cloud and Azure PlayFab are important parts of Microsoft Game Dev. Not only are more and more modern games requiring multiplayer support but it is also becoming more common for single-player games to store player data in the cloud. Therefore, the cloud is becoming more and more important in game development.

At the Game Developers Conference in March 2022, Microsoft announced a new program, **ID@Azure**, designed to help game developers develop games using the Microsoft Azure cloud and Azure PlayFab services. Any game developer can apply to join the program, whether they are an independent game developer or a game studio. After joining the program, you can get up to $5,000 in Azure credits, so you can access many cloud services, get a free Azure PlayFab Standard Plan, get expert support, and so on.

---

**Note**

If you are interested in the ID@Azure program, you can find more information at `https://aka.ms/idazure`.

---

Now that you have an understanding of what Microsoft Game Dev is, let's move on to exploring what the Microsoft Azure cloud and Azure PlayFab are.

## Microsoft Azure Cloud

Microsoft Azure is a cloud computing service platform where you can find the following services:

- **Cloud computing** services, such as Azure App Service, Azure Functions, and Azure Virtual Machines

- **Database** services, such as Cosmos DB, Azure SQL Database, and Azure Cache for Redis

- **Storage** services, such as an Azure Storage account and Data Lake Storage

- **Networking** services, such as Azure Application Gateway, Azure Firewall, and Azure Load Balancer

- **Analytics** services, such as Azure Data Factory and Azure Synapse Analytics

- **Security** services, such as Azure Defender and Azure DDoS Protection

- **AI** services, such as Azure Cognitive Services and Azure Bot Service

In the game industry, game servers are usually deployed in data centers as close as possible to players, which not only reduces network latency but also meets data sovereignty requirements in certain countries and regions:

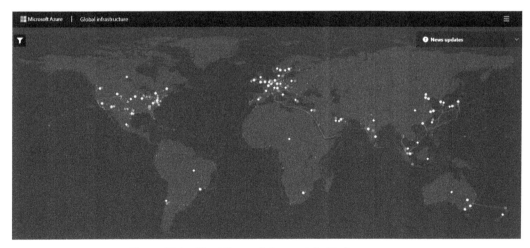

Figure 11.2 – The Microsoft Azure global infrastructure

According to data from Microsoft, the Microsoft Azure cloud covers 140 countries and regions around the world, and the number of available areas is more than any other cloud platform. The huge global coverage helps game developers quickly deploy game services for target countries or regions.

> **Note**
> You can find more information about the Microsoft Azure Global Infrastructure at `https://infrastructuremap.microsoft.com/explore`.

In addition to using Azure data centers to host games, game developers can also develop games using Azure virtual machines on the Microsoft Azure cloud. A new Azure Game Development Virtual Machine was announced at the Game Developers Conference in March 2022, which is customized for game developers and pre-installed with tools such as the Microsoft Game Development Kit, Visual Studio 2019 Community Edition, and Blender to enable game production on the cloud.

> **Note**
> If you are interested in the Azure Game Development Virtual Machine, you can find more information at `https://aka.ms/gamedevvmdocs`.

## Azure PlayFab

PlayFab is a complete backend service for building and operating real-time games. In early 2018, Microsoft acquired PlayFab. Now, PlayFab has joined the Azure family and changed its name to Azure PlayFab, becoming a part of Azure. Azure PlayFab combines the Azure cloud with PlayFab; the Azure cloud brings reliability, global-scale accessibility, and enterprise-grade security, while PlayFab provides game developers with a complete game backend service.

As a complete backend service solution, Azure PlayFab mainly provides the following functions for game developers to develop games:

- Built-in authentication that game developers can use to enable player registration, login, and even track players across devices

- The ability to create dynamically scaled multiplayer servers and manage player data on the cloud

- The ability to easily implement a leaderboard on the backend server

> **Note**
> Azure PlayFab also provides other services for maintaining and operating games, such as **Liveops** (short for **Live Operations**) and data analytics services, which can be used to manage game content, such as making updates to a game without releasing a new version, and reporting and analyzing game data daily. They are beyond what we need here, but if you're interested, you can find out more at `https://docs.microsoft.com/en-us/gaming/playfab`.

In the rest of this chapter, we will integrate Azure PlayFab into a Unity project to implement player registration, login, data saving, loading, and a leaderboard.

Let's move on!

# Setting up Azure PlayFab for a Unity project

In this example, we will add player registration, login, data saving, loading, and leaderboard functions to a *Flappy Bird*-style game in Unity:

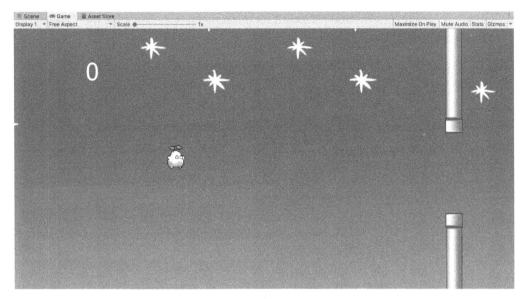

Figure 11.3 – The Unity project

Next, we will first create a new Azure PlayFab account, set up a game studio and a game title in Azure PlayFab, and then set up the Azure PlayFab SDK in this Unity project.

# Creating a new Azure PlayFab account

First of all, we need a new Azure PlayFab account. To create a new Azure Playfab account, let's perform the following steps:

1.  Visit the home page of Microsoft Azure PlayFab at `https://playfab.com/` and click the **SIGN UP** button at the upper-right corner to open the sign-up page:

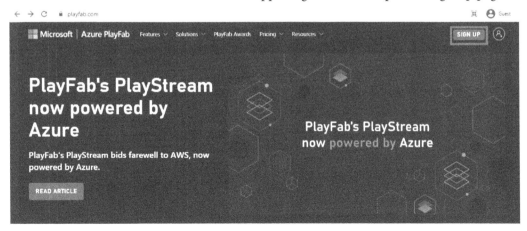

Figure 11.4 – The home page of Azure PlayFab

2.  Enter your email address and password on the sign-up page and click the **Create a free account** button:

Figure 11.5 – Creating a free account

3.  You will then receive a verification email from Azure PlayFab to verify your email address; click **VERIFY YOUR EMAIL ADDRESS**:

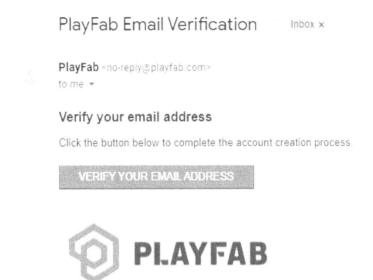

Figure 11.6 – An email address verification from Azure PlayFab

4.  After the email address verification is complete, you can log in with the Azure PlayFab account you just created, and you can see your game studio and a game title already set up in the Azure PlayFab developer portal, also known as **Game Manager** in Azure PlayFab:

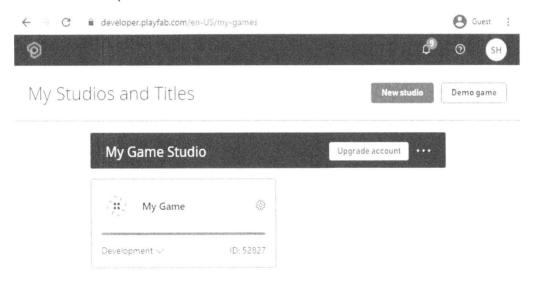

Figure 11.7 – My Game Studio in Azure PlayFab

Now that we have created a new Azure PlayFab account, we can start looking at how to set up a game studio and a game title in Azure PlayFab.

# Setting up a game studio and a game title in Azure PlayFab

After creating an Azure PlayFab account, the next task is to set up your own game studio and game title:

1.  The default game studio is called **My Game Studio**, which doesn't make much sense, so you can click **...** | **Studio settings** on the right to open the **Edit Studio** page:

Figure 11.8 – Opening the Edit Studio page

2.  On the **Edit Studio** page, change the game studio name to UnityBook and click the **Save studio** button to save:

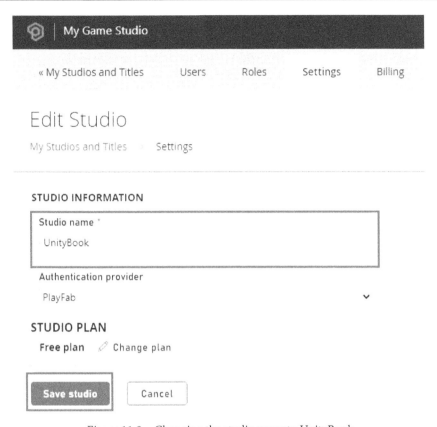

Figure 11.9 – Changing the studio name to UnityBook

3.   Similarly, the default game title is **My Game**, which also doesn't make much sense. As shown in the following figure, you can click the gear button and then **Edit title info** to open the **Edit Title** page:

Figure 11.10 – Opening the Edit Title page

4.  On the **Edit Title Information** page, you can set various information about the title here, such as name, genre, and player modes. Let's change the title name to `Chapter11-AzurePlayfabAndUnity` and click **Save title** to save:

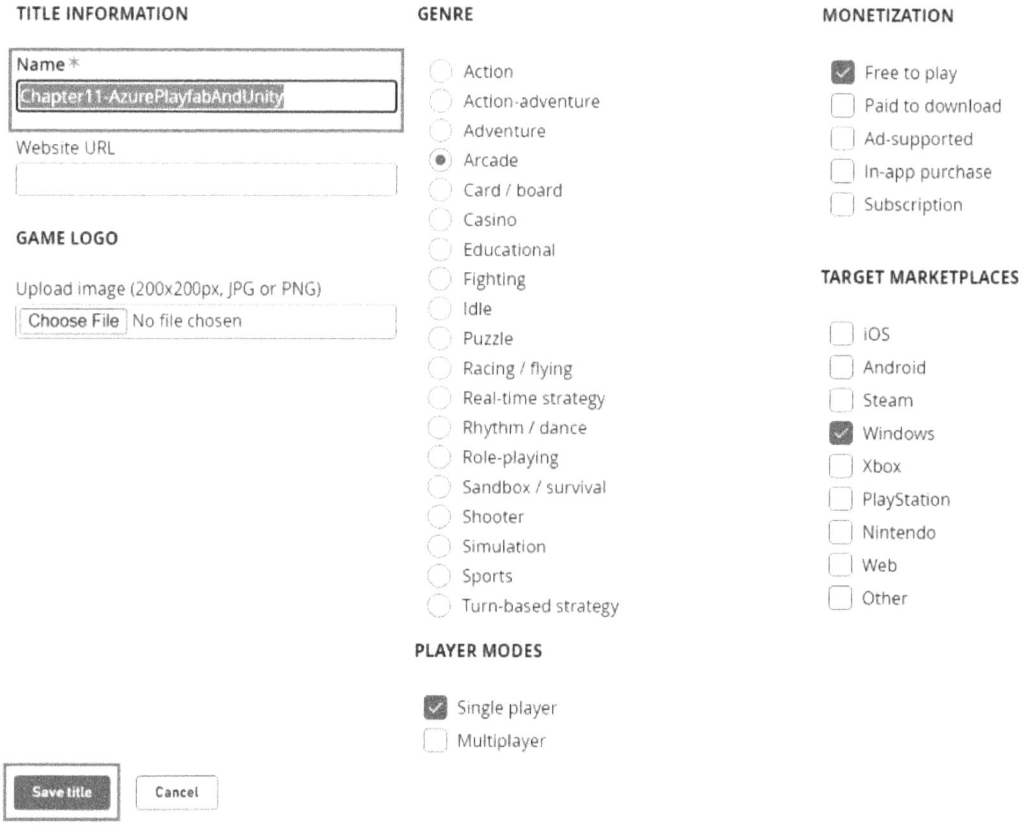

TITLE INFORMATION

Name *

Chapter11-AzurePlayfabAndUnity

Website URL

GAME LOGO

Upload image (200x200px, JPG or PNG)

Choose File   No file chosen

GENRE

○ Action
○ Action-adventure
○ Adventure
◉ Arcade
○ Card / board
○ Casino
○ Educational
○ Fighting
○ Idle
○ Puzzle
○ Racing / flying
○ Real-time strategy
○ Rhythm / dance
○ Role-playing
○ Sandbox / survival
○ Shooter
○ Simulation
○ Sports
○ Turn-based strategy

PLAYER MODES

☑ Single player
☐ Multiplayer

MONETIZATION

☑ Free to play
☐ Paid to download
☐ Ad-supported
☐ In-app purchase
☐ Subscription

TARGET MARKETPLACES

☐ iOS
☐ Android
☐ Steam
☑ Windows
☐ Xbox
☐ PlayStation
☐ Nintendo
☐ Web
☐ Other

Save title    Cancel

Figure 11.11 – Changing the title name

Now that we have set up a game studio and a game title in Azure PlayFab, let's turn our attention to setting up the Azure PlayFab SDK in the Unity project!

## Setting up the Azure PlayFab SDK in the Unity project

In order to access the API in Azure PlayFab from Unity, we need to import the Azure PlayFab SDK into the Unity project first:

1.  You can find the Azure PlayFab SDK at `https://docs.microsoft.com/en-us/gaming/playfab/sdks/unity3d/`. Here, you can also find the link to the Unity PlayFab SDK GitHub repository, as shown in *Figure 11.12*:

Figure 11.12 – The Azure PlayFab SDK download links

2.  Drag and drop the `UnitySDK` package you just downloaded into the Unity Editor.
    The **Import Unity Package** window will pop up, where you can preview the
    contents of the package, and then click the **Import** button to import it into
    this Unity project:

Figure 11.13 – Importing the Azure PlayFab SDK

3.  Once the SDK has been imported, you will find the PlayFab menu in the Unity Editor toolbar. Then, you can click on **PlayFab** > **MakePlayFabSharedSettings** to open the **PlayFabSharedSettings** window, where you need to configure settings to connect this Unity project to the game title in Azure PlayFab:

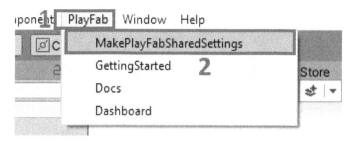

Figure 11.14 – The PlayFab SDK has been imported

4.  In the **Play Fab Shared Settings** window, you should provide the game title ID and the developer secret key of your game title, as shown in *Figure 11.15*:

Figure 11.15 – The Play Fab Shared Settings window

5.  In order to find out the game title ID and the developer secret key, you need to go back to the developer portal of Azure PlayFab, where you can find the game title ID on the game title item:

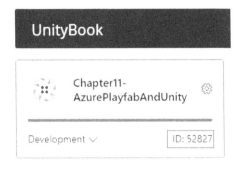

Figure 11.16 – The game title ID

6.  The developer secret key is tightly coupled to the game title in Azure PlayFab, so on the developer portal, you first need to click on the title item to open the overview page of the game title. As shown in the following figure, you need to click on the gear button and then click **Title settings** to open the settings page of the game title:

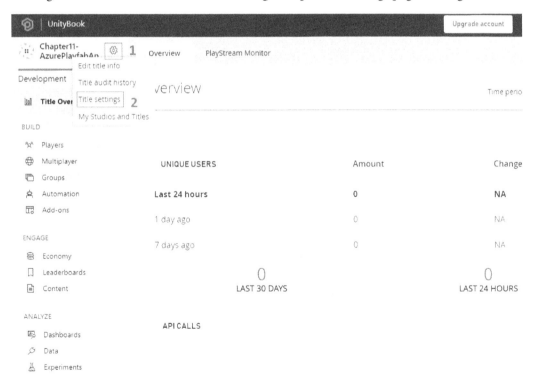

Figure 11.17 – The Overview page of the game title

7.  In the settings page of the game title, select the **Secret Keys** tab to switch to the secret keys settings, where you can find the default developer secret key:

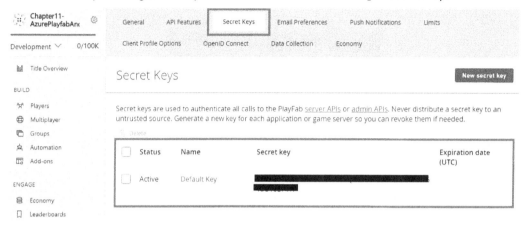

Figure 11.18 – The Secret Keys page

8.  Go back to the **Play Fab Shared Settings** window in Unity, and use the title ID and the developer secret key you just got from the Azure PlayFab developer portal to set the title ID and the developer secret key.

Now that we have set up the Azure PlayFab SDK for this Unity project, you should now have an understanding of Azure PlayFab, including the Azure PlayFab developer portal, (which is also called Game Manager), how to set up a game studio and a game title, and how to import Azure PlayFab's SDK into a Unity project and connect the game title in Azure PlayFab to the project. Next, let's move on to exploring how to register and log in a player via Azure PlayFab.

# Signing up and logging in players using Azure PlayFab in Unity

In the demo project mentioned in the *Technical requirements* section, you can find the signup and login UI panel in **AzurePlayFabIntegration folder | StartScene**, which we will use to implement the signup and login functionality:

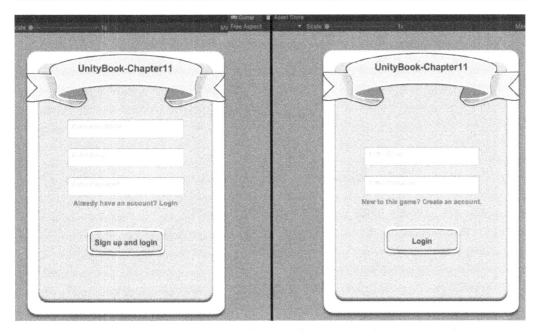

Figure 11.19 – The signup tab (left) and the login tab (right) on the UI panel

As shown in *Figure 11.19*, like many common signup and login pages, the signup and login UI panels in our example also have two tabs, namely the signup tab and the login tab, which can be switched by clicking the red reminder text on the panel. The signup tab requires the player to provide a username, email, and password to create a new player account in Azure PlayFab, while the login tab only requires the player to provide the email and password to log in.

## Signing up players in Azure PlayFab

Next, let's take a look at how to implement the signup function first:

1. Create a new folder in the `AzurePlayFabIntegration` folder and name it `Scripts`:

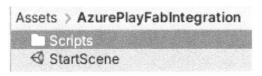

Figure 11.20 – Creating a Scripts folder

2.  Create a new C# script in the `Scripts` folder, name it
    `AzurePlayFabAccountManager`, and add the following code:

```csharp
using System.Text;
using System.Security.Cryptography;
using UnityEngine;
using UnityEngine.UI;
using PlayFab;
using PlayFab.ClientModels;

public class AzurePlayFabAccountManager :
  MonoBehaviour
{
    [SerializeField]
    private InputField _userName, _email, _password;

    [SerializeField]
    private Text _message;

    public void OnSignUpButtonClick()
    {
        var userRequest = new
          RegisterPlayFabUserRequest
        {
            Username = _userName.text,
            Email = _email.text,
            Password = Encrypt(_password.text)
        };

        PlayFabClientAPI.RegisterPlayFabUser(userRequest,
          OnRegisterSuccess, OnError);
    }

  public void
    OnRegisterSuccess(RegisterPlayFabUserResult result)
    {
        _message.text = "created a new account!";
```

```
        var displayNameRequest = new
          UpdateUserTitleDisplayNameRequest
        {
            DisplayName = result.Username
        };
            PlayFabClientAPI.
            UpdateUserTitleDisplayName(display
            NameRequest, OnUpdateDisplayNameSuccess,
            OnError);
    }

    public void OnError(PlayFabError error)
    {
        _message.text = error.ErrorMessage;
    }

    private static string Encrypt(string input)
    {
        var md5 = new MD5CryptoServiceProvider();
        var bytes = Encoding.UTF8.GetBytes(input);
        bytes = md5.ComputeHash(bytes);
        return Encoding.UTF8.GetString(bytes);
    }
}
```

This is quite a long script; let's break down the code as follows:

- We add `System.Security.Cryptography` and the `System.Text` namespace with the `using` keyword to encrypt the password in the `Encrypt` method.

- We add `PlayFab` and the `PlayFab.ClientModels` namespace with the `using` keyword to access the API that Azure PlayFab offers.

- In the `fields` section, we reference three `InputField` UI elements to provide the username, email address, and password. Also, we get a reference to the `Text` UI element to display the message from Azure PlayFab.

- We create a new instance of `RegisterPlayFabUserRequest` and call `PlayFabClientAPI.RegisterPlayFabUser` to register this user in Azure PlayFab.

- We also have two callbacks – `OnRegisterSuccess`, which is called when the result is received, and `OnError`, which is called when an error occurs.

- In `OnRegisterSuccess`, we create a new instance of `UpdateUserTitleDisplayNameRequest` and call `PlayFabClientAPI.UpdateUserTitleDisplayName` to update the user's display name with the username at registration; otherwise, the user's display name is an empty string by default. Also, you can use this method to allow the user to change the account's display name in the future.

3. Drag and drop `AzurePlayFabAccountManager.cs` onto the **SignupAndLogin** GameObject in the scene, and assign the UI elements to the corresponding fields, as shown in *Figure 11.21*:

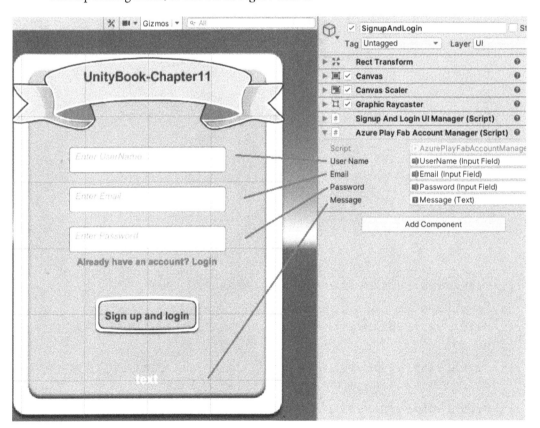

Figure 11.21 – Setting up AzurePlayFabAccountManager

4. Select the **Sign up and log in** button in the UI panel to open the **Inspector** window of this button. In the **Inspector** window, first, click the + button at the bottom of the **On Click()** section, then select the GameObject that AzurePlayFabAccountManager is attached to, and finally, select the method defined in the AzurePlayFabAccountManager class that will be called when the button is clicked, namely OnSignUpButtonClick, as shown in *Figure 11.22*:

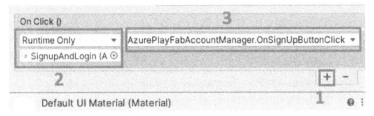

Figure 11.22 – Setting up the sign-up button

5. Run the game, and enter the username, email address, and password in the **Sign up and login** UI panel, and then click the **Sign up and login** button to send a register user request to Azure PlayFab. As shown in the following figure, a new account is created:

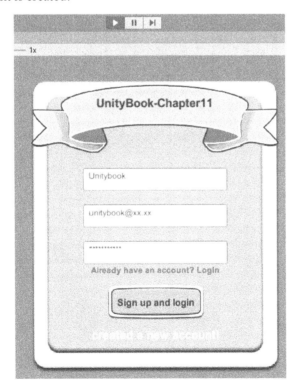

Figure 11.23 – A new account is created

6.  Let's go back to the dashboard of the game title in Azure PlayFab. In the dashboard, you can see that there is a new API call and a new user has been created. Then, we can also click the **Players** button to open the **Players** page for more information:

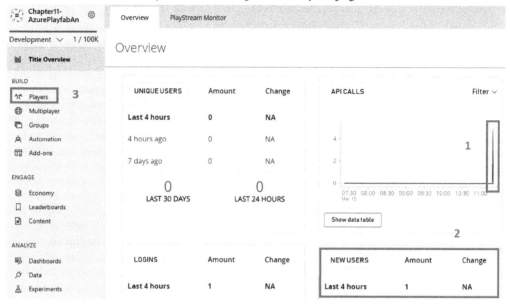

Figure 11.24 – The dashboard of the game title

7.  Take a look at the player list on the **Players** page; you can see the new account we just created. There is also some information about the account, such as the last login time, the time the account was created, and which country the player logged in from:

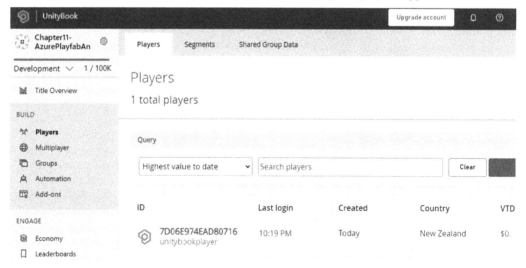

Figure 11.25 – The Players page

Now that we've implemented the registration function, it's time to implement the login function for the players who have accounts.

# Logging in players in Azure PlayFab

By taking the following steps, we will require players to provide an email and password to log in, and if the login is successful, they will jump to our *Flappy Bird*-style game scene:

1. Go back to the `AzurePlayFabAccountManager` script and add the following code:

```
// ... pre-existing code ...

using UnityEngine.SceneManagement;

//... pre-existing code ...

    public void OnLoginButtonClick()
    {
        var userRequest = new
          LoginWithEmailAddressRequest
        {
            Email = _email.text,
            Password = Encrypt(_password.text)
        };
        PlayFabClientAPI.
          LoginWithEmailAddress(userRequest,
    OnLoginSuccess, OnError);
    }

    public void OnLoginSuccess(LoginResult result)
    {
        _message.text = "login successful!";
        StartGame();
    }

    private static void StartGame()
    {
```

```
                    SceneManager.LoadScene(1);
        }
```

Let's break down the newly added code as follows:

- First of all, the `UnityEngine.SceneManagement` namespace is added with the `using` keyword. This is because if the player logs in successfully, we need to switch the scene from the login scene to the game scene, and the logic related to scene loading is defined in this namespace.

- We create a new instance of `LoginWithEmailAddressRequest` and call `PlayFabClientAPI.LoginWithEmailAddress` to log the player into Azure PlayFab.

- In addition to using email to log in, Azure PlayFab also offers multiple login methods, such as calling `PlayFabClientAPI .LoginWithFacebook` to log in with a Facebook access token, and calling `PlayFabClientAPI. LoginWithGameCenter` to log in with an iOS Game Center player identifier.

- The `SceneManager.LoadScene` method will be called when the player logs in successfully. The `SceneManager.LoadScene` method takes an `int` parameter, which is the index of the target scene.

- There are two scenes in this example – the first one is `StartScene`, with an index of `0`, which allows players to register or log in here; and the second is GameScene, with an index of `1`, which allows players to play the game, so we use an index of `1` to switch from `StartScene` to `GameScene`.

2. Select the **Login** button in the UI panel to open the **Inspector** window of this button. In the **Inspector** window, first, click the + button at the bottom of the **On Click()** section, then select the same GameObject that `AzurePlayFabAccountManager` is attached to, and finally, select the method defined in the `AzurePlayFabAccountManager` class that will be called when the **Login** button is clicked; this time, it is the `OnLoginButtonClick` method, as shown in *Figure 11.26*:

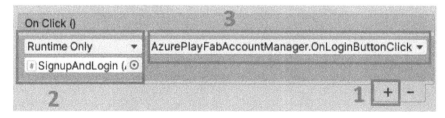

Figure 11.26 – Setting up the Login button

3. Run the game, switch to the login tab, enter the email address and password, and then click the **Login** button to send a login user request to Azure PlayFab, as shown in the following figure:

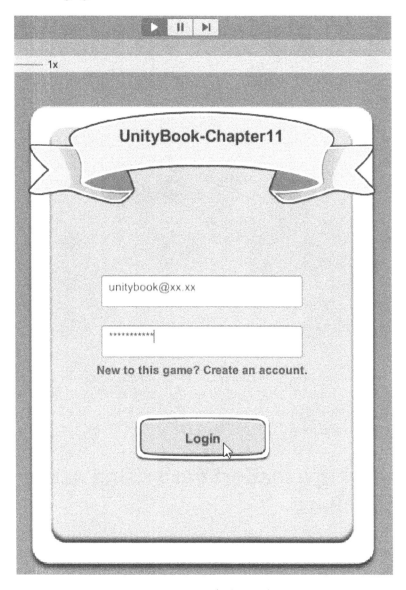

Figure 11.27 – The login tab

4.   Then, if the player logs in successfully, the game scene will be loaded and the game will start, as shown in the following figure:

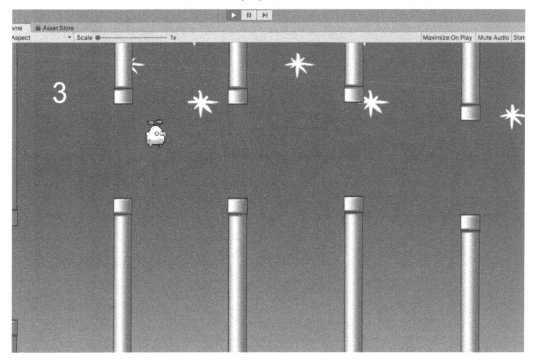

Figure 11.28 – Playing the game

In this section, you learned how to use the Azure PlayFab API to register a user, how to update a user's display name in Azure PlayFab via the Azure PlayFab API, and how to use that to log into Azure PlayFab from a Unity game. Next, we will explore how to implement a leaderboard in a Unity game using Azure PlayFab.

# Implementing a leaderboard using Azure PlayFab in Unity

Most games today use leaderboards, which indicate who is the best performer in a game and increase gamer engagement with a game. In this section, we will be exploring how to implement a leaderboard using Azure PlayFab in our Unity project.

# Setting up a leaderboard in Azure PlayFab

In order to use Azure PlayFab's leaderboard feature, we first need to set up a leaderboard in the developer portal of Azure PlayFab:

1.  Go back to the dashboard of the game title in Azure PlayFab. In the dashboard, you will find the **Leaderboards** option in the left column; click it to open the **Leaderboards** page:

Figure 11.29 – The Leaderboards option

2.  As shown in the following figure, no leaderboard has been created yet, so click the **New leaderboard** button to create a new leaderboard in Azure PlayFab:

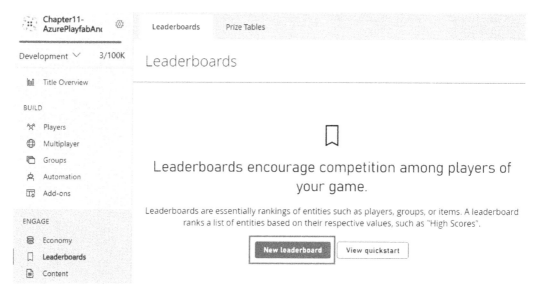

Figure 11.30 – Creating a new leaderboard

3.  In the **New Leaderboard** settings panel, we will set up three properties for this leaderboard, namely **Statistic name**, **Reset frequency**, and **Aggregation method**, as shown in *Figure 11.31*:

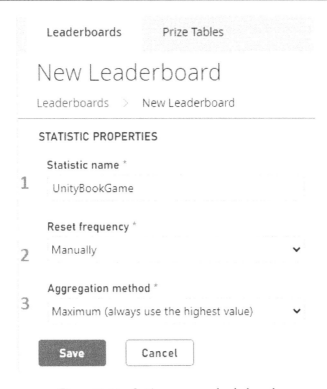

Figure 11.31 – Setting up a new leaderboard

Let's explain these three properties one by one:

- **Statistic name** is the name of this leaderboard; we've named it `UnityBookGame` in this example.

- **Reset frequency** determines how often the leaderboard should be reset. There are five options:

  - **Manually**: This is the default value, and we leave the reset frequency set to this so that the leaderboard doesn't reset automatically.

  - **Hourly**: Automatically resets the leaderboard every hour.

  - **Daily**: Automatically resets the leaderboard every day.

  - **Weekly**: Automatically resets the leaderboard every week.

  - **Monthly**: Automatically resets the leaderboard every month.

- **Aggregation method** determines how the scores from the players are saved. There are four options:

  + **Last**: This is the default option; it is always updated with a new value, regardless of whether it is higher or lower than the existing value.

  + **Minimum**: Always use the lowest value.

  + **Maxmum**: Always use the highest value. We choose this option to save the highest score of a player in our game.

  + **Sum**: Add this value to the existing value.

4.  Click the **Save** button in the **New Leaderboard** settings panel; then, we have an empty leaderboard set up in Azure PlayFab:

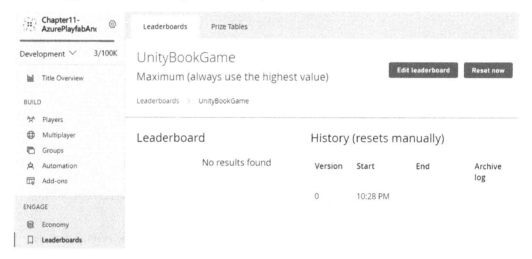

Figure 11.32 – A new empty leaderboard

5.  In order to allow the Unity game to post player statistics requests to Azure PlayFab, we also need to enable the **Allow client to post player statistics** option in Azure PlayFab. So, click the gear icon at the upper-left corner first, and then select the **Title settings** option to open the settings page of the game title, as shown in the following figure:

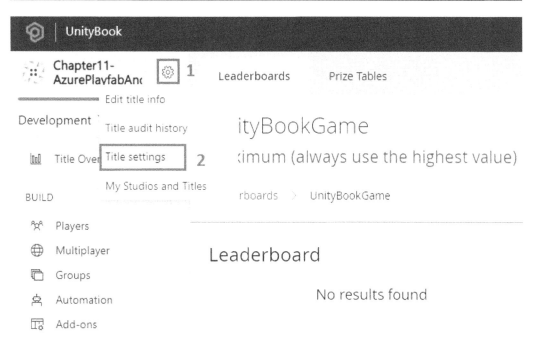

Figure 11.33 – Open the Title settings page

6. On the settings page, click on the **API Features** tab to switch to the **API Features** settings:

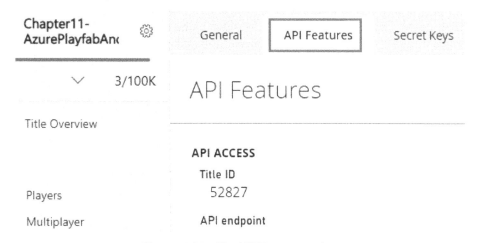

Figure 11.34 – The API Features settings

7. Scroll down to the **ENABLE API FEATURES** section, enable the **Allow client to post player statistics** option, and save, as shown in *Figure 11.35*:

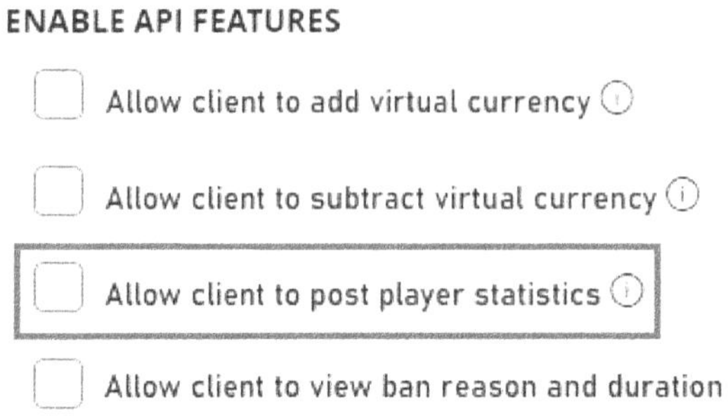

**ENABLE API FEATURES**

☐ Allow client to add virtual currency ⓘ

☐ Allow client to subtract virtual currency ⓘ

☐ Allow client to post player statistics ⓘ

☐ Allow client to view ban reason and duration

Figure 11.35 – Enable the Allow client to post player statistics option

Now that we've created and set up a leaderboard in Azure PlayFab, let's move on to explore how to update the score of a player from Unity using the Azure PlayFab API.

## Updating the score of a player from Unity using the Azure PlayFab API

When a player completes a game and has a higher score than before, we want to update the player's score on the leaderboard in Azure PlayFab. Let's perform the following steps to implement it:

1. Create a new C# script in the Scripts folder, name it AzurePlayFabLeaderboardManager, and add the following code:

```csharp
using System.Collections.Generic;
using UnityEngine;
using PlayFab;
using PlayFab.ClientModels;

public class AzurePlayFabLeaderboardManager :
  MonoBehaviour
{
    public void UpdateLeaderboardInAzurePlayFab(int
      score)
    {
        var scoreUpdate = new StatisticUpdate
        {
```

```
        StatisticName = "UnityBookGame",
        Value = score
    };

    var scoreUpdateList = new
      List<StatisticUpdate> { scoreUpdate };

    var scoreRequest = new
      UpdatePlayerStatisticsRequest
      {
          Statistics = scoreUpdateList
      };

    PlayFabClientAPI.UpdatePlayerStatistics
      (scoreRequest, OnUpdateSuccess, OnError);
}

public void OnUpdateSuccess
  (UpdatePlayerStatisticsResult result)
{
    Debug.Log("Update Success!");
}

public void OnError(PlayFabError error)
{
    Debug.LogError(error.ErrorMessage);
}
}
```

Let's break down the code, as follows:

- In the `UpdateLeaderboardInAzurePlayFab` method, we create a new instance of the `StatisticUpdate` class, which encapsulates the data that needs to be updated for the leaderboard. Here, we provide the name of the leaderboard and the player's score.

- Then, we create a list of `StatisticUpdate` and add the instance of `StatisticUpdate` that we just created to it.

- After that, we create a new instance of the `UpdatePlayerStatisticsRequest` class, which encapsulates the `StatisticUpdate` list we used to update the leaderboard, and call `PlayFabClientAPI.UpdatePlayerStatistics` to update the leaderboard in Azure PlayFab.

- We also have two callbacks – `OnUpdateSuccess`, which is called when the result is received, and `OnError`, which is called when an error occurs.

2.   Then, we need to make sure that the `UpdateLeaderboardInAzurePlayFab` method will be called when the game is over. So, let's open the example project's **Game** scene in the **BasicGame | Scenes** folder, as shown in *Figure 11.36*:

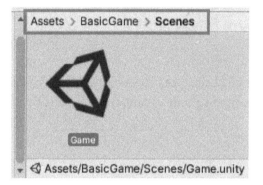

Figure 11.36 – The example game scene

3.   Create a new GameObject in the **Game** scene, name it `AzurePlayFabManager`, and then drag and drop `AzurePlayFabLeaderboardManager.cs` onto it, as shown in *Figure 11.37*:

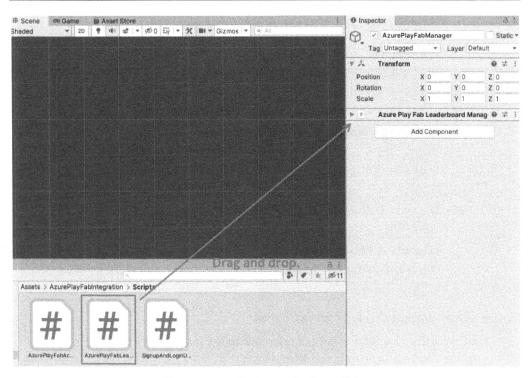

Figure 11.37 – Setting up the GameObject

4.  Next, we need to modify an existing C# script in the example project. You can find the `ExampleGameManager.cs` file in the **BasicGame** > **Scripts** folder; double-click it to open it, as shown in *Figure 11.38*:

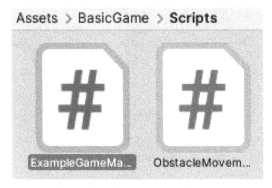

Figure 11.38 – The ExampleGameManager.cs file

5.  Add the following code to the ExampleGameManager class:

```
// ... pre-existing code ...

[SerializeField]
private AzurePlayFabLeaderboardManager
  _azurePlayFabLeaderboardManager;

// ... pre-existing code ...

public void GameOver()
  {
  _azurePlayFabLeaderboardManager.
    UpdateLeaderboardInAzurePlayFab(score);
  }
```

Let's break down the added code, as follows:

- First, we add a new field to get the reference to the instance of AzurePlayFabLeaderboardManager.

- Then, we call the UpdateLeaderboardInAzurePlayFab method in GameOver to update the leaderboard.

6.  Remember to assign the reference to the AzurePlayFabLeaderboardManager instance to the field we just added:

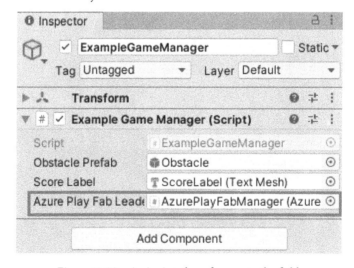

Figure 11.39 – Assigning the reference to the field

7. Let's go back to the **Start** scene and run the game in the Editor, using the player account we created in the previous section to log in and play. In the following screenshot, we have **4** points at the end of the game:

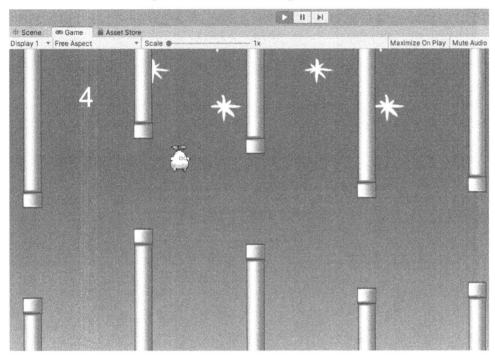

Figure 11.40 – We have 4 points in the game

8. At the same time, go to the **Leaderboard** dashboard of Azure PlayFab, where you can see that the player's highest score is **4** points and they rank first:

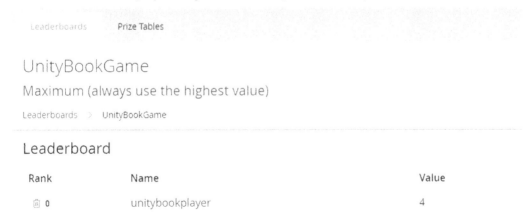

Figure 11.41 – The leaderboard in Azure PlayFab

We have called the API to update the score of a player on the leaderboard in Azure PlayFab from Unity. Our next challenge will be to load the leaderboard data from it.

# Loading the leaderboard data from Azure PlayFab in Unity

In the example project, you also can find the leaderboard panel in **BasicGame | Scenes | GameScene**, which we will use to display the top 10 players on the leaderboard loaded from Azure PlayFab. By default, this UI panel is not activated, so now, we can't see it in the **Game** view:

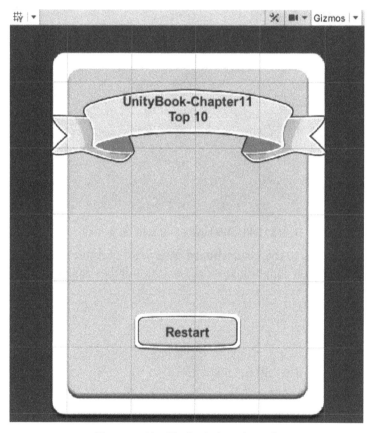

Figure 11.42 – The leaderboard panel

Before we start exploring how to load leaderboard information from Azure PlayFab, let's register more players and add more items to the **UnityBookGame** leaderboard, as shown in the following figure:

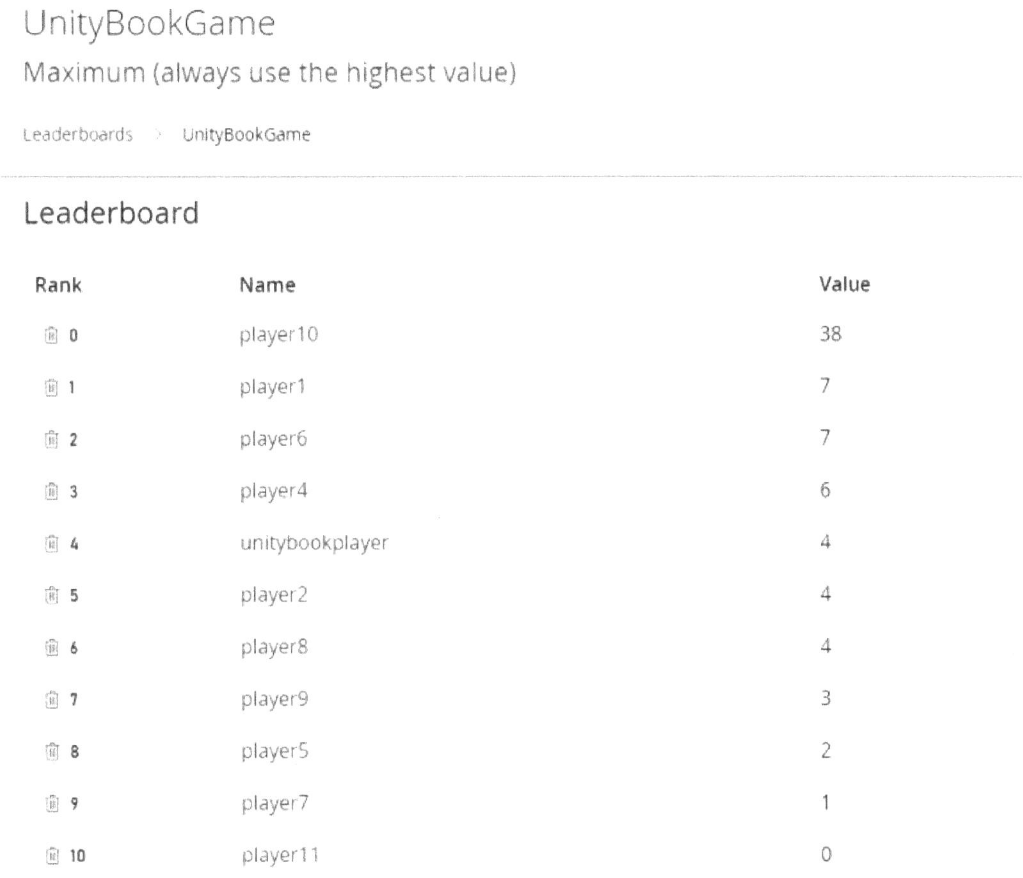

## UnityBookGame

Maximum (always use the highest value)

Leaderboards  >  UnityBookGame

## Leaderboard

| Rank | Name | Value |
| --- | --- | --- |
| 0 | player10 | 38 |
| 1 | player1 | 7 |
| 2 | player6 | 7 |
| 3 | player4 | 6 |
| 4 | unitybookplayer | 4 |
| 5 | player2 | 4 |
| 6 | player8 | 4 |
| 7 | player9 | 3 |
| 8 | player5 | 2 |
| 9 | player7 | 1 |
| 10 | player11 | 0 |

Figure 11.43 – The UnityBookGame leaderboard

Then, our first task is to call the Azure PlayFab API in Unity to get leaderboard data. To do so, proceed as follows:

1.  Go back to the `AzurePlayFabLeaderboardManager` script and add the code, as follows:

```
// ... pre-existing code ...

    public void LoadLeaderboardDataFromAzurePlayFab()
    {
        var loadRequest = new GetLeaderboardRequest
        {
            StatisticName = "UnityBookGame",
            StartPosition = 0,
            MaxResultsCount = 10
        };

        PlayFabClientAPI.GetLeaderboard(loadRequest,
          OnLoadSuccess, OnError);
    }


// ... pre-existing code ...

    public void OnLoadSuccess(GetLeaderboardResult
      result)
    {
        Debug.Log("Load Success!");
    }
```

Let's break down the added code, as follows:

- We create a new method and name it `LoadLeaderboardDataFromAzurePlayFab`.
- In the `LoadLeaderboardDataFromAzurePlayFab` method, we create a new instance of `GetLeaderboardRequest` and call `PlayFabClientAPI.GetLeaderboard` to retrieve up to `10` entries from the `UnityBookGame` leaderboard, starting at index `0`.
- We also add a new callback, `OnLoadSuccess`, which is called when the result is received to print `"Load Success!"` in the console window.

2.  Then, go back to `ExampleGameManager.cs` and update the code as follows:

```
public void GameOver()
{
  _azurePlayFabLeaderboardManager.
    UpdateLeaderboardInAzurePlayFab(score);
  _azurePlayFabLeaderboardManager.
    LoadLeaderboardDataFromAzurePlayFab();
}
```

The newly added code will get leaderboard information from Azure PlayFab when the game is over.

3.  Run the game and take a look at the console window when the game is over; we successfully load leaderboard data from Azure PlayFab, as shown in the following screenshot:

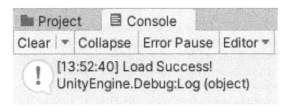

Figure 11.44 – Load Success!

Now that we've received successful results from Azure PlayFab, our next task is display this data in the leaderboard UI panel in our Unity project:

1.  Go back to the `AzurePlayFabLeaderboardManager` script again and update the code, as follows:

```
// ... pre-existing code ...
    [SerializeField]
    private GameObject _leaderboardUIPanel;
    [SerializeField]
    private List<Text> _itemsText;


// ... pre-existing code ...
    public void OnLoadSuccess(GetLeaderboardResult
      result)
    {
```

```
            _leaderboardUIPanel.SetActive(true);
            CreateRankingItemsInUnity(result.Leaderboard);
    }


    private void CreateRankingItemsInUnity
        (List<PlayerLeaderboardEntry> items)
    {
        foreach(var item in items)
        {
            var itemText = _itemsText[item.Position];
            itemText.text = $"{item.Position + 1}:
            {item.Profile.DisplayName} –
              {item.StatValue}";
        }
    }
```

Let's break down the newly added code, as follows:

- In the `fields` section, we reference the `Leaderboard` GameObject in the scene; this is because the leaderboard panel is not activated by default, and we want to activate it at the end of the game to display leaderboard information from Azure PlayFab. Additionally, we get a reference to the list of `Text` UI elements used to display each item on the leaderboard.

- In `OnLoadSuccess`, we activate the `Leaderboard` GameObject in the scene and receive the leaderboard information from Azure PlayFab. Then, call the `CreateRankingItemsInUnity` method, which takes `List<PlayerLeaderboardEntry>` as a parameter.

- In `CreateRankingItemsInUnity`, we update the `Text` UI elements to display information about each item, including the player's rank, display name, and score.

2.  Don't forget to assign these UI elements to these newly added fields accordingly in the `AzurePlayFabManager` GameObject in the **Game** scene, as shown in *Figure 11.45*:

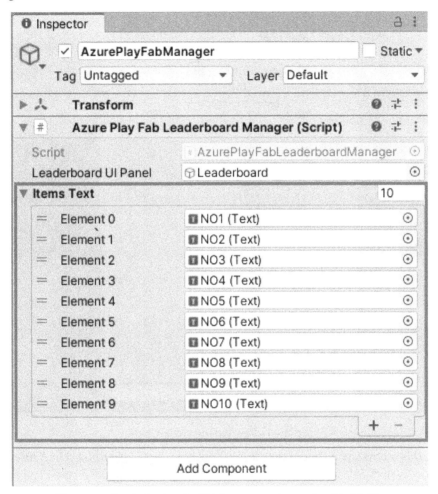

Figure 11.45 – Assigning the UI elements to the newly added fields

3.  Let's go back to the **Start** scene and run the game. In the following screenshot, we can see that the leaderboard UI panel appears at the end of the game and displays the ranking information for the top 10 players:

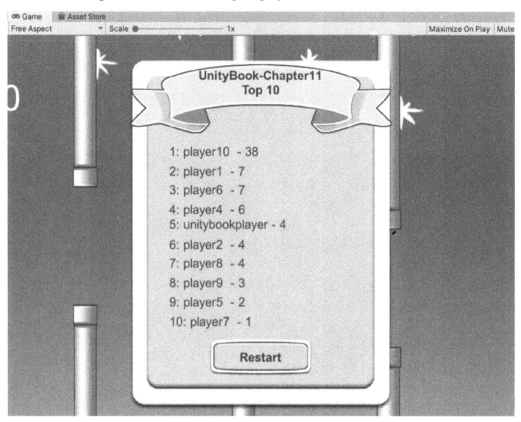

Figure 11.46 – The leaderboard in Unity

By reading this section, you should now know how to set up a leaderboard in Azure PlayFab, how to update leaderboard data from a Unity game using the Azure PlayFab API, and how to use the Azure PlayFab API to get leaderboard data from Azure PlayFab and display it in Unity.

# Summary

This chapter is the last chapter of the book and the final level of our long adventure. Along the way, you've learned a lot of different topics on how to develop games with the Unity game engine, which may include areas you're familiar with, such as how to implement UI in Unity using the MVVM architectural pattern, or there may be content that you have never touched upon before, such as rendering pipelines and related mathematical knowledge. I hope you enjoyed the journey and are ready for the new challenges ahead.

In addition to the Unity engine itself, this chapter also focuses on Microsoft Game Dev, the Microsoft Azure cloud, and Azure PlayFab. We discussed what they are and why we should consider using them in game development. Then, we used an example project to demonstrate how to create a new Azure PlayFab developer account, set up the Azure PlayFab SDK in the Unity project, and implement the registration, login and leaderboard functions in Unity through the Azure PlayFab API.

By reading this chapter and this book, I hope that you now understand that the knowledge required for game development is not limited to how to use a game engine; it also involves knowledge of programming, computer graphics, and even cloud services.

While this chapter is the end of the book, this book is only the beginning of your game developer journey. Keep learning, and keep growing!

# Index

Hi!

I'm Jiadong Chen, author of *Game Development with Unity for .NET Developers*. I really hope you enjoyed reading this book and found it useful for increasing your productivity and efficiency in developing games with Unity and Microsoft Azure cloud.

It would really help us (and other potential readers!) if you could leave a review on Amazon sharing your thoughts on *Game Development with Unity for .NET Developers*.

Go to the link below or scan the QR code to leave your review:

`https://packt.link/r/1801078076`

Your review will help us to understand what's worked well in this book, and what could be improved upon for future editions, so it really is appreciated.

Best wishes,

*Jiadong Chen*

Packt.com

Subscribe to our online digital library for full access to over 7,000 books and videos, as well as industry leading tools to help you plan your personal development and advance your career. For more information, please visit our website.

## Why subscribe?

- Spend less time learning and more time coding with practical eBooks and Videos from over 4,000 industry professionals

- Improve your learning with Skill Plans built especially for you

- Get a free eBook or video every month

- Fully searchable for easy access to vital information

- Copy and paste, print, and bookmark content

Did you know that Packt offers eBook versions of every book published, with PDF and ePub files available? You can upgrade to the eBook version at packt.com and as a print book customer, you are entitled to a discount on the eBook copy. Get in touch with us at customercare@packtpub.com for more details.

At www.packt.com, you can also read a collection of free technical articles, sign up for a range of free newsletters, and receive exclusive discounts and offers on Packt books and eBooks.

# Other Books You May Enjoy

If you enjoyed this book, you may be interested in these other books by Packt:

**Hands-on Game Development with Unity 2018.1**

Raymundo Barrera

ISBN: 978-1-78646-543-6

- Develop a foundation in game development in Unity by using industry-standard techniques

- Design and implement prototypes to iterate quickly

- Build a reusable framework to make development smoother

- Master Unity's latest features to stay ahead of the curve

- Master the best practices and techniques you need to know to develop professional games in Unity from beginning concept to launch

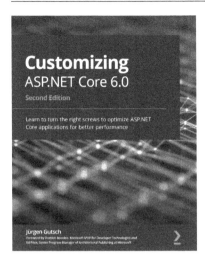

**Customizing ASP.NET Core 6.0**

Jürgen Gutsch

ISBN: 978-1-80323-360-4

- Explore various application configurations and providers in ASP.NET Core 6
- Enable and work with caches to improve the performance of your application
- Understand dependency injection in .NET and learn how to add third-party DI containers
- Discover the concept of middleware and write your middleware for ASP.NET Core apps
- Create various API output formats in your API-driven projects
- Get familiar with different hosting models for your ASP.NET Core app

# Packt is searching for authors like you

If you're interested in becoming an author for Packt, please visit authors.packtpub. com and apply today. We have worked with thousands of developers and tech professionals, just like you, to help them share their insight with the global tech community. You can make a general application, apply for a specific hot topic that we are recruiting an author for, or submit your own idea.

www.ingramcontent.com/pod-product-compliance
Lightning Source LLC
Chambersburg PA
CBHW081450050326
40690CB00015B/2743